Structural Stability in Engineering Practice

Also Available from E & FN Spon

Aluminium Design and Construction
J.B. Dwight

Applied Structural and Mechanical Vibrations
P. Gatti and V. Ferrari

Bridge Deck Behaviour
E.C. Hambly

Construction Methods and Planning
J.R. Illingworth

Design of Structural Elements
C. Arya

Designer's Guide to the Dynamic Response of Structures
A.P. Jeary

Earthquake Engineering
Y.X. Hu, S.C. Liu and W. Dong

Earthquake Resistant Concrete Structures
G. Penelis and A.J. Kappos

Explosive Loading of Engineering Structures
P.S. Bulson

High Performance Concrete
P-C Aitcin

Steel Structures
T.J. MacGinley

Structural Assessment
Edited by K.S. Virdi, F.K. Garas, J.L. Clarke, G.S.T. Armer

Structural Mechanics
A. Carpinteri

Silos
Edited by C.J. Brown and J. Neilsen

For information on these and any other books on the subject please contact The Marketing Department, E & FN Spon, 11 New Fetter Lane, London EC4P 4EE . Tel:+44 (0)171 842 2001, Fax:+44 (0)171 842 2298, www.efnspon.com

Structural Stability in Engineering Practice

Edited by
Lajos Kollár
Technical University Budapest

CRC Press
Taylor & Francis Group
Boca Raton London New York

CRC Press is an imprint of the
Taylor & Francis Group, an **informa** business

A SPON PRESS BOOK

CRC Press
Taylor & Francis Group
6000 Broken Sound Parkway NW, Suite 300
Boca Raton, FL 33487-2742

First issued in paperback 2019

© 1999 by Taylor & Francis Group, LLC
CRC Press is an imprint of Taylor & Francis Group, an Informa business

No claim to original U.S. Government works

ISBN-13: 978-0-419-23790-7 (hbk)
ISBN-13: 978-0-367-44755-7 (pbk)

British Library Cataloguing in Publication Data
A catalogue record for this book is available from the British Library

Library of Congress Cataloging in Publication Data
Structural stability in engineering practice / edited by L. Kollár.
     p. cm.
   Includes bibliographical references.
   1. Structural stability. 2. Load factor design. I. Kollár, Lajos
TA656.S778 1999
624. 1'71--dc21   99-14295
                                   CIP

Visit the Taylor & Francis Web site at
http://www.taylorandfrancis.com

and the CRC Press Web site at
http://www.crcpress.com

Contents

Preface   xi

1 Loss of stability and post-buckling behaviour *Lajos Kollár*   1
  1.1 The main kinds of loss of stability of 'centrally' loaded (geometrically perfect)
       structures   2
       1.1.1 Symmetric stable bifurcation   2
       1.1.2 Symmetric unstable bifurcation   3
       1.1.3 Asymmetric bifurcation   4
       1.1.4 The degenerating case of the symmetric bifurcation   5
       1.1.5 General remarks   6
  1.2 The imperfection sensitivity of the structures   7
  1.3 Loss of stability with a limit point (divergence of equilibrium; snapping through)   11
  1.4 The influence of plasticity   13
  1.5 Some practical points of view for estimating the post-critical behaviour   16
  1.6 Evaluation of buckling experiments by the generalized Southwell plot   20

2 Summation theorems concerning critical loads of bifurcation *Tibor Tarnai*   23
  2.1 On the summation theorems   23
  2.2 The *Southwell* theorem   26
  2.3 *Dunkerley* type theorems and formulae   30
       2.3.1 The *Dunkerley* theorem   30
       2.3.2 The *Föppl-Papkovich* theorem   35
       2.3.3 The *Kollár* conjecture   43
       2.3.4 The *Melan* theorem   51
       2.3.5 The *Rankine* formula   55
  2.4 Conclusions   56

3 Interaction of different buckling modes in the post-buckling range *Lajos Kollár*   59
  3.1 Description of the phenomenon   59
  3.2 The post-buckling load-bearing capacity of a braced column   60
  3.3 The post-buckling load-bearing capacity of the ribbed plate   69
  3.4 The post-buckling load-bearing behaviour of the box bar   74
  3.5 The interaction of the buckling modes of cylindrical shells   80
       3.5.1 Nonlinear shell equations   80

3.5.2 The eigenfunctions of the cylindrical shell 82
3.5.3 The post-buckling behaviour of the shell 84

4 Stability of elastic structures with the aid of the catastrophe theory *Zsolt Gáspár* 88
4.1 Statement of the problem 88
4.2 Definitions 89
4.3 Thom's theorem 90
4.4 The cuspoid catastrophes 92
    4.4.1 The fold catastrophe 92
    4.4.2 The cusp catastrophe 93
    4.4.3 The swallowtail catastrophe 95
    4.4.4 The butterfly catastrophe 96
4.5 The umbilic catastrophes 97
    4.5.1 The elliptic umbilic 97
    4.5.2 The hyperbolic umbilic 99
4.6 Imperfection-sensitivity of structures 100
    4.6.1 What kind of catastrophes arise? 100
    4.6.2 The method 100
    4.6.3 The fold catastrophe 102
    4.6.4 The cusp catastrophe 106
    4.6.5 Higher order cuspoid catastrophes 115
    4.6.6 The umbilic catastrophes 117
4.7 Probability of the instability 127

5 Buckling of frames *Josef Appeltauer, Lajos Kollár* 129
5.1 General theory of frame buckling *Josef Appeltauer* 129
    5.1.1 Description of the phenomena 129
    5.1.2 Mechanical models describing the various kinds of loss of stability 130
    5.1.3 Bifurcation 135
    5.1.4 Divergence 145
    5.1.5 Snapping through 153
    5.1.6 Practical applications 155
    5.1.7 Conclusions 161
5.2 Approximate stability analysis of frames by the buckling analysis of the individual columns *Lajos Kollár* 161
    5.2.1 Basic principles of the method 161
    5.2.2 Stability investigation of braced frames 167
    5.2.3 Stability investigation of unbraced frames 175

6 Application of the sandwich theory in the stability analysis of structures *István Hegedűs* and *László P. Kollár* 187
6.1 Assumptions, definitions 188
6.2 Sandwich beam with thin faces (Timoshenko-beam) 190
6.3 Sandwich beam with thick faces 195
    6.3.1 Incompressible core 195

6.3.2   Compressible core   199
6.4 Models based on the sandwich beam with thick faces   204
    6.4.1   Beams with flexural deformations only ($S = \infty$ or $S = 0$)   205
    6.4.2   Beams with shear deformations only ($D_0 = \infty$ and $D_l = 0$)   205
    6.4.3   Sandwich beam with thin faces ($D_l = 0$)   207
    6.4.4   Beam on an elastic foundation which restrains the rotations –
           Csonka-beam ($D_0 = \infty$)   207
    6.4.5   Sandwich beam with thick faces on an elastic foundation which restrains
           the rotations   209
    6.4.6   Sandwich beam with thick faces on an elastic foundation which restrains
           the displacements   209
    6.4.7   Isotropic sandwich plate   210
    6.4.8   Orthotropic sandwich plate   213
    6.4.9   Orthotropic shallow sandwich shell   215
    6.4.10 Multi-layered sandwich cantilever beam   216
6.5 Approximate expressions for the calculation of the buckling load   217
    6.5.1   Parallel and serial connections of beams (Föppl-Papkovich's and
           Southwell's theorem)   217
    6.5.2   Cantilever beams on elastic foundation which restrains the rotation   221
    6.5.3   Multi-layered sandwich beam   222
6.6 Some applications of the sandwich theory in structural engineering   223
    6.6.1   Discrete structures with regular built-up   223
    6.6.2   Exact analysis of discrete structures using the theory of difference equa-
           tions   223
    6.6.3   Trusses   226
    6.6.4   Laced (Vierendeel) column   230
    6.6.5   Frames and shear walls   235
    6.6.6   Combined torsional and in-plane buckling of multistorey buildings   238

7 Bracing of building structures against buckling Lajos Kollár and Károly Zalka   242
7.1 Basic principles   242
7.2 The necessary stiffness of the bracing core   243
    7.2.1   The bending stiffness of the bracing core   243
    7.2.2   The torsional stiffness of the bracing core   245
    7.2.3   Generalization of the results. Spatial behaviour   251
7.3 The necessary strength of the bracing core   256
7.4 Bracing system of shear walls and cores   258
    7.4.1   The equivalent column   259
    7.4.2   Uniformly distributed load over the height   262
    7.4.3   Concentrated load at top floor level   265
    7.4.4   Supplementary remarks   265
7.5 Stability analysis of the columns of the building   268
    7.5.1   Sway critical loads   269
    7.5.2   Sway versus nonsway critical loads   272
    7.5.3   Conclusions   273

8 Buckling of arches and rings *Lajos Kollár*   276
  8.1 Buckling of bars with curved axis (arches) in their own plane   277
       8.1.1 Buckling of rings and arches with circular axis   277
       8.1.2 Arches with noncircular axes.   285
       8.1.3 Snapping through of flat arches   291
       8.1.4 Buckling of arches with thin-walled, open cross sections   299
       8.1.5 Buckling of arches with hangers or struts   308
  8.2 Lateral buckling of rings and arches   321
       8.2.1 Lateral buckling of centrally compressed arches with circular axis 321
       8.2.2 Buckling of centrally compressed arches with axes other than circular 327
       8.2.3 Lateral buckling of centrally compressed arches loaded by hangers or
              struts   329
       8.2.4 Post-critical behaviour of laterally buckling arches.   335
       8.2.5 Lateral buckling of arches bent in the plane of the arch   335

9 Special stability problems of beams and trusses *Lajos Kollár*   339
  9.1 Problems of lateral stability of beams   339
       9.1.1 The governing differential equations of lateral buckling and some
              conclusions   340
       9.1.2 The energy method for determining the critical loads of suspended
              beams   344
       9.1.3 Determination of the critical load by the summation theorem   349
  9.2 Lateral stability of the nodes of plane trusses   353
  9.3 Snapping through of shell-beams in the plane of bending   356

10 Stability of viscoelastic structures *György Ijjas*   358
  10.1 Introduction   358
       10.1.1 General remarks   358
       10.1.2 Material properties   358
  10.2 Various kinds of creep buckling   359
  10.3 Structures made of fluid-type material, exhibiting symmetric unstable
       post-critical behaviour   363
       10.3.1 Description of the phenomenon   363
       10.3.2 The pseudo-equilibrium surface   365
       10.3.3 The total potential energy   368
       10.3.4 Supplementary remarks   371
  10.4 Structures made of solid-type material, exhibiting symmetric unstable
       post-critical behaviour   371
       10.4.1 Description of the phenomenon   371
       10.4.2 The pseudo-equilibrium surface   373
       10.4.3 The total potential energy   375
       10.4.4 Supplementary remarks   376
  10.5 Structures made of *Dischinger*-type material, exhibiting symmetric
       unstable post-critical behaviour   376
       10.5.1 Description of the phenomenon   376

10.5.2 The pseudo-equilibrium surface 378
10.5.3 The total potential energy 379
10.5.4 Supplementary remarks 380
10.6 Structures made of fluid-type material, exhibiting symmetric stable post-critical
behaviour 380
10.6.1 Description of the phenomenon 380
10.6.2 The pseudo-equilibrium surface 382
10.6.3 The total potential energy 382
10.6.4 Supplementary remarks 383
10.7 Structures made of solid-type material, exhibiting symmetric stable post-critical
behaviour 384
10.7.1 Description of the phenomenon 384
10.7.2 The pseudo-equilibrium surface 384
10.7.3 The total potential energy 386
10.8 Creep of the dashpot 386
10.9 Two remarks about the problems appearing in the literature 387
10.9.1 Importance of the degree of approximations 387
10.9.2 Importance of the elastic behaviour 388

11 Buckling under dynamic loading *Lajos Kollár* 389
11.1 Description of the dynamic loading process 389
11.2 Buckling of an initially curved bar under a falling load 391
11.3 Generalizations 395

12 Stability paradoxes *Lajos Kollár* 398
12.1 Structures behaving differently from common engineering sense 398
12.1.1 Instability of blown-up rubber balloons 398
12.1.2 The buckling length in the case of a load of varying direction, passing
through a fixed point 402
12.1.3 Instability of a bar in tension 403
12.1.4 Structures with infinitely great critical forces 404
12.1.5 Structures with abruptly changing rigidity characteristics 405
12.2 Destabilizing by stiffening and stabilizing by softening 407
12.2.1 Stabilizing by increasing the length 407
12.2.2 The destabilizing effect of an additional support 410
12.2.3 Paradoxes with torsional buckling 411
12.2.4 The destabilizing effect of damping in the case of nonconservative
forces 412

References 415
Author Index 443
Subject Index 449

Preface

Many books have been published on structural stability. Most of them, intended mainly to be used as textbooks, treat the subject by explaining the phenomena and presenting derivations to obtain the results. There are other books, intended to be used by practicing engineers, which collect ready-to-use formulas, but generally do not explain their background.

The aim of our book is somewhat different. We try, on the one hand, to elucidate the – sometimes complicated – phenomena of buckling in a clear engineering way, but mostly without presenting mathematical derivations for which we refer to the literature. We think this useful for design engineers, but also researchers may find the explanations helpful because they can see the gaps and unsolved problems needing research. On the other hand, we collect formulas, diagrams, tables which can be used in practice, and give their explanations and range of validity, so that design engineers can judge in which cases the use of the formulas is justified. We tried to formulate the explanations in a visual way, clearly understandable for engineers working in practice. We mainly covered problems which are either seldom treated, or cannot be found easily in the literature. We also show the application of generally valid approximate methods (as the summation theorems of *Southwell* and *Dunkerley*, seldom used in practice) in several fields, enabling the engineer to solve complex problems in a simple way.

We think that the general use of computers in engineering practice renders clear, simple considerations and approximate methods even more important than before, because they yield the only reliable control of computer results. In addition, they are indispensable in preliminary design where a clear insight into static behaviour is essential to choose the appropriate structure. We hope that our book can help the engineer also in this respect.

Like with other books on structural stability, the results published here can be applied in practice with due consideration to the relevant Building Codes. So the critical loads derived on the basis of the elastic theory should be reduced due to the effects of accidental imperfections, plasticity, creep and, in the case of reinforced concrete structures, of cracking.

1

Loss of stability and post-buckling behaviour

Lajos KOLLÁR

In this book we mainly deal with the loss of stability of elastic structures, caused by the critical load. To understand the phenomena involved and to judge the safety of structures we have to know the post-critical behaviour of the structure. Thus in the following sections, relying on the works of KOITER [1945, 1963], CROLL and WALKER [1972], and THOMPSON and HUNT [1973], we show with simple models various kinds of loss of stability and of post-critical behaviour. (In Chapter 4 we also find all these derived from the general theory.)

In this chapter we will investigate the characteristics of the *initial* post-buckling behaviour, in accordance with the approximate theory of *Koiter*. This is in most cases sufficient to obtain all the information needed for the dimensioning of the structure. There are, however, complex cases of stability when the initial character of post-buckling behaviour later fundamentally changes. We show examples of this in Section 1.4 and Chapter 3.

In accordance with the descriptive character of this chapter, we are content to state that the state of equilibrium is *stable* if the deformation, measured from the geometrically perfect state, increasing in the absolute sense, needs an increasing load (in this case the diagram of the load plotted against the displacement, i.e. the 'equilibrium path', is ascending). The state of equilibrium is *unstable* if the (in the absolute sense) increasing deformation requires a decreasing load (the equilibrium path is descending). Where the tangent of the equilibrium path is horizontal, there is a transition between the stable and unstable states. This transitory point (or section) is called *critical*; there the state of equilibrium can be stable, indifferent (neutral) or unstable. The vertical sections need special consideration.

1.1 THE MAIN KINDS OF LOSS OF STABILITY OF 'CENTRALLY' LOADED (GEOMETRICALLY PERFECT) STRUCTURES

1.1.1 Symmetric stable bifurcation

Let us consider the rigid bar shown in *Fig. 1-1a*, supported at its lower end by a hinge and against rotation by a spring with the spring constant c. The spring be unstressed when the bar is vertical.

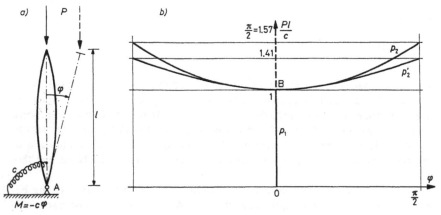

Fig. 1-1 Symmetric stable bifurcation. *a*) Model, *b*) equilibrium paths

Loading the bar by a force P and rotating it by the angle φ we can write the equilibrium of the moments around point A:

$$Pl \sin\varphi - c\varphi = 0. \qquad (1\text{-}1)$$

A possible solution of Eq. (1-1) is $\varphi = 0$. The corresponding 'first equilibrium path' is to be seen in *Fig. 1-1b*, denoted by p_1. The second possibility is $\varphi \neq 0$. In this case P is unequivocally determined by φ:

$$\frac{Pl}{c} = \frac{\varphi}{\sin\varphi}. \qquad (1\text{-}2)$$

The corresponding 'second equilibrium path' p_2 is shown by the upwards curved symmetric curve in *Fig. 1-1b*.

The stable, indifferent (neutral, critical) or unstable character of the equilibrium states can be determined by the equilibrium method [CROLL and WALKER, 1972], (see also Chapter 4). According to what has been said earlier, we will draw conclusions on stability from the ascending or descending character of the equilibrium path.

The section OB of path p_1 is stable, because here the structure has 'no other possibility' than to get along this path to B. Here the equilibrium path 'bifurcates': the state of equilibrium becomes critical. The path p_2 is stable, since it is ascending, but the further section of p_1 is unstable, since from here the structure can 'jump' to p_2 which represents a larger deformation.

The path p_2 shows that the structure has an increasing load-bearing capacity in the post-buckling range.

For the following discussion, we have to know above all the 'initial' post-buckling behaviour, i.e. that during small displacements. To this purpose – following KOITER – we develop the sine function appearing in (1-2) into a power series:

$$\frac{\varphi}{\sin\varphi} \approx 1 + \frac{\varphi^2}{6}. \tag{1-3}$$

The approximate equilibrium path p_2' thus obtained

$$\frac{Pl}{c} = 1 + \frac{\varphi^2}{6} \tag{1-4}$$

has been also plotted in *Fig. 1-1b*.

Both p_2 and p_2' equilibrium paths are symmetric to the vertical axis, i.e. they are symmetric in φ. This is the reason why we call this kind of bifurcation symmetric.

In *Fig. 1-1* and in the subsequent similar figures, we will denote the stable branches of the equilibrium paths by solid line, the unstable branches by dashed line, and the indifferent ones by dotted line.

1.1.2 Symmetric unstable bifurcation

Let us investigate the structure depicted in *Fig. 1-2*, which differs from that of *Fig. 1-1* only in the supporting spring. We assume that the spring always remains horizontal, and that it takes tension and compression as well.

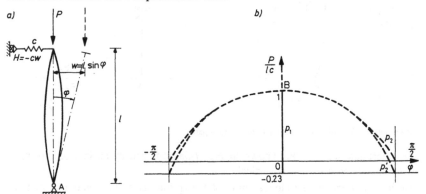

Fig. 1-2 Symmetric unstable bifurcation. *a*) Model, *b*) equilibrium paths

The moment equilibrium equation around point A is

$$Pl \sin\varphi - cl \sin\varphi \cdot l \cos\varphi = 0,$$

or

$$\left(\frac{P}{lc} - \cos\varphi\right) \sin\varphi = 0. \tag{1-5}$$

One solution is again $\varphi = 0$, giving the equilibrium path p_1, see *Fig. 1-2b*. The second solution is

$$\frac{P}{lc} = \cos\varphi \qquad (1\text{-}6)$$

yielding the equilibrium path p_2, curved downwards, in *Fig. 1-2b*.

Of the path p_1 we can state the same as in the case of *Fig. 1-1b*. At point B the equilibrium again bifurcates, but the path p_2 is now – with the exception of the critical point B – unstable all along, since it is everywhere descending. Hence this structure has a decreasing post-buckling load-bearing capacity.

For the investigation of the initial post-buckling behaviour we develop the cosine function appearing in (1–6) into a power series:

$$\frac{P}{lc} \approx 1 - \frac{\varphi^2}{2}. \qquad (1\text{-}7)$$

The corresponding approximate equilibrium path p_2' is plotted in *Fig. 1-2b*.

Both paths p_2 and p_2' are symmetric in φ.

1.1.3 Asymmetric bifurcation

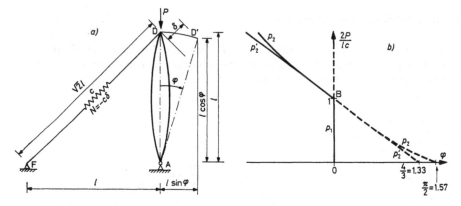

Fig. 1-3 Asymmetric bifurcation. *a*) Model, *b*) equilibrium paths

Let us now investigate the structure shown in *Fig. 1-3a*, stiffened by an inclined spring. The spring be unstressed if the bar is vertical.

First we have to compute the elongation δ of the spring due to the rotation φ of the bar. Writing *Pythagoras'* theorem for the distance FD' we obtain:

$$\left(\sqrt{2}l + \delta\right)^2 = (l + l\sin\varphi)^2 + (l\cos\varphi)^2 = 2l^2(1 + \sin\varphi),$$

from which

$$\delta = \sqrt{2}l\left(\sqrt{1 + \sin\varphi} - 1\right). \qquad (1\text{-}8)$$

We resolve the force $N = (-)c\delta$ acting at the point D' into a horizontal (N_H) and a vertical (N_V) component:

$$N_H = N\frac{l(1 + \sin\varphi)}{\sqrt{2}l + \delta} = N\frac{\sqrt{1 + \sin\varphi}}{\sqrt{2}}, \tag{1-9a}$$

$$N_V = N\frac{l\cos\varphi}{\sqrt{2}l + \delta} = N\frac{\cos\varphi}{\sqrt{2}\sqrt{1 + \sin\varphi}}. \tag{1-9b}$$

The moment equilibrium equation around A gives:

$$Pl\sin\varphi = N_H l\cos\varphi - N_V l\sin\varphi, \tag{1-10}$$

whose one solution is $\varphi = 0$; since then, according to (1-8), $\delta = 0$, and thus $N_H = N_V = 0$. The corresponding equilibrium path is p_1 in *Fig. 1-3b*.
The other solution is (with $\varphi \neq 0$):

$$\frac{2P}{lc} = 2\cos\varphi(\sqrt{1 + \sin\varphi} - 1)\left[\frac{\sqrt{1 + \sin\varphi}}{\sin\varphi} - \frac{1}{\sqrt{1 + \sin\varphi}}\right]$$

$$= 2\cot\varphi\left(1 - \frac{1}{\sqrt{1 + \sin\varphi}}\right), \tag{1-11}$$

yielding the equilibrium path p_2 (*Fig. 1-3b*).

Hence the equilibrium again bifurcates at point B, but the left-hand side of p_2 is unstable, while the right-hand side is stable. The behaviour of the structure is thus asymmetric with respect to φ: the post-buckling load-bearing capacity is decreasing at positive φ, and increasing at negative φ.

Due to the complicated structure of Eq. (1-11), the approximate equilibrium path p_2, characterizing the initial post-buckling behaviour, becomes even more important. p_2 can be constructed by developing the trigonometric and square root functions into power series.

$$\cot\varphi \approx \frac{1}{\varphi} - \frac{\varphi}{3},$$

$$\frac{1}{\sqrt{1 + \sin\varphi}} \approx 1 - \frac{\varphi}{2} + \frac{3}{8}\varphi^2.$$

By so doing, and neglecting the powers of φ, we obtain

$$\frac{2P}{lc} = 1 - \frac{3}{4}\varphi, \tag{1-12}$$

which is represented by the inclined straight line p'_2 in *Fig 1-3b*.

1.1.4 The degenerating case of the symmetric bifurcation

GÁSPÁR [1984] has proven that the post-buckling load-bearing capacity of the structure shown in *Fig. 1-4a* is exactly constant. That is, the moment equilibrium equation around A on a bar rotated by the angle φ gives:

$$Pl\sin\varphi - cl(1 - \cos\varphi)l\sin\varphi - cl\sin\varphi \cdot l\cos\varphi = 0, \tag{1-13a}$$

or

$$P\sin\varphi = cl\sin\varphi. \tag{1-13b}$$

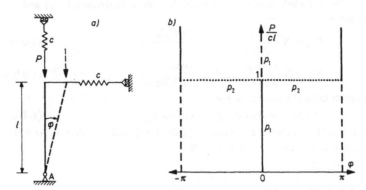

Fig. 1-4 The degenerating case of the symmetric bifurcation. *a*) Model, *b*) equilibrium paths

One solution is $\varphi = 0$ (path p_1 in *Fig. 1-4b*), and the other is $P = cl$, defining the horizontal path p_2 which again bifurcates at $\varphi = \pm\pi$ into a stable and an unstable branch.

1.1.5 General remarks

It is worthwhile to observe that, in the cases of symmetric bifurcation, when developing the deformation into series, we had to take terms of the second degree in φ into account, because the linear terms disappeared, and the post-critical behaviour has been determined by the second degree term. With the asymmetric bifurcation, however, the term linear in φ became dominant, and it was not necessary to take further terms into account.

If we further simplified the description of the phenomenon and neglected the first (linear or quadratic) term containing φ, then the formulas would give a constant post-critical load-bearing capacity in all three cases (*Fig. 1-4*), that is, we would not obtain any information on the post-critical behaviour.

On the other hand, it would not make much sense to take, in addition to the terms appearing in the expressions of the approximate equilibrium paths p'_2, further terms of higher degree into account, because they would not basically modify the *initial* character of the post-critical behaviour. However, in exceptional cases, it may happen that from the expression of the post-critical equilibrium path, developed into a power series, the linear, quadratic, etc., terms are missing. In such cases the first nonzero term containing φ on a higher power (and its sign) decides the character of the initial post-critical behaviour.

1.2 THE IMPERFECTION SENSITIVITY OF THE STRUCTURES

All cases treated so far referred to the so-called central compression, i.e. the structure had no initial displacement (deformation) in the direction of buckling, and no such displacement came about under loads inferior to the critical one. The three kinds of post-critical behaviour of centrally compressed structures, shown in *Figs. 1-1, 1-2* and *1-3*, however, yield valuable information on the question of whether the initial imperfections (geometric inaccuracies of the structure, eccentricity of the load), always present in practice, influence the maximum load to be carried by the structure, and if so, to what extent?

To investigate the 'eccentically loaded' structures we choose the structure to be seen in *Fig. 1-4a*, which behaves under central load according to *Fig. 1-4b*. It is well known from the linear (elementary) theory of stability [TIMOSHENKO and GERE, 1961] that the initial amplitude w_0 (and any of the characteristic data) of a structure, having an initial imperfection of the same shape as its buckling shape, increases under the influence of a compressive force P, according to the curve of *Fig. 1-5*, to the value w:

$$w = \frac{w_0}{1 - \dfrac{P}{P_{cr}}} \qquad (1\text{-}14a)$$

with P_{cr} as the critical force. Expressing P/P_{cr} we obtain:

$$\frac{P}{P_{cr}} = 1 - \frac{w_0}{w}. \qquad (1\text{-}14b)$$

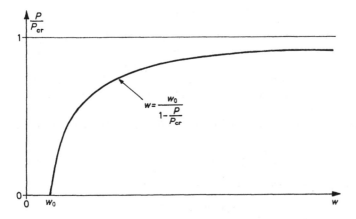

Fig. 1-5 The behaviour of the structure of Fig. 1-4 under eccentric loading

Instead of an exact derivation we assume that the 'eccentric' behaviour of the 'centrally loaded' structures of *Figs. 1-1b, 1-2b* and *1-3b* can be obtained by distorting the characteristic curve of *Fig. 1-5* in the same ratio as the diagram describing the post-critical behaviour of the centrally loaded structure differs from the horizontal straight line of *Fig. 1-4* or *1-5*. In other words: the curves of *Figs. 1-1b, 1-2b* and *1-3b*, referring to perfect structures, have to be multiplied by the expression appearing at the right-hand side of Eq. (1-14b). By so doing we obtain the same results as with the derivations to be found in KOITER [1945] or CROLL and WALKER [1972], (see also Chapter 4).

Let us begin the investigation with the structure behaving as shown in *Fig. 1-1b*. Let the equation of the approximate equilibrium path p'_2 be (writing w instead of φ):

$$\frac{P}{P_{cr}} = 1 + k_1 w^2. \qquad (1\text{-}15)$$

Multiplying the right-hand side by $(1 - w_0/w)$:

$$\frac{P}{P_{cr}} = \left(1 - \frac{w_0}{w}\right)\left(1 + k_1 w^2\right) = 1 - \frac{w_0}{w} - k_1 w_0 w + k_1 w^2. \qquad (1\text{-}16)$$

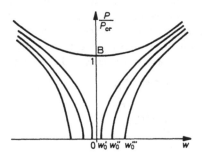

Fig. 1-6 The behaviour of the structure of Fig. 1-1a under eccentric loading

We plotted the curves corresponding to (1-16) for various w_0-s in *Fig. 1-6*. As can be seen, the curves have no peak points, but they monotonically increase with increasing (absolute value of) amplitude w. That is, the structure has – when maintaining the assumption of a perfectly elastic material – no 'maximum load-bearing capacity', but the load, which the structure is able to carry, can increase without any limits. Consequently, the structure is not sensitive to initial imperfections.

Let us investigate structures behaving as shown in *Fig. 1-2b*. The equation of the equilibrium path p'_2 is now:

$$\frac{P}{P_{cr}} = 1 - k_2 w^2. \qquad (1\text{-}17)$$

Multiplying this by the right-hand side of Eq. (1-14b) we obtain:

$$\frac{P}{P_{cr}} = 1 - \frac{w_0}{w} + k_2 w_0 w - k_2 w^2. \qquad (1\text{-}18)$$

The corresponding curves are shown in *Fig. 1-7a*. Every curve has a peak point, and these lie lower than point B, corresponding to the critical load of the perfect structure. Hence the initial imperfection reduces the maximum load P_{max}, which the structure is able to carry: the structure is sensitive to imperfections.

Let us determine how P_{max} depends on the value w_0 characterizing the imperfection. We set the first derivative of the right-side expression of (1-18) equal to zero:

$$\frac{w_0}{w^2} + k_2 w_0 - 2k_2 w = 0,$$

that is

$$2k_2\frac{w^3}{w_0} - k_2w^2 - 1 = 0. \tag{1-19}$$

The value of w pertaining to $P_{\max}$ is greater than w_0, since w has to increase considerably until $P_{\max}$ is reached. Assuming that $w(P_{\max}) \gg w_0$, the second term is much smaller than the first one, so that it can be neglected.

The w pertaining to $P_{\max}$ will thus be:

$$w = \sqrt[3]{\frac{w_0}{2k_2}}. \tag{1-20}$$

Introducing this into (1-18) we obtain $P_{\max}$ as a function of w_0:

$$\frac{P_{\max}}{P_{cr}} = 1 - 3\sqrt[3]{\frac{k_2}{4}}w_0^{\frac{2}{3}} + 2\left(\sqrt[3]{\frac{k_2}{4}}w_0^{\frac{2}{3}}\right)^2. \tag{1-21}$$

The last term is proportional to the square of the second one. We are interested above all in the 'initial' decrease of the load-bearing capacity, and, since this means small w_0-s, we can assume that $k_2w_0^2 \ll \sqrt{k_2}w_0$. Hence the third term can be neglected, and we arrive at the following relation:

$$\frac{P_{\max}}{P_{cr}} = 1 - 1.890\sqrt[3]{k_2}w_0^{\frac{2}{3}}. \tag{1-22}$$

The shape of the corresponding curve has been plotted in *Fig. 1-7b*.

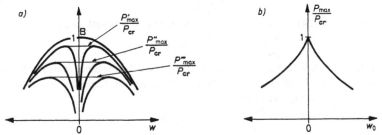

Fig. 1-7 The behaviour of the structure of Fig. 1-2a under eccentric loading. *a*) Equilibrium paths, *b*) imperfection sensitivity

Finally let us investigate the structure which behaves according to *Fig. 1-3b*. The equation of the path p_2' is

$$\frac{P}{P_{cr}} = 1 - k_3w. \tag{1-23}$$

Multiplying this by the right-hand side of Eq. (1-14*b*):

$$\frac{P}{P_{cr}} = 1 - \frac{w_0}{w} + k_3w_0 - k_3w. \tag{1-24}$$

In *Fig. 1-8a* we plotted some curves corresponding to (1-24) pertaining to different w_0-s. The curves on the left-hand side of the figure have the same character as those of *Fig. 1-6*, i.e. they have no peak points and ascend monotonically. The curves of the right-hand side are similar to those of *Fig. 1-7a*. We will determine their peak points.

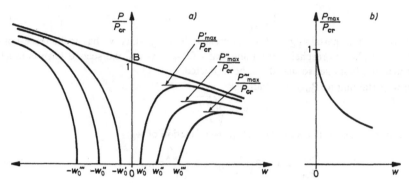

Fig. 1-8 The behaviour of the structure of Fig. 1-3a under eccentric loading. *a*) Equilibrium paths, *b*) imperfection sensitivity

Setting the first derivative of the right-hand side of Eq. (1-24) with respect to w equal to zero we obtain

$$\frac{w_0}{w^2} - k_3 = 0,$$

or

$$w = \sqrt{\frac{w_0}{k_3}}. \tag{1-25}$$

Introducing this into (1-24) we obtain P_{max} corresponding to the peak point:

$$\frac{P_{max}}{P_{cr}} = 1 - 2\sqrt{k_3}\sqrt{w_0} + k_3 w_0. \tag{1-26}$$

The third term is proportional to the square of the second one. Since we investigate in the first place the influence of small w_0-s, the third term can be neglected in comparison to the second one. So, finally, we obtain:

$$\frac{P_{max}}{P_{cr}} = 1 - 2\sqrt{k_3}\sqrt{w_0}, \tag{1-27}$$

see the curve plotted in *Fig. 1-8b*. It is worth while to note that this curve starts, due to the expression $\sqrt{w_0}$, with a vertical tangent, so that the drop in the load-bearing capacity is, at the beginning, very sharp.

It is obvious from the foregoing that from the diagram describing the post-buckling load-bearing capacity of the 'centrally' compressed, perfect structure it is possible to conclude on the degree of its imperfection sensitivity. Structures with a symmetric unstable bifurcation are less sensitive to imperfections than those with an asymmetric bifurcation, and the latter are the more sensitive, the steeper the straight line describing the decrease of the load-bearing capacity of the perfect structure descends (i.e. the greater the coefficient k_3).

1.3 LOSS OF STABILITY WITH A LIMIT POINT
(DIVERGENCE OF EQUILIBRIUM; SNAPPING THROUGH)

We still have to describe a fourth kind of the loss of stability, that is, of the post-buckling load-bearing capacity, which does not come about by bifurcation, but by a so-called limit point.

This phenomenon is very well described by the expression 'divergence of equilibrium' according to which, after reaching the limit point, the internal resistance of the structure grows slower than the influence of the external load: the two 'diverge' (*Fig. 1-9a*). In a broader sense, all structures belong here whose load-displacement diagram begins to descend after reaching a peak point. An example is the over-reinforced concrete beam, in which the gradual crushing of the concrete in the compressed zone results in the load-deflection diagram shown in *Fig. 1-9a*. This kind of failure is, however, not called 'loss of stability', because the basic feature of the instability phenomena, the influence of the deformation on the internal forces, is absent. That is, the 'real' kinds of loss of stability are always caused by the modification of the internal forces by the deformation, which is mathematically expressed by the fact that the equilibrium equations have to be written for the deformed, buckled shape of the structure. (The equations written for the undeformed state would always show a stable state of equilibrium.) For the bent reinforced concrete beam, however, we can write the equations for the undeformed shape, and still obtain the diagram of *Fig. 1-9a*. Hence the failure is not caused by the influence of the deformation on the internal forces.

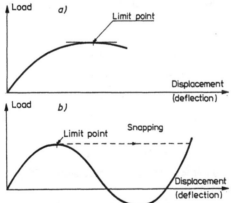

Fig. 1-9 Loss of stability with a limit point. *a*) Divergence of equilibrium, *b*) snapping through

A special case of the loss of stability with a limit point is when the post-critical load-bearing curve, after having a descending section, again begins to ascend. This means that the structure, reaching the limit point, jumps over ('snaps through') to the ascending, stable section of the curve (*Fig. 1-9b*). This phenomenon can be studied on the model consisting of two rigid bars and two springs, shown in *Fig. 1-10a*. The structure must be sufficiently flat ($f \ll 2l$) in order to make the corresponding approximations acceptable.

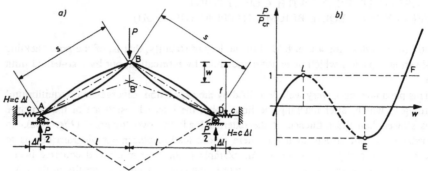

Fig. 1-10 Snapping through. *a*) Model, *b*) equilibrium path

The length of the inclined bars is:

$$s = \sqrt{l^2 + f^2} \approx l \left(1 - \frac{1}{2}\frac{f^2}{l^2}\right). \tag{1-28}$$

It follows from Eq. (1-28) that the maximum value of the horizontal displacement Δl of the hinges A and D is $f^2/(2l)$, which is negligible in comparison with l, according to our asssumption on the flatness of the structure.

The relation between the vertical displacement of the middle hinge B and the horizontal displacements Δl is given by *Pythagoras'* theorem written for the changed shape:

$$(l + \Delta l)^2 + (f - w)^2 = s^2 = l^2 + f^2$$

or

$$2fw - w^2 = 2l \cdot \Delta l + (\Delta l)^2. \tag{1-29}$$

Here we can neglect $(\Delta l)^2$ in comparison with $2l \cdot \Delta l$, and thus:

$$\Delta l = \frac{f}{l}w - \frac{w^2}{2l}. \tag{1-30}$$

The moment equation of the forces acting on the left half of the structure with respect to the displaced hinge B' is:

$$\frac{P}{2}(l + \Delta l) - H(f - w) = 0. \tag{1-31}$$

Introducing here the relation $H = c \cdot \Delta l$ and the expression (1-30), we obtain the load P as a function of the displacement w:

$$P = c \left(\frac{2f^2}{l^2}w - \frac{3f}{l^2}w^2 + \frac{w^3}{l^2}\right). \tag{1-32}$$

To obtain its maximum, we differentiate P with respect to w, from which

$$w\,(P_{\max}) = f\left(1 - \frac{1}{\sqrt{3}}\right) = 0.423f \tag{1-33a}$$

is obtained. Introducing this into (1-32) yields:

$$P_{\max} = 0.385c\frac{f^3}{l^2}. \tag{1-33b}$$

We have plotted the curve $P(w)$ in *Fig. 1-10b*. After reaching the critical load P_{cr}, the structure 'snaps through' along the dashed line into the lower position B (provided the load decreases accordingly); or if the load remains constant, the structure follows the horizontal dashed line and jumps into the position F. This structure essentially behaves as the 'imperfect' structures of *Figs. 1-7* and *1-8*. There is, however, a fundamental difference between them: the structure of *Fig. 1-10* has no 'central' (or 'perfect') case. Although we can speak of imperfection sensitivity (e.g. the original position of the hinge B may be lower, thus reducing the load causing snapping through), but we cannot refer the imperfection itself to the 'central' case; we can thus only speak of a sensitivity to the *increment* of imperfection.

1.4 THE INFLUENCE OF PLASTICITY

Although in this book we mainly deal with elastic structures, nevertheless we find it expedient to present the influence of plasticity on the loss of stability.

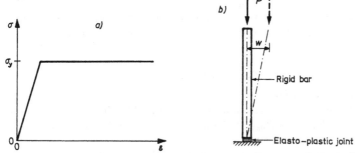

Fig. 1-11 The influence of plasticity on the loss of stability. *a*) Elastic–perfectly plastic material model, *b*) the structure investigated

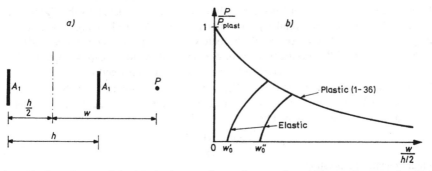

Fig. 1-12 The influence of plasticity in the case of a two-flange section. *a*) The two-flange section, *b*) the reducing effect of plasticity on the load-bearing capacity

For the sake of simplicity we confine ourselves to the elastic-plastic material model corresponding to *Fig. 1-11a*, and we assume that the yielding capability of the material is infinite (elastic–perfectly plastic material).

The structure investigated in *Fig. 1-11b* consists of a rigid bar which is supported by a very short joint made of elastic-perfectly plastic material. This joint be first a 'two-flange' section depicted in *Fig. 1-12a*, whose flanges are thin enough to assume a constant normal stress distribution all over the thickness; and whose web does not take any normal stresses parallel to the bar axis, it only serves to rigidly connect the two flanges (the cross section of *van der Neut*, see also in Sect. 3.4). The behaviour of this section is characterized, for our purposes, by the fact that if the normal stress σ, parallel to the bar axis, reaches in one of the flanges the yield stress σ_y, the section becomes plastic at once, without any transition.

Let us write the moment equilibrium equation for the left flange, assuming that in the right flange $\sigma = \sigma_y$ arises:

$$P\left(w + \frac{h}{2}\right) = A_1 \sigma_y h. \tag{1-34}$$

Introducing the notation

$$P_{\text{plast}} = 2A_1\sigma_y, \tag{1-35}$$

representing the load-bearing capacity of the cross section under central compression, the maximum load to be carried by the cross section, as a function of imperfection w, can be written as

$$\frac{P}{P_{\text{plast}}} = \frac{1}{1 + \frac{w}{h/2}}. \tag{1-36}$$

This relation is shown in *Fig. 1-12b*.

Hence the behaviour of the perfect structure is described by the vertical straight line starting from O, until P reaches P_{plast}. From here on the descending curve according to (1-36) represents the equilibrium path.

If the structure of *1-11b* has an initial eccentricity w_0, then the curve $P(w)$ starts according to the diagram shown in *Fig. 1-6* (instead of which we may take also that of *Fig. 1-5*, as an approximation), but when this intersects the plastic load-bearing curve (1-36), from this point on, this latter determines the load-bearing capacity, which is decreasing.

Hence the structure does not behave exactly as the curves depicted on the right-hand side of *Fig. 1-8a* predict, but similarly. The phenomenon sketched in *Fig. 1-12b* is essentially independent of whether the elastic behaviour of the structure corresponds to *Fig. 1-5, 1-6, 1-7a* or *1-8a*. Plasticity renders the post-critical load-bearing capacity decreasing in every case, i.e. the structure becomes sensitive to imperfections.

Let us now investigate the model of *Fig. 1-11b* having a joint with a 'solid' cross section (*Fig. 1-13a*). First we determine the value P_{elast} of P, which causes just the yield stress $\sigma = \sigma_y$ in one of the extreme fibres (*Fig. 1-13b*):

$$\sigma_y = \frac{P_{\text{elast}}}{bh} + \frac{P_{\text{elast}}w}{bh^2/6}. \tag{1-37}$$

Introducing again the notation

$$P_{\text{plast}} = bh\sigma_y, \tag{1-38}$$

we obtain the relation sought for:

$$\frac{P_{\text{elast}}}{P_{\text{plast}}} = \frac{1}{1 + 3\dfrac{w}{h/2}},$$

(1-39)

see the lower curve in *Fig. 1-13d*.

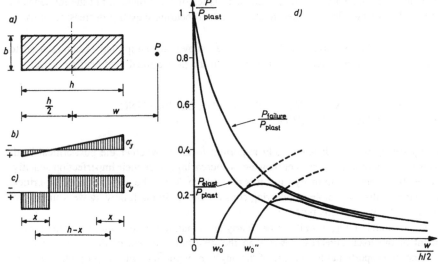

Fig. 1-13   Influence of plasticity in the case of a solid cross section. *a*) The solid section investigated, *b*) the stress distribution causing yield stress in the extreme fibre, *c*) the stress distribution causing plastification of the whole cross section, *d*) the reducing effect of plasticity on the load-bearing capacity

Let us now determine the value P_{failure} of P, which makes, with a given eccentricity w, the cross section entirely plastic (*Fig. 1-13c*). The projection equation is:

$$P_{\text{failure}} = b(h - 2x)\sigma_y,$$

(1-40)

and the moment equation with respect to the centroid of the cross section is:

$$P_{\text{failure}}w = bx\sigma_y(h - x).$$

(1-41)

Eliminating x and using Eq. (1-38), we obtain:

$$\frac{P_{\text{failure}}}{P_{\text{plast}}} = \sqrt{\left(\frac{w}{h/2}\right)^2 + 1} - \frac{w}{h/2}.$$

(1-42)

See the upper curve of *Fig. 1-13d*.

The gradual plastification of the cross section occurs at P values between the two curves.

If the structure has some initial eccentricity w_0 and we load it gradually, then it deforms according to a curve pertaining to one of the elastic cases treated earlier (*Fig. 1-13d*). When this curve intersects the curve $P_{\text{elast}}/P_{\text{plast}}$, plastification begins; so that the load-bearing capacity becomes less than if the structure remained elastic. It is also obvious that the load-bearing capacity cannot be higher than P_{failure}. The curve describing the behaviour of the structure thus deviates from the 'elastic' curve (indicated by dashed line) when reaching

P_{elast}, and asymptotically approaches the limit curve $P_{failure}$. The behaviour of the structure will thus be similar to the right-hand side of *Fig. 1-8a*: the load-bearing capacity becomes decreasing, and the imperfection sensitivity is approximately described by the curve of *Fig. 1-8b*.

Both cases shown belong to the category of composite stability phenomena: the *initial* behaviour is determined by the elastic characteristics of the structure, but the influence of plasticity, becoming effective *later*, renders the load-bearing capacity decreasing in every case.

We would like to emphasize that both cases shown are simplified examples used to illustrate the decreasing effect of the plasticity on the post-critical load-bearing capacity.

1.5 SOME PRACTICAL POINTS OF VIEW FOR ESTIMATING THE POST-CRITICAL BEHAVIOUR

For engineering practice it is basically important to know whether the post-critical load-bearing capacity of a structure is decreasing or increasing, since their imperfection sensitivity depends on this fact. In many cases it is possible to decide the nature of the post-critical behaviour visually, by simple engineering considerations. In the following we want to give some points of view to this purpose.

On the model of *Fig. 1-1a* it can be clearly seen that, during the rotation φ, the turning moment of the load increases slower than the restoring moment of the spring. That is, the former is proportional to the sine of the angle φ, while the latter is proportional to the angle φ itself. Consequently, the post-buckling load-bearing capacity must be of increasing nature.

In the case of *Fig. 1-2a* the situation is opposite: here the turning moment grows faster than the restoring one, since the former is proportional to $\sin\varphi$, while the latter to $\sin\varphi \cdot \cos\varphi$. The post-critical load-bearing capacity will thus be of decreasing nature.

In the case of elastic bar structures the post-buckling load-bearing capacity is – with a good approximation – constant. The restoring moment is exerted by the internal bending moment of the bars, due to their bending rigidity. If this rigidity is reduced by some effect, the *post-buckling load-bearing capacity* becomes *decreasing*. Such an effect is the *plasticity* of the material, which causes the internal bending moment to increase with increasing curvature slower than in the elastic range or not at all. This was shown in Sect. 1.4.

Further examples for the decreasing post-buckling load-bearing capacity are the shell structures. In Sect. 3.5 we will investigate the post-critical behaviour of the axially compressed cylindrical shells.

In complex structures the interaction of local and global buckling also renders the post-critical load-bearing capacity in most cases decreasing. We will show three examples for this in Sects 3.2–3.4.

Let us now show some examples for the *increasing post-buckling load-bearing capacity*. Such are, as a rule, *the plane plates*. We will investigate how this increase in the load-bearing capacity comes about.

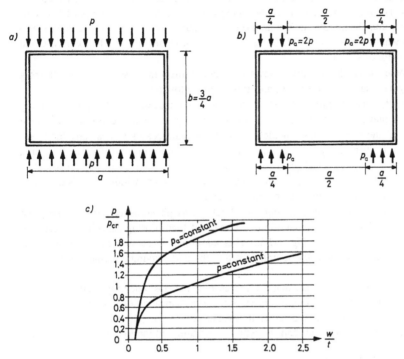

Fig. 1-14 Investigation of the causes of the increasing post-buckling load-bearing capacity. *a*) The plane
plate investigated, *b*) the modified load arrangement, *c*) the behaviour of the initially curved plate

Let us consider the plane plate shown in *Fig. 1-14a*, compressed in one direction, and
simply supported all along its four sides. It will be assumed that these hinged supports
allow the plate edges to shift freely parallel to the edge.

On the basis of the computations of SÄTTELE, RAMM and FISCHER [1981], performed
with the nonlinear finite element method, we show in *Fig. 1-14c* the load-deflection diagram of the initially curved plate of the side ratio $b/a = 3/4$. We have plotted on the
horizontal axis the ratio of the deflection w of the middle point of the plate to the plate
thickness, and on the vertical axis the ratio of the uniformly distributed load p taken by the
plate to the classical critical (bifurcation) load p_{cr}.

From the two curves, that denoted by p gives the behaviour under the load uniformly
distributed according to *Fig. 1-14a*, and the other curve denoted by p_a represents the behaviour under the modified load arrangement shown in *Fig. 1-14b*, which is that the load
p is taken off the middle section $a/2$ of the plate and transferred to the two side sections
$a/4 - a/4$; the load intensity thus increases here to $p_a = 2p$.

From *Fig. 1-14c* we can read off the following.

The curve p shows an increase in the load-bearing capacity of about 60% as compared
with the bifurcation load of the perfectly plane plate. (From here on the plasticity, also

taken into account, renders the load-bearing capacity decreasing.) This increase can be explained by the '*membrane action*' of the plate: due to the deflection, the plate becomes doubly curved, thus the middle surface undergos elongations and distortions, causing membrane forces which hinder the further increase of deflection, see e.g. in [TIMOSHENKO and WOINOWSKI-KRIEGER, 1959].

On the other hand, the p_a curve shows that the (further) increase of the load-bearing capacity may have another reason too, and this is the following.

According to TIMOSHENKO and GERE [1961] the maximum post-buckling load-bearing capacity of the plane plate shown in *Fig. 1-14a* can be approximately determined from the assumption that the central, most buckled, part of the plate ceases to bear any load, and only the two side strips carry the total load. The load-bearing capacity of these strips are, however, greater than the critical (bifurcation) load of the whole plate (this is the so-called *paradox of plate buckling*). Thus, by appropriately choosing the width of the side strips, we can determine a 'maximum' load-bearing capacity.

The load arrangement shown in *Fig. 1-14b* takes this phenomenon into account, and this is also the explanation of the additional increase of load-bearing capacity corresponding to the curve p_a, shown in *Fig. 1-14c*. This additional increase can, however, come about only if the load can actually be transferred to the parts having a greater load-bearing capacity.

In summary we can thus state that the increase of the post-buckling load-bearing capacity can be caused by two circumstances: first, a certain additional stiffness becomes active in the structure itself during larger deformations; second, the structure has certain parts which can carry more load than the whole structure, and the load can be transferred to these parts.

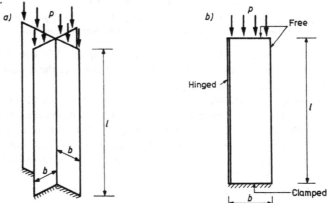

Fig. 1-15 Example of the investigation of increasing post-buckling load-bearing capacity. *a*) The bar investigated, *b*) one 'wing' of the bar

On the basis of these principles we can easily decide whether the post-critical load-bearing capacity of the thin-walled bar shown in *Fig. 1-15a* increases during torsional buckling or not.

During torsional buckling the four 'wings' of the bar buckle as long plates simply supported along the bar axis, while their outer edges are free. One of these plates is depicted in *Fig. 1-15b*. Its critical (bifurcation) load is inversely proportional to the width b (paradox of plate buckling), see TIMOSHENKO and GERE [1961]. That is, if the load can be rearranged in such a way that it 'comes closer' to the middle point of the cross section, then the plates – having an 'effective' width smaller than b – will be able to carry more load, i.e. the post-buckling load-bearing capacity will be increasing. In addition, the original plane plates become doubly curved during buckling, causing in-plane deformations and pertaining membrane forces in the middle surfaces of the plates, hindering the buckling deformation. Thus the bar has an additional load-bearing capacity independently of the rearranging of the load.

According to the exact analysis [TIMOSHENKO and GERE, 1961] the common bars show a slight increase of load-bearing capacity after buckling, but this is so small that it can be neglected, and we can consider their post-buckling load-bearing capacity practically as *constant.*

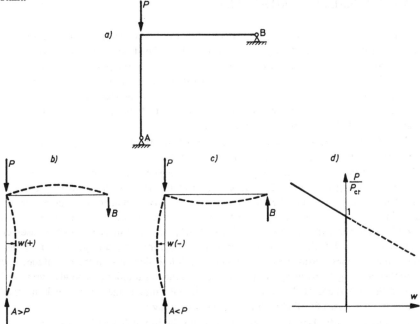

Fig. 1-16 Example of the 'asymmetric' post-critical behaviour. *a*) The frame investigated, *b*) the buckling shape pertaining to decreasing post-buckling load-bearing capacity, *c*) the buckling shape pertaining to increasing post-buckling load-bearing capacity, *d*) the post-critical behaviour of the frame

Let us investigate the frame structure shown in *Fig. 1-16a*. It is very easy to understand without any computation that *its post-critical behaviour is asymmetric.*

If the frame buckles according to *Fig. 1-16b*, then at point B – according to the bent shape of the beam – a support force pointing downwards is needed. Thus at point A a support force greater than P arises $(A = P + B)$, and this will also be the compressive force in the column. Assuming that the post-critical load-bearing capacity of the column is constant, then after buckling P must decrease in order to keep A constant (see the dashed line on the right-hand side of *Fig. 1-16d*).

If, however, the frame buckles as shown in *Fig. 1-16c*, then the support force B points upwards and, consequently, $A < P$ $(A = P - B)$, so that P must increase during buckling in order to keep A (i.e. the compressive force in the column) constant. Hence the post-buckling load-bearing capacity of the frame is now increasing, as shown by the solid line on the left-hand side of *Fig. 1-16d*.

The detailed investigation of the frame structure shown in *Fig. 1-16a* can be found in [KOITER, 1967], [THOMPSON and HUNT, 1973], [BRUSH and ALMROTH, 1975], [KOUNADIS, GIRI and SIMITSES, 1977].

1.6 EVALUATION OF BUCKLING EXPERIMENTS BY THE GENERALIZED SOUTHWELL PLOT

The critical load of structures (i.e. the maximum load they can carry) is determined in many cases by model tests, during which we may load the structure up to failure, but it is important to know, besides the critical (failure) load, whether the post-critical load-bearing capacity of the structure is constant, increasing or decreasing, because the safety factor can only be properly assessed with knowledge of all these. Also the requirement may arise not to load the model – at least in some loading cases – up to failure, in order to save money, but nevertheless to be able to establish these data.

Since the real models are never perfect, we obtain from the loading test one of the curves $P(w)$ of *Figs. 1-5, 1-6, 1-7, 1-8, 1-9b* or *1-12b*.

These curves, however, cannot give a reliable answer to the foregoing questions, because the ascending sections of the curves $P(w)$ of the various cases are very similar to each other, so that we cannot be sure which one we have obtained and, moreover, their peak point, i.e. the critical load, cannot be reliable extrapolated. This problem is solved by *Southwell*'s plot [TIMOSHENKO and GERE, 1961], which transforms, in the cases corresponding to *Fig. 1-5*, the curved load-deflection diagrams of the imperfect structures to straight lines. In these cases the buckling deformation w depends on the amplitude of initial crookedness w_0 as given by Eq. (1-14). In this, the assumption, made also previously, was used that the shape of the initial curvature is the same as that of the buckling mode, which is in most cases fulfilled with a sufficient accuracy.

Introducing the displacement δ measured from the initial (imperfect) shape:

$$w - w_0 = \delta \tag{1-43}$$

Eq. (1-14) can be rearranged in the following form:

$$\frac{\delta}{P} = \frac{\delta + w_0}{P_{cr}}. \tag{1-44}$$

This is the equation of a straight line in the co-ordinate system δ, δ/P (*Fig. 1-17*). That is, if we measure δ-s belonging to several values of P, and plot these data in the co-ordinate system δ, δ/P, we obtain a straight line, the cotangent of which yields the critical load. The inclination of a straight line can be determined with a greater accuracy than an extrapolated peak point, so that *Southwell*'s plot makes the determination of the critical load more reliable.

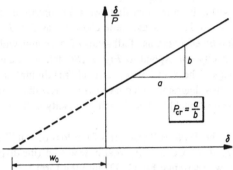

Fig. 1-17 The *Southwell* diagram valid for structures with constant post-buckling load-bearing capacity

Let us investigate now which shape the diagram of *Southwell* will assume in the case of structures having a nonconstant post-critical load-bearing capacity [KOLLÁR, 1972].

The $P(w)$ diagram of a structure having an *increasing* post-buckling load-bearing capacity according to *Fig. 1-6* is described by Eq. (1-16). Rearranging this according to (1-14), making use of (1-43), we arrive at the relation

$$\frac{\delta}{P} = \frac{\delta + w_0}{P_{cr}} - \frac{k_1}{P} (w_0 + \delta)^2 \delta. \tag{1-45}$$

All elements of the second term of the right-hand side are positive, that is, with increasing δ we have to subtract more and more from the equation of *Southwell*'s straight line (1-44), so that we arrive at a diagram *curved downwards* (*Fig. 1-18a*).

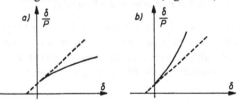

Fig. 1-18 The shape of the *Southwell* diagram in the cases of varying post-buckling load-bearing capacity.
a) Increasing load-bearing capacity, *b*) decreasing load-bearing capacity

The equation (1-18) of the $P(w)$ diagram of structures with a *decreasing* post-buckling load-bearing capacity according to *Fig 1-7a* can be rearranged into the form

$$\frac{\delta}{P} = \frac{\delta + w_0}{P_{cr}} + \frac{k_2}{P} (w_0 + \delta)^2 \delta. \tag{1-46}$$

Here the second term of the right-hand side increases the first term corresponding to *Southwell*'s straight line, so that we obtain a diagram *curved upwards* (*Fig. 1-18b*).

The equation (1-24) of the right-hand side of the diagram of a structure with asymmetric behaviour corresponding to *Fig. 1-8a*, representing a decreasing post-buckling load-bearing capacity, can be rearranged into the form:

$$\frac{\delta}{P} = \frac{\delta + w_0}{P_{cr}} + \frac{k_3}{P}(w_0 + \delta)\delta. \tag{1-47}$$

This differs from Eq. (1-46) only in the fact that $(w_0 + \delta)$ appears in it to the first power, but it also *curves upwards*, that is, it has the same character as *Fig. 1-18b*.

As we can see, the *Southwell* diagrams of all kinds of structures with a decreasing post-buckling load-bearing capacity correspond to *Fig. 1-18b*, but we cannot decide to which group the structure belongs. The value of the critical (maximum) load can also not be determined reliably from this diagram. The curved *Southwell* diagrams can thus be used only to decide whether the post-buckling load-bearing capacity of the structure is increasing or decreasing.

We can also construct the *Southwell* diagram of the loss of stability with a limit point (snapping through, see *Fig. 1-9*). Considering that now $w \equiv \delta$ (the displacement measured from the initial position), we rearrange Eq. (1-32) into the form

$$\frac{\delta}{P} = \frac{l^2}{c\left(2f^2 - 3f\delta + \delta^2\right)}. \tag{1-48}$$

Up to the end of the snapping through process $\delta/f < 2$, so that the denominator is always positive, consequently we obtain an ascending curve, see. *Fig. 1-19*.

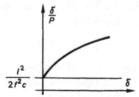

Fig. 1-19 The *Southwell* diagram of the loss of stability with a limit point

We still have to consider the cases of *Figs. 1-12b* and *1-13b*. It may happen that the *initial* section of the $P(w)$ diagram corresponds to *Fig. 1-5* (or *1-6*), i.e. to an increasing post-buckling load-bearing capacity, although the *plastification, occurring later*, causes the load-bearing capacity to decrease. However, the *Southwell* diagram gives no information on this phenomenon.

2

Summation theorems concerning critical loads of bifurcation

TIBOR TARNAI

2.1 ON THE SUMMATION THEOREMS

Summation formulae are used in the theory of elastic stability so that approximate estimates of the critical load factors of a complex problem are obtained by combining the load factors of subproblems in different ways. If the critical load factors are directly added, then the formula is called a *Southwell* type formula. If the reciprocals of the critical load factors are added, then the formula is called a *Dunkerley* type formula. First we consider examples of the Southwell and Dunkerley formulae.

Example 2-1: Consider the torsional buckling of a thin-walled bar of height H with the lower end built-in and the upper end free, subjected to a concentrated vertical force P at the free end. The equilibrium of the deformed bar can be described by the differential equation

$$EI_\omega \Phi'''' + \left(Pi_p^2 - GI_t \right) \Phi'' = 0, \qquad (2\text{-}1)$$

with the boundary conditions

$$\Phi(H) = 0, \quad \Phi'(H) = 0, \quad \Phi''(0) = 0, \quad GI_t\Phi'(0) - EI_\omega\Phi'''(0) = Pi_p^2\Phi'(0). \qquad (2\text{-}2)$$

Here Φ is the rotation of the cross section, i_p is the radius of gyration, and prime denotes differentiation along the axis of the bar. The torsional stiffness of the cross section is composed from two parts: the warping stiffness EI_ω and the *Saint-Venant* torsional stiffness GI_t. Let us consider first the case where the bar has warping stiffness only ($GI_t = 0$). In this case the critical load is [TIMOSHENKO and GERE, 1961]:

$$P_{cr,1} = \frac{1}{i_p^2} \frac{\pi^2 EI_\omega}{4H^2}.$$

Let us consider now the case where the bar has *Saint-Venant* torsional stiffness only ($EI_\omega = 0$). In this case the critical load is

$$P_{cr,2} = \frac{GI_t}{i_p^2}.$$

The *Southwell* summation approximates the critical load of the original bar by the expression

$$P_S = P_{cr,1} + P_{cr,2}. \qquad (2\text{-}3)$$

The exact value of P_{cr} is given by TIMOSHENKO and GERE [1961]:

$$P_{cr} = \frac{1}{i_p^2}\left(GI_t + \frac{\pi^2 EI_\omega}{4H^2}\right).$$

It can be seen that in this example $P_S = P_{cr}$, that is, the *Southwell* formula (2-3) provides the exact value of the critical load.

Example 2-2: Fig. 2-1 shows a simply supported thin-walled bar of doubly symmetric ($I_x > I_y$) open cross section, subjected to an axial force N and a couple M whose plane is perpendicular to the x axis. We want to determine the values of the N, M pairs by the *Dunkerley* formula under which the bar comes into the critical state. Consider first the case where the bar is subject to the force N only ($M = 0$). Suppose that the critical load for torsional buckling is greater than the critical load for flexural buckling about the y axis. In this case the bar buckles in the weak direction, that is, about the y axis, and the *Euler* critical load is

$$N_{y,cr} = \frac{\pi^2 EI_y}{l^2}.$$

Let us consider now the case where the bar is subjected to the couple M only ($N = 0$). The critical value of the couple can be obtained from TIMOSHENKO and GERE [1961]:

$$M_{cr} = \pm\sqrt{\frac{\pi^2 EI_y}{l^2}}\sqrt{GI_t + \frac{\pi^2 EI_\omega}{l^2}}.$$

Since N and M are of different dimension, the *Dunkerley* formula should be written in dimensionless form as follows:

$$\frac{N}{N_{y,cr}} + \frac{M}{M_{cr}} = 1, \tag{2-4}$$

where M and M_{cr} should be considered with the same sign. To assess the accuracy of the *Dunkerley* formula (2-4), we have to determine the exact relationship between the critical pair N, M by solving the set of differential equations

$$EI_y u'''' + M\Phi'' + Nu'' = 0,$$

$$EI_\omega \Phi'''' - GI_t \Phi'' + Mu'' + Ni_p^2\Phi'' = 0$$

with the boundary conditions

$$u(0) = u(l) = u''(0) = u''(l) = \Phi(0) = \Phi(l) = \Phi''(0) = \Phi''(l) = 0$$

where u is the displacement of the centroid of the cross section in the x direction. If $M_{\Phi,cr}$ denotes the critical load of the bar for pure torsional buckling then the desired expression has the form

$$\left(1 - \frac{N}{N_{y,cr}}\right)\left(1 - \frac{N}{N_{\Phi,cr}}\right) - \frac{M^2}{M_{cr}^2} = 0 \tag{2-5}$$

[TIMOSHENKO and GERE, 1961] where

$$N_{\Phi,cr} = \frac{1}{i_p^2}\left(GI_t + \frac{\pi^2 EI_\omega}{l^2}\right).$$

Let $N \le N_{y,cr} < N_{\Phi,cr} < \infty$, then in the co-ordinate system M/M_{cr}, $N/N_{y,cr}$, Eq. (2-4) represents a straight line and Eq. (2-5) represents a hyperbola, only one of whose branches should be considered (*Fig. 2-1b*). In the interval $0 < M/M_{cr} < 1$, points of the *Dunkerley* line (2-4) lie below the curve (2-5) inside the stability domain.

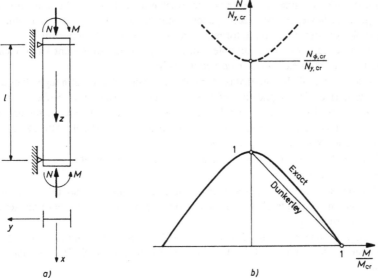

Fig. 2-1 Lateral buckling of a bar subjected to an axial force and a couple at its ends. *a*) The layout of the loads, *b*) diagram of the loads and its approximation by the *Dunkerley* line

Therefore the *Dunkerley* formula is on the safe side for any value of M/N but subject to the constraint $N > 0$. If N and M denote the exact values of the normal force and bending moment in a critical state, then instead of (2-4) the inequality below will be valid

$$\frac{N}{N_{y,cr}} + \frac{M}{M_{cr}} \geq 1. \tag{2-6}$$

If $N_{\Phi,cr} = \infty$, then the hyperbola is transformed into a parabola and the *Dunkerley* formula conservatively approximates with a very large error. If $N_{\Phi,cr} = N_{y,cr}$, then the hyperbola degenerates into two crossing straight lines and the *Dunkerley* formula is exact. [The case $N_{\Phi,cr} < N_{y,cr}$ was excluded.]

When the ratio of M to N is fixed, for instance, force N acts with an excentricity e ($M = eN$), the loading force system depends only on one parameter: N. From (2-6) for N_{cr} we obtain

$$\frac{1}{N_{y,cr}} + \frac{1}{\frac{1}{e}M_{cr}} \geq \frac{1}{N_{cr}}, \tag{2-7}$$

which is a better known form of the *Dunkerley* formula. It is seen that the summation applies to the reciprocals of the critical loads.

The practical advantage of the summation formulae is that there exist solutions to the subproblems, or it is easy to determine them, which can be used to obtain an approximate solution for the original problem which would be difficult to obtain directly.

With the development of computer technology, the significance of the summation formulae in high accuracy calculations has decreased but, in everyday design practice, their significance has not decreased at all. These formulae are widely used in practice.

In both examples here we could say whether the *Southwell* and the *Dunkerley* formulae are or are not on the safe side, as we know the exact critical loads of the structures in question. In general, however, the exact solution is not known. Therefore it is useful to know the conditions under which these formulae are conservative. The aim of this chapter is to present these conditions where possible. For some problems we do not know them; and for some types of problems *no such conditions exist.*

The algebraic forms of the summation formulae are very simple, although they are often difficult to derive. The application of functional analysis (theory of Hilbert spaces and linear operators defined in them) can be effective in this respect. The mathematical base of these formulae can be found, for example, in MIKHLIN [1970], AKHIEZER and GLAZMAN [1977], and WEINBERGER [1974, 1995].

Most of the mathematical terms we need were defined by WEINBERGER [1995]. Here we introduce some additional ones.

Definition: The set X is dense in the set Y if any arbitrarily small neighbourhood of any element of Y contains elements of X. (The set of rational numbers is dense in the set of real numbers, for instance.)

Definition: Consider a linear operator S defined in a Hilbert space H. Let (u,v), u, $v \in H$ denote the scalar product in H, and D_S the domain of definition of S. We say that the linear operator S is

(*i*)   symmetric, if $(Su,v) = (u,Sv)$ for all u, $v \in D_S$ and D_S is dense in H;

(*ii*) positive, if $(Su,u) > 0$ for all $u \neq 0$, $u \in D_S$;

(*iii*) positive definite, if there exists a number $\alpha > 0$ such that $(Su,u) \geq \alpha (u,u)$, $u \in D_S$.

Definition: The spectrum of a linear operator A is the set of numbers λ such that the operator $A - \lambda I$ has no unique bounded inverse. (In a general case, the identity operator I is replaced by a linear operator B and we refer to the spectrum of an operator pair.) An operator (or operator pair) comprises a discrete spectrum if the spectrum consists only of eigenvalues.

Theorem: Let A and B be symmetric linear operators in H such that (Bu,u) is completely continuous with respect to (Au,u) (the definition of the term 'completely continuous' is given by WEINBERGER [1974, 1995]), then the operator pair $A - \lambda B$ is of discrete spectrum. (We do not prove this here.)

The proofs of the forthcoming theorems will not be presented here. They can be found in the Hungarian edition of this book [KOLLÁR, 1991], or in TARNAI [1995].

2.2 THE SOUTHWELL THEOREM

In the theory of structural stability, the theorem of SOUTHWELL [1922] is stated as: If the stiffness of a structure is composed of parts, then the smallest critical load parameter of the structure is not less than the sum of the smallest critical load parameters corresponding to the partial stiffnesses. The theorem can be stated purely in mathematical terms as follows.

Theorem 2-1: Let A and B symmetric linear operators in a Hilbert space H such that A is positive definite, B is positive and $D_A \subset D_B \subset H$. Suppose (Bu,u) is completely continuous with respect to (Au, u). Let $A = \sum_{i=1}^{n} A_i$ where the operators A_i $(i = 1, 2, \ldots, n)$ have the same properties as A. If λ_0 and λ_i denote the smallest eigenvalues of the eigenvalue problem

$$(A - \lambda B)u = 0 \tag{2-8}$$

and

$$(A_i - \lambda B)\, u = 0, \quad i = 1, 2, \ldots, n, \tag{2-9}$$

respectively, then we have

$$\lambda_0 \geq \sum_{i=1}^{n} \lambda_i. \tag{2-10}$$

Remarks: It is important in the theorem that B is positive. If B is not positive then the the eigenvalue problems may have also negative eigenvalues and so the statement in the theorem is not valid.

The theorem remains valid also for a degenerate case where (Bu, u) is not completely continuous with respect to $(A_i u, u)$ but $(A_i u, u)/(Bu, u) > 0$, $u \in D_{A_i}$ and λ_i is defined as the infimum of the *Rayleigh* quotient $(A_i u, u)/(Bu, u)$. This occurs, for instance, if for some i the spectrum of (2-9) is only one point, but it is an eigenvalue with infinite multiplicity as occurs in Example 2-1.

Example 2-3: Consider a column of variable cross section with the lower end built-in and the upper end free. Let the column be subject to the action of a distributed axial load given by the equation $q = q_1 (l - z)/l$ (*Fig. 2-2*). Let the variation of the second moment of area of the cross section be given by the equation

$$bI = I_1 + I_2 \frac{l-z}{l} + I_3 \left(\frac{l-z}{l}\right)^2 + I_4 \left(\frac{l-z}{l}\right)^3.$$

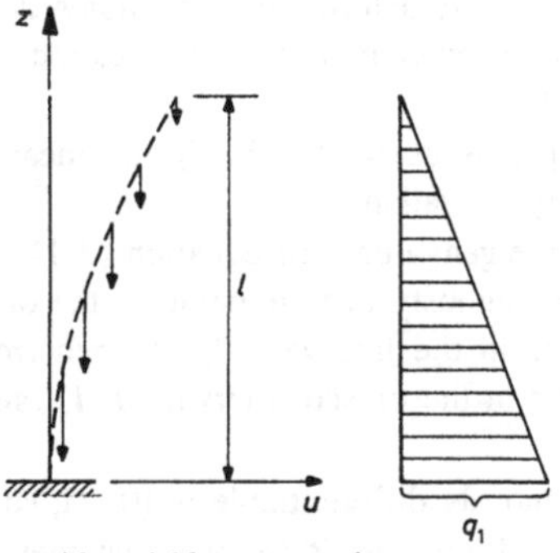

Fig. 2-2 Buckling of a cantiliver with variable cross section

The equilibrium of the bar can be described by the equation

$$\left(EIu''\right)'' + q_1 \left[\frac{(l-z)^2}{2l}u'\right]' = 0 \tag{2-11}$$

with the boundary conditions

$$u(0) = u'(0) = u''(l) = u'''(l) = 0. \tag{2-12}$$

Let us introduce the following notation:

$$A_1 u = \left(E I_1 u'' \right)'', \tag{2-13a}$$

$$A_2 u = \left(E I_2 \frac{l-z}{l} u'' \right)'', \tag{2-13b}$$

$$A_3 u = \left[E I_3 \left(\frac{l-z}{l} \right)^2 u'' \right]'', \tag{2-13c}$$

$$A_4 u = \left[E I_4 \left(\frac{l-z}{l} \right)^3 u'' \right]'', \tag{2-13d}$$

$$A u = \sum_{i=1}^{4} A_i u, \tag{2-13e}$$

$$B u = - \left[\frac{(l-z)^2}{2l} u' \right]', \tag{2-13f}$$

$$\lambda = q_1. \tag{2-13g}$$

Differential expressions (2-13a to f) with the boundary conditions (2-12) define differential operators satisfying the conditions of Theorem 2-1, and the eigenvalue problem (2-11), (2-12) may be written in the form $Au - \lambda Bu = 0$, and the eigenvalue problems $A_i u - \lambda Bu = 0$ ($i = 1, \ldots, 4$) can be defined. If λ_0 denotes the least critical value of the load parameter of the original bar, and λ_i denotes the least critical load parameter of the bar with bending rigidity $EI_i \, [(l - z)/l]^{i-1}$, then from the *Southwell* Theorem 2-1 we have

$$\lambda_0 \geq \sum_{i=1}^{4} \lambda_i. \tag{2-14}$$

The values λ_i ($i = 1, \ldots, 4$) can be determined by *Tables 2-14* on page 131 of TIMOSHENKO and GERE [1961], and a lower bound of the desired value of the critical load parameter λ_0 obtained from (2-14).

As an aid to understanding it is useful to clarify the meaning of the conditions in Theorem 2-1 and show how they are satisfied.

We are looking for the least eigenvalue λ of equation (2-11) in the following class of functions u: u is four times continuously differentiable in the open interval $\{0, l\}$, three times continuously differentiable in the interval $\{0, l]$ open from left and closed from right, continuously differentiable in the closed interval $[0, l]$, and satisfies the boundary conditions (2-12).

Functions u are once continuously differentiable in $[0, l]$, so they are continuous and consequently bounded in $[0, l]$. From this it follows that they are square integrable in $[0, l]$. For an arbitrary pair u, v of such functions there exists the integral

$$(u, v) = \int_0^l uv \, dz$$

which has the properties of the scalar product. These functions are therefore elements of $L^2 [0, l]$, that is, the space of square integrable functions in the interval $[0, l]$. The space $L^2 [0, l]$ is a *Hilbert* space.

Consider the operator

$$A = \frac{d^2}{dz^2} \left(EI \frac{d^2}{dz^2} \right)$$

defined on the class of functions u. Because of the linear properties of differentiation, A is a linear operator. A is symmetric since for, arbitrary functions u, $v \in D_A$ with integration in parts we obtain

$$(Au, v) = \int_0^l \left(EIu'' \right)'' v\,dz = \left[\left(EIu'' \right)' v \right]_0^l - \left[EIu'' v' \right]_0^l$$

$$+ \left[u' EIv'' \right]_0^l - \left[u \left(EIv'' \right)' \right]_0^l + \int_0^l u \left(EIv'' \right)'' dz = (u, Av).$$

As a consequence of the boundary conditions (2-12) all the integrated parts are equal to zero. *In statics, symmetry of operator A means the equality of the internal 'alien' works, that is, if there is a set of stress and a set of strain independent of each other, then their internal work is equal to the internal work of the set of strain and the set of stress corresponding to the given sets of stress and strain due to Hooke's law.*

D_A is dense in $L^2 [0, l]$. It comes from the fact that any square integrable function in $[0, l]$ can be approximated almost everywhere in $[0, l]$ with arbitrary exactness by functions which are four times continuously differentiable in $[0, l]$, three times continuously differentiable in $[0, l]$ and once continuously differentiable in $[0, l]$ (for instance, by polynomials).

It is also possible to show that the operator A is positive. Let $u \in D_A$, $u \neq 0$ be arbitrary then, integrating by parts, we obtain

$$(Au, u) = \int_0^l \left(EIu'' \right)'' u\,dz$$

$$= \left[\left(EIu'' \right)' u \right]_0^l - \left[EIu'' u' \right]_0^l + \int_0^l EIu'' u'' dz = \int_0^l EIu''^2 dz > 0$$

because $EI > 0$ and the integrated parts are equal to zero in consequence of the boundary conditions (2-12). *In statics, positivity of operator A means the positivity of the strain energy.* (The scalar product is twice the strain energy.)

It is possible to show that A is positive definite but we omit this now.

Introduce the notation $g(z) = (l - z)^2 / (2l)$ by which operator B in (2-13f) can be written as

$$B = -\frac{d}{dz} \left[g(z) \frac{d}{dz} \right].$$

This operator is defined on the class of functions twice continuously differentiable in the open interval $\{0, l\}$, once continuously differentiable in the interval $[0, l\}$ closed from left open from right, vanishing at point $z = 0$. As three times and four times differentiability is not required, it follows that the domain of definition of B contains that of A, that is, $D_A \subset D_B$.

Operator B is linear because of the linear properties of differentiation. B is symmetric since for any $u, v \in D_B$ by integration by parts it is obtained:

$$(Bu, v) = - \int_0^l \left[g(z) u' \right]' v \, dz$$

$$= - \left[g(z) u' v \right]_0^l + \left[u g(z) v' \right]_0^l - \int_0^l u \left[g(z) v' \right]' dz = (u, Bv).$$

The integrated parts are equal to zero because of the boundary conditions $u(0) = 0$, $u'(0) = 0$ in (2-12) and $g(l) = 0$. *In statics, symmetry of operator B means the equality of the external 'alien' works, that is, the validity of the Maxwell-Betti theorem.*

D_B is dense in $L^2 [0, l]$. This can be shown similar to the case of operator A.

Operator B is positive because for any $u \in D_B$, $u \neq 0$ integration by parts yields:

$$(Bu, u) = - \int_0^l \left[g(z) u' \right]' u \, dz$$

$$= - \left[g(z) u' u \right]_0^l + \int_0^l g(z) u' u' dz = \int_0^l g(z) u'^2 dz > 0.$$

This holds as $g(z) \geq 0$, $z \in [0, l]$, and the integrated parts are equal to zero in consequence of the boundary conditions $u(0) = 0$, $u'(0) = 0$ in (2-12) and $g(l) = 0$. *In statics, positivity of operator B means the positivity of external work.* (The scalar product is twice the external work.)

To show that (Bu, u) is completely continuous with respect to (Au, u) is a larger task that we omit here. *In statics, complete continuity results in that the load has denumerable (infinitely many) critical values. These are well separated, and a definite buckling form belongs to each of them.*

We remark here that the *Southwell* theorem can be generalized also for quadratic eigenvalue problems [TARNAI, 1980].

2.3 DUNKERLEY TYPE THEOREMS AND FORMULAE

2.3.1 The Dunkerley theorem

In the theory of stability the classical theorem of DUNKERLEY [1894] is the following: The reciprocal of the least critical load parameter of an elastic structure subjected to a complex load system is not greater than the sum of the reciprocals of the least critical load parameters of the same structure subjected to subsystems of the load. In this theorem mathematically it is supposed that all the operators are positive definite or positive, that is, where all the eigenvalues (critical load parameters) are positive. The theorem, however, can be generalized for cases where there are both positive and negative eigenvalues. Now we formulate the *Dunkerley* theorem mathematically for this generalized case.

Theorem 2-2: Let A and B be symmetric linear operators in a *Hilbert* space H such that A is positive, and there exists a constant c such that

$$|(Bu, u)| \leq c\,(Au, u)\,, \; u \in D_A \subset D_B \subset H. \tag{2-15}$$

Suppose that the quadratic functional (Bu, u) is completely continuous with respect to (Au, u). Let $B = \sum\limits_{i=1}^{n} B_i$ where the operators B_i $(i = 1, 2, \ldots, n)$ have the same properties as B, and let there exist constants c_i associated to B_i to satisfy (2-15). If λ_0 and λ_i denote the least positive eigenvalues of the eigenvalue problem

$$(A - \lambda B)\,u = 0 \tag{2-16}$$

and

$$(A - \lambda B_i)\,u = 0, \; i = 1, 2, \ldots, n, \tag{2-17}$$

respectively, then we have

$$\frac{1}{\lambda_0} \leq \sum_{i=1}^{n} \frac{1}{\lambda_i}. \tag{2-18}$$

Remark: If for some i, $(A - \lambda B_i)\,u = 0$ has no positive eigenvalues then

$$\sup_{u \in D_A} \frac{(B_i u, u)}{(Au, u)} = 0,$$

and in (2-18), $1/\lambda_i$ should be replaced by 0.

In the following, some examples of the application of the *Dunkerley* theorem will be presented for lateral buckling of beam-columns. As shown in Chapter 9, equilibrium of thin-walled bars with symmetric open cross section subjected to both axial and transverse forces in a buckled state can be described by the differential equations below if the normal force N is constant:

$$EI_y u_T'''' + (M_x \Phi)'' + N\left(u_T'' + y_0 \Phi''\right) = 0, \tag{2-19}$$

$$EI_\omega \Phi'''' - GI_t \Phi'' + M_x u_T'' + t M_x'' \Phi - \beta_1 \left(M_x \Phi'\right)' + N\left(y_0 u_T'' + i_{pT}^2 \Phi''\right) = 0. \tag{2-20}$$

Here u_T is the lateral displacement of the shear centre of the cross section, t is the distance to the point of application of the transverse load q from the shear centre, and β_1, y_0, i_{pT}^2 are cross sectional constants. Suppose we have a one-parameter load system. Let λ denote the load parameter, M_{x1}, N_1, q_1 the reference values of the functions M_x, N, q. So,

$$M_x = \lambda M_{x1}, \; N = \lambda N_1, \; q = \lambda q_1.$$

Introduce the following notation:

$$A_{11} u_T = EI_y u_T'''', \tag{2-21a}$$

$$A_{22} \Phi = EI_\omega \Phi'''' - GI_t \Phi'', \tag{2-21b}$$

$$B_{11} u_T = -N_1 u_T'', \tag{2-21c}$$

$$B_{12}^{(1)} \Phi = -y_0 N_1 \Phi'', \tag{2-21d}$$

$$B_{12}^{(2)} \Phi = -(M_{x1} \Phi)'', \tag{2-21e}$$

$$B_{21}^{(1)} u_T = -y_0 N_1 u_T'', \tag{2-21f}$$

$$B_{21}^{(2)} u_T = -M_{x1} u_T'', \tag{2-21g}$$

$$B_{22}^{(1)} \Phi = -i_{pT}^2 N_1 \Phi'', \tag{2-21h}$$

$$B_{22}^{(2)} \Phi = \beta_1 \left(M_{x1} \Phi' \right)', \tag{2-21i}$$

$$B_{22}^{(3)} \Phi = -t M_{x1}'' \Phi, \tag{2-21j}$$

$$B_{12} = B_{12}^{(1)} + B_{12}^{(2)}, \tag{2-22a}$$

$$B_{21} = B_{21}^{(1)} + B_{21}^{(2)}, \tag{2-22b}$$

$$B_{22} = B_{22}^{(1)} + B_{22}^{(2)} + B_{22}^{(3)}. \tag{2-22c}$$

Differential expressions $(2-21a$ to $j)$ with given homogeneous boundary conditions denote differential operators. In this way equations (2-19 and 20) with given boundary conditions can be written in the form

$$\begin{bmatrix} A_{11} & 0 \\ 0 & A_{22} \end{bmatrix} \begin{bmatrix} u_T \\ \Phi \end{bmatrix} - \lambda \begin{bmatrix} B_{11} & B_{12} \\ B_{21} & B_{22} \end{bmatrix} \begin{bmatrix} u_T \\ \Phi \end{bmatrix} = \begin{bmatrix} 0 \\ 0 \end{bmatrix}, \tag{2-23}$$

or in a shorter form $(A - \lambda B)w = 0$ by introducing notation

$$A = \begin{bmatrix} A_{11} & 0 \\ 0 & A_{22} \end{bmatrix}, \qquad B = \begin{bmatrix} B_{11} & B_{12} \\ B_{21} & B_{22} \end{bmatrix}, \qquad w = \begin{bmatrix} u_T \\ \Phi \end{bmatrix}. \tag{2-24}$$

In most practical cases the boundary conditions result in operators A and B satisfying the conditions of Theorem 2-2.

Example 2-4: Let operator B be of the form

$$B = B_1 + B_2 \tag{2-25}$$

where

$$B_1 = \begin{bmatrix} B_{11} & B_{12}^{(1)} \\ B_{21}^{(1)} & B_{22}^{(1)} \end{bmatrix}, \qquad B_2 = \begin{bmatrix} 0 & B_{12}^{(2)} \\ B_{21}^{(2)} & B_{22}^{(2)} + B_{22}^{(3)} \end{bmatrix} \tag{2-26}$$

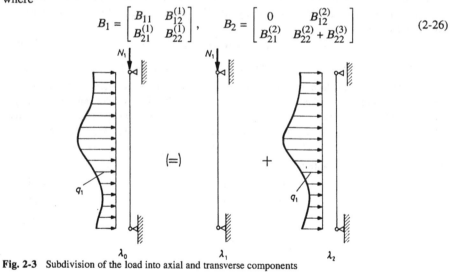

Fig. 2-3 Subdivision of the load into axial and transverse components

This means that the system of external forces is divided into two parts: (*a*) axial and (*b*) transverse loads (*Fig. 2-3*). Due to Theorem 2-2 a lower bound can be given to the least positive critical load parameter by the least critical load parameter for torsional buckling and the least positive critical load parameter for lateral buckling. This is a traditional way of applying the *Dunkerley* theorem.

Example 2-5: Consider the problem of lateral buckling of the bar without axial forces ($N_1 = 0$):

$$\begin{bmatrix} A_{11} & 0 \\ 0 & A_{22} \end{bmatrix} \begin{bmatrix} u_T \\ \Phi \end{bmatrix} - \lambda \begin{bmatrix} 0 & B_{12}^{(2)} \\ B_{21}^{(2)} & B_{22}^{(2)} + B_{22}^{(3)} \end{bmatrix} \begin{bmatrix} u_T \\ \Phi \end{bmatrix} = \begin{bmatrix} 0 \\ 0 \end{bmatrix}. \tag{2-27}$$

Let us subdivide operator B_2 in (2-26) into two parts:

$$B_2 = B_2^{(1)} + B_2^{(2)} \tag{2-28}$$

such that

$$B_2^{(1)} = \begin{bmatrix} 0 & B_{12}^{(2)} \\ B_{21}^{(2)} & B_{22}^{(2)} \end{bmatrix}, \qquad B_2^{(2)} = \begin{bmatrix} 0 & 0 \\ 0 & B_{22}^{(3)} \end{bmatrix}. \tag{2-29}$$

Let operator $B_{22}^{(3)}$ in (2-21j) be positive, that is, let the point of application of the load be above the shear centre. In this case the conditions of Theorem 2-2 are satisfied by the operators A, B_2, $B_2^{(1)}$, $B_2^{(2)}$ defined by (2-24), (2-28), (2-29), and to the least positive critical load parameter we can give an approximate value in the following way (*Fig. 2-4*).

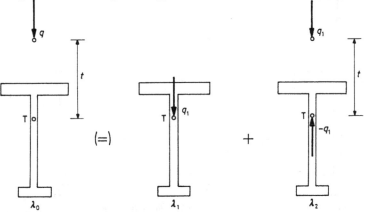

Fig. 2-4 Grouping of loads

First we determine the least positive critical load parameter for the case where the point of application of the load is the shear centre $\left(B_2^{(2)} = 0 \right)$. This is λ_1. Then we determine the critical load of the structure for the case where additionally to the original load with original point of application, another load is applied whose intensity is the same as that of the original, but its direction is the opposite, and its point of application is the shear centre $\left(B_2^{(1)} = 0 \right)$. This is equivalent to the case where the bar is simply supported along an axis passing through the shear centre. This support allows the cross sections to rotate. The critical load parameter in this case is λ_2. The theorem yields the inequality

$$\frac{1}{\lambda_0} \leq \frac{1}{\lambda_1} + \frac{1}{\lambda_2}.$$

Example 2-6: Another way of solving the problem (2-27) can be if the operator B_2 is divided into two parts as

$$B_2 = \tilde{B}_2^{(1)} + \tilde{B}_2^{(2)} \tag{2-30}$$

such that

$$\tilde{B}_2^{(1)} = \begin{bmatrix} 0 & B_{12}^{(2)} \\ B_{21}^{(2)} & B_{22}^{(3)} \end{bmatrix}, \qquad \tilde{B}_2^{(2)} = \begin{bmatrix} 0 & 0 \\ 0 & B_{22}^{(2)} \end{bmatrix}. \tag{2-31}$$

Operators A, B_2, $\tilde{B}_2^{(1)}$, $\tilde{B}_2^{(2)}$ defined by (2-24), (2-30), (2-31) satisfy the conditions of Theorem 2-2, therefore

$$\frac{1}{\lambda_0} \leq \frac{1}{\lambda_1} + \frac{1}{\lambda_2},$$

where λ_1 is the least positive critical load parameter of such a bar whose cross section is doubly symmetric and whose stiffness properties are the same as those of the original bar, and where the distance between the point of application of the transverse load and the shear centre is the same as that of the original bar; λ_2 is the least positive critical load parameter of the original bar subjected to the load at the shear centre ($t = 0$) when the lateral displacement of the shear centre is prevented (*Fig. 2-5*).

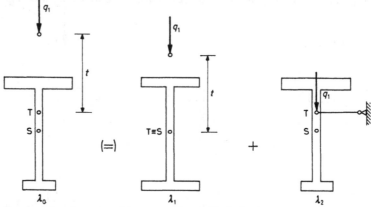

Fig. 2-5 Imaginary transformation of the structure for the *Dunkerley* theorem

The Strigl formula: There are known investigations to sharpen the Dunkerley inequality (2-18) in the Theorem 2-2, that is, to bring the inequality closer to equality (*Fig. 2-6*). These investigations led to the nonlinear versions of the *Dunkerley* formula.

Let A and B be symmetric linear operators in a *Hilbert* space H such that A is positive definite and B is positive. In order to have a discrete spectrum, let us suppose that the quadratic functional (Bu, u) is completely continuous with respect to (Au, u). Let $B = B_1 + B_2$ where the operators B_1 and B_2 have the same properties as B. If λ_0 denotes the least eigenvalue of the eigenvalue problem

$$(A - \lambda B)u = 0, \tag{2-32}$$

and λ_1, u_1 and λ_2, u_2 denote the least eigenvalues and the corresponding eigenelements of the eigenvalue problems

$$(A - \lambda B_1)u = 0 \tag{2-33}$$

and

$$(A - \lambda B_2)u = 0, \tag{2-34}$$

respectively, then according to STRIGL [1955], a very good approximation is provided by the quadratic *Dunkerley* formula

$$\frac{\lambda_0}{\lambda_1} + \frac{\lambda_0}{\lambda_2} - \frac{\lambda_0^2}{\lambda_1 \lambda_2}\left[1 - \frac{(Au_1,\, u_2)^2}{(Au_1,\, u_1)(Au_2,\, u_2)}\right] \approx 1 \tag{2-35}$$

which can be written in the form

$$\begin{vmatrix} (Au_1, u_1)\left(1 - \frac{\lambda_1}{\lambda_0}\right) & (Au_1, u_2) \\ (Au_1, u_2) & (Au_2, u_2)\left(1 - \frac{\lambda_2}{\lambda_0}\right) \end{vmatrix} \approx 0. \tag{2-36}$$

Relationship (2-36) is the simplest case of a highly nonlinear interaction formula derived by STRIGL [1955]. He has also applied formula (2-35) with success for cases where operator B was not positive but only symmetric, and so the problem also had negative eigenvalues. Practical application of formula (2-35) is somewhat difficult since not only the least positive eigenvalues of the subproblems are required but also the corresponding eigenelements (eigenfunctions).

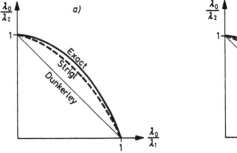
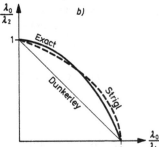

Fig. 2-6 Approximation more exact than the Dunkerley straight line: *a*) On the safe side, *b*) partly on the safe side, partly on the unsafe side

Numerical experiments by *Strigl* have shown that formula (2-35) can lead to approximations both on the safe side (*Fig. 2-6a*) and the unsafe side (*Fig. 2-6b*).

It should be noted that the Dunkerley theorem can be generalized also for quadratic eigenvalue problems [TARNAI, 1980].

2.3.2 The Föppl-Papkovich theorem

The algebraic form of the *Föppl-Papkovich* formula is the same as that of the *Dunkerley* formula, but its physical background is different [FÖPPL, 1933], [PAPKOVICH, 1963]. Its essence is the following. Let a structure be characterized by n stiffness parameters. Consider an imaginary structure obtained from the original, all of whose stiffness parameters are increased up to infinity except the ith one which is kept unchanged. Let the least critical load parameter of the structure obtained in this way be denoted by λ_i. Let this procedure be done for $i = 1, 2, \ldots, n$. If λ_0 denotes the least critical load parameter of the original structure, then for its reciprocal a good approximation is given by the reciprocals of the least critical load parameters of the imaginary rigidized structures:

$$\frac{1}{\lambda_0} \approx \sum_{i=1}^{n} \frac{1}{\lambda_i}. \tag{2-37}$$

However, it is not clear whether or not the approximation is conservative. In the following we will give conditions under which formula (2-37) results in approximation on the safe side.

Definition: Let A be a symmetric linear operator in *Hilbert* space H. We say that elements u and v $(u, v \in D_A)$ are A-orthogonal, if $(Au, v) = 0$. Let X and Y be subspaces in D_A. We say that X and Y are A-orthogonal, if for every $u \in X$ and $v \in Y$ we have $(Au, v) = 0$. Let Z be a subspace in D_A. We say that Z is A-orthogonal direct sum of X and Y, denoted by $Z = X \oplus Y$, if every $w \in Z$ can be written in the form $w = u + v$ such that $u \in X$, $v \in Y$ and $(Au, v) = 0$.

Theorem 2-3: Let A and B be symmetric linear operators in *Hilbert* space H such that A is positive definite, B is positive and $D_A \subset D_B \subset H$. Let the quadratic functional (Bu, u) be completely continuous with respect to (Au, u). Let $D_1, D_2, \ldots, D_n$ be pairwise A-orthogonal subspaces in D_A such that $D_A = D_1 \oplus D_2 \oplus \ldots \oplus D_n$. If λ_0 and λ_i denotes the least eigenvalue of the eigenvalue problem

$$(A - \lambda B)u = 0, \qquad u \in D_A \tag{2-38}$$

and

$$(A - \lambda B)u = 0, \qquad u \in D_i, \qquad i = 1, 2, \ldots, n, \tag{2-39}$$

respectively, then we have

$$\frac{1}{\lambda_0} \leq \sum_{i=1}^{n} \frac{1}{\lambda_i}. \tag{2-40}$$

Remarks: In Theorem 2-3 it is important that B is positive. Conservative approximation is therefore guaranteed only for problems having positive eigenvalues. In the Theorem it is not required that (Bu, u) is completely continuous with respect to (Au, u) in every subproblem $(u \in D_i, \quad i = 1, 2, \ldots, n)$. Thus, the Theorem is valid for degenerate subproblems whose spectrum has only one point and it is an eigenvalue with infinite multiplicity, or whose spectrum is continuous. In the case of continuous spectrum, however, we cannot talk about the least eigenvalue but the infimum of the *Rayleigh* quotient. Such problems can arise for buckling of bars where the effect of shearing forces on the deflection is taken into consideration.

There are problems where by application of the *Föppl* method (partial rigidizing), the conditions of the *Dunkerley* theorem are satisfied, and (2-40) is due to the *Dunkerley* theorem. In such a case the *Föppl-Papkovich* formula is identical to the *Dunkerley* formula. There are, however, cases for which the *Föppl-Papkovich* formula is different from the *Dunkerley* formula, and the *Föppl-Papkovich* formula approximates the exact value with a smaller error than the *Dunkerley* formula. We will show an example for each of the two cases.

Example 2-7: Consider the buckling of a bar of constant cross section with hinged ends, subjected to compressive force N at its ends. The differential equation and the boundary conditions of the problem are:

$$EI_y u'''' + Nu'' = 0, \qquad (2\text{-}41)$$

$$u(0) = u''(0) = u(l) = u''(l) = 0. \qquad (2\text{-}42)$$

By the notation $Au = EI_y u''''$, $\lambda N_1 = N$, $Bu = -N_1 u''$ the eigenvalue problem (2-41), (2-42) can be written in the form $(A - \lambda B)u = 0$. Eq. (2-41), however, can be written in the form

$$u'''' + \frac{N}{EI_y} u'' = 0, \qquad (2\text{-}43)$$

which with the notation $\bar{A}u = u''''$, $\lambda N_1 = N$, $\bar{B}u = -N_1 u'' / (EI_y)$ reads $(\bar{A} - \lambda \bar{B})u = 0$. Subdivide the length of the bar into n parts with points $0 = z_0 < z_1 < z_2 < \ldots < z_{n-1} < z_n = 1$ (*Fig.* 2-7). The value of the stiffness parameter on every segment is EI_y. Introduce the following notation:

$$K_i = \begin{cases} EI_y & \text{if } z_{i-1} < z \le z_i, \\ \infty, & \text{if } z \le z_{i-1} \text{ or } z > z_i, \end{cases}$$

$$\bar{B}_i u = -N_1 \frac{u''}{K_i}, \qquad i = 1, 2, \ldots, n.$$

It is clear that $\bar{B} = \sum_{i=1}^{n} \bar{B}_i$, and the conditions of the *Dunkerley* theorem (Theorem 2-2) are satisfied. Therefore, if λ_i denotes the least eigenvalue of the eigenvalue problem $(\bar{A} - \lambda \bar{B}_i) u = 0$, then (2-18) holds, but considering the physical background it expresses the inequality (2-40), as by partial rigidizing the bar we have defined A-orthogonal subspaces of displacement functions u. This can be shown as follows. The ith A-orthogonal subspace D_i consists of the four times differentiable functions u satisfying boundary conditions (2-42) such that – because of rigidizing – u is linear outside the interval $z_{i-1} < z \le z_i$. (We note here that for $i = 2, 3, \ldots, n-1$ the boundary conditions $u''(0) = u''(l) = 0$ are automatically satisfied.) Let $u \in D_i$, $v \in D_j$ $(i \ne j)$ be arbitrary elements, then by integration by parts

$$(Au, v) = \int_0^l EI_y u'''' v \, dz = \left[EI_y u''' v \right]_0^l - \left[EI_y u'' v' \right]_0^l + \int_0^l EI_y u'' v'' \, dz.$$

In this expression the integrated parts vanish in consequence of the boundary conditions. The second derivative of the function u and v vanishes outside the interval $z_{i-1} < z \le z_i$ and $z_{j-1} < z \le z_j$, respectively. Since these intervals are disjoint, it follows that $u'' v''$ is identically equal to zero. Therefore $(Au, v) = 0$, that is, u and v are A-orthogonal; consequently D_i and D_j are A-orthogonal also. So, the conditions of the *Föppl-Papkovich* theorem (Theorem 2-3) are satisfied, too. By writing the *Rayleigh* quotient we also can show that the least eigenvalues of the eigenvalue problems $(\bar{A} - \lambda \bar{B}_i) u = 0$, $u \in D_A$ and $(A - \lambda B) u = 0$, $u \in D_i$ are the same: λ_i. In this example, therefore, the *Dunkerley* theorem and the *Föppl-Papkovich* theorem result in the same formula.

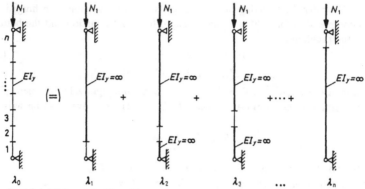

Fig. 2-7 Partial rigidizing of the hinged bar

Example 2-8: Consider torsional buckling of a bar of height l with the lower end built-in and the upper end free. The equilibrium equation and the boundary conditions with the notation used in Section 2.3.1 are

$$\begin{bmatrix} A_{11} & 0 \\ 0 & A_{22} \end{bmatrix} \begin{bmatrix} u_T \\ \Phi \end{bmatrix} - \lambda \begin{bmatrix} B_{11} & B_{12}^{(1)} \\ B_{21}^{(1)} & B_{22}^{(1)} \end{bmatrix} \begin{bmatrix} u_T \\ \Phi \end{bmatrix} = 0, \tag{2-44}$$

$$u_T(0) = u_T''(0) = u_T'(l) = u_T'''(l) = 0, \tag{2-45a}$$

$$\Phi(0) = \Phi''(0) = \Phi'(l) = \Phi'''(l) = 0. \tag{2-45b}$$

Eq. (2-44) with notation (2-24) and (2-26) takes the form

$$Aw - \lambda B_1 w = 0. \tag{2-46}$$

Introduce the notation

$$B_1^{(1)} = \begin{bmatrix} B_{11} & 0 \\ 0 & 0 \end{bmatrix}, \ B_1^{(2)} = \begin{bmatrix} 0 & 0 \\ 0 & B_{22}^{(1)} \end{bmatrix}, \ B_1^{(3)} = \begin{bmatrix} 0 & B_{12}^{(1)} \\ B_{21}^{(1)} & 0 \end{bmatrix}.$$

It is clear that $B_1 = B_1^{(1)} + B_1^{(2)} + B_1^{(3)}$. Let $\lambda_1, \lambda_2, \lambda_3$ denote the least eigenvalues of the eigenvalue problems

$$\left(A - \lambda B_1^{(1)} \right) w = 0,$$

$$\left(A - \lambda B_1^{(2)} \right) w = 0,$$

$$\left(A - \lambda B_1^{(3)} \right) w = 0,$$

respectively. If λ_0 denotes the least eigenvalue of the problem (2-46), then due to the *Dunkerley* theorem we obtain:

$$\frac{1}{\lambda_0} \leq \frac{1}{\lambda_1} + \frac{1}{\lambda_2} + \frac{1}{\lambda_3}.$$

If EI_y and GI_t are the two stiffness parameters in the *Föppl* method, then the *Föppl-Papkovich* theorem results in the inequality

$$\frac{1}{\lambda_0} \leq \frac{1}{\lambda_1} + \frac{1}{\lambda_2}.$$

It can be seen that in this case the *Föppl-Papkovich* formula and the *Dunkerley* formula are not the same, and the *Föppl-Papkovich* formula is more accurate. (This is obvious, as λ_1 and λ_2 in the *Föppl-Papkovich* formula are identical to λ_1 and λ_2 in the *Dunkerley* formula, but there is the term $1/\lambda_3$ in the *Dunkerley* formula that is not present in the *Föppl-Papkovich* formula.)

Example 2-9: (The *Plantema* paradox) In the field of sandwich structures the *Föppl* method is known as the *method of split rigidities*. In many problems the method of split rigidities provides exact results or conservative approximations. PLANTEMA [1952], however, found an example where the result of this method was *not* on the safe side. Now we will investigate the reason of this paradoxical behaviour.

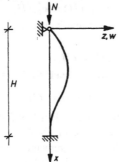

Fig. 2-8 Buckling of a thin-faced sandwich bar with the lower end built-in and the upper end hinged

Consider the buckling of a bar of height H with the lower end built-in and the upper end hinged, subjected to a concentrated vertical force N at the hinged end (*Fig. 2-8*). Suppose that the bar undergoes both bending and shearing deformations. The displacement w of a point of the axis of the bar is composed of two parts: displacement w_D due to the bending deformation and displacement w_S due to the shearing deformation, that is,

$$w = w_D + w_S.$$

The equilibrium of the bar – as is usual in the literature [TIMOSHENKO and GERE, 1961] – can be expressed by using w_D only. If D_0 and S denote the bending and the shear rigidities, and $\lambda = N$ the equilibrium equation takes the form

$$w_D'''' - \lambda \left(\frac{w_D''''}{S} - \frac{w_D''}{D_0} \right) = 0 \tag{2-47}$$

and as shown in Chapter 6 the boundary conditions are

$$w_D(0) = w_D''(0) = w_D(H) - \frac{D_0}{S} w_D''(H) = w_D'(H) = 0. \tag{2-48}$$

The eigenvalue problem (2-47), (2-48) for $S = \infty$ leads to the eigenvalue problem

$$w_D'''' + \lambda \frac{w_D''}{D_0} = 0, \tag{2-49}$$

$$w_D(0) = w_D''(0) = w_D(H) = w_D'(H) = 0, \tag{2-50}$$

and for $D_0 = \infty$ leads to the eigenvalue problem

$$w_D'''' - \lambda \frac{w_D''''}{S} = 0, \tag{2-51}$$

$$w_D(0) = w_D''(0) = w_D''(H) = w_D'(H) = 0. \tag{2-52}$$

Let $\lambda_0, \lambda_1, \lambda_2$ denote the least eigenvalue of the eigenvalue problem (2-47), (2-48), (2-49), (2-50), (2-51), (2-52), respectively. Introduce the notation

$$\alpha^2 = \frac{N}{D_0\left(1 - \frac{N}{S}\right)}. \tag{2-53}$$

As $N = \lambda$ it is easy to show that the reciprocal of the least eigenvalue of the eigenvalue problem

$$w_D'''' - \lambda\left(\frac{w_D''''}{S} - \frac{w_D''}{D_0}\right) = 0, \tag{2-54}$$

$$w_D(0) = w_D''(0) = w_D(H) = w_D'(H) = 0 \tag{2-55}$$

is equal to $1/\lambda_1 + 1/\lambda_2$. This eigenvalue comes from the least positive solution of the characteristic equation

$$\tan \alpha H = \alpha H. \tag{2-56}$$

The eigenvalue problem (2-47), (2-48), however, leads to the characteristic equation

$$\left(1 + \frac{D_0}{S}\alpha^2\right)\tan \alpha H = \alpha H \tag{2-57}$$

where the coefficient of $\tan \alpha H$ is a number greater then 1. It follows that the least positive root of (2-57) is less than the least positive root of (2-56), and due to (2-56) the less α is, the less N (or λ) is; we obtain that the least eigenvalue λ_0 of the eigenvalue problem (2-47), (2-48) is less than the least eigenvalue $(1/\lambda_1 + 1/\lambda_2)^{-1}$ of the eigenvalue problem (2-54), (2-55)

$$\frac{1}{\lambda_0} > \frac{1}{\lambda_1} + \frac{1}{\lambda_2},$$

that is, the *Föppl* method is on the unsafe side. Looking at the problem (2-47), (2-48), however, the explanation is not at all obvious. (*A*-orthogonality of the bending mode and the shearing mode does not occur, the operators in the problem are not symmetric, and the boundary conditions change with partial rigidizing of the bar.)

To resolve the paradox situation let us describe the problem using both the bending displacement w_D and the shearing displacement w_S explicitly. The equilibrium of the bar in a buckled shape can be written by the system of differential equations

$$D_0 w_D'''' + \lambda\left(w_D'' + w_S''\right) = 0, \tag{2-58a}$$

$$-Sw_S'' + \lambda\left(w_D'' + w_S''\right) = 0, \tag{2-58b}$$

and the boundary conditions

$$w_D(0) = 0, \quad w_D''(0) = 0, \quad w_D(H) - \frac{D_0}{S}w_D''(H) = 0, \quad w_D'(H) = 0, \tag{2-59a}$$

$$w_S(0) = 0, \qquad w_S(H) + w_D(H) = 0. \tag{2-59b}$$

By introducing the notation

$$A_{11}w_D = D_0 w_D'''', \tag{2-60a}$$

$$A_{22}w_S = -Sw_S'', \tag{2-60b}$$

$$B_{11}w_D = -w_D'', \tag{2-60c}$$

$$B_{22}w_S = -w_S'', \tag{2-60d}$$

we obtain the eigenvalue problem

$$\begin{bmatrix} A_{11} & 0 \\ 0 & A_{22} \end{bmatrix} \begin{bmatrix} w_D \\ w_S \end{bmatrix} - \lambda \begin{bmatrix} B_{11} & B_{22} \\ B_{11} & B_{22} \end{bmatrix} \begin{bmatrix} w_D \\ w_S \end{bmatrix} = 0, \tag{2-61}$$

which with the notation

$$A = \begin{bmatrix} A_{11} & 0 \\ 0 & A_{22} \end{bmatrix}, \qquad B = \begin{bmatrix} B_{11} & B_{22} \\ B_{11} & B_{22} \end{bmatrix}, \qquad w = \begin{bmatrix} w_D \\ w_S \end{bmatrix} \tag{2-62}$$

can be written in the form $(A - \lambda B)w = 0$. It can be established that the bending displacement w_D and the shearing displacement w_S as vectors

$$\begin{bmatrix} w_D \\ 0 \end{bmatrix}, \qquad \begin{bmatrix} 0 \\ w_S \end{bmatrix},$$

are A-orthogonal. Operator B is symmetric and positive, and (Bw, w) is completely continuous with respect to (Aw, w). *Operator A, however, is not symmetric!* This, however, is not the only reason of the paradoxical behaviour. The question of why the bound goes the wrong way was answered by WEINBERGER (private communication): Since $D_0 w_D'''' + S w_S'' = 0$ and $w_D''(0) = w_S(0) = 0$, we must have $D_0 w_D'' + S w_S = cx$ for some constant c. Since $w_S(H) = -w_D(H)$, the second boundary condition in (2-59b) says that $c = 0$. The *Föppl* bound appears to come from splitting the space of displacement pairs (w_D, w_S) into the subspaces of pairs $(u, 0)$ and $(0, v)$. To satisfy the imposed conditions, one must have $u(0) = u(H) = u'(H) = 0$ and $v(0) = v(H) = 0$. Then any linear combination of such pairs satisfies the additional constraint $w_D(H) = 0$. In other words, the bound given by the method of split rigidities is a lower bound for the critical buckling load of a more constrained problem, but not for the lower buckling load of the original problem. *In mathematical terms, the two subspaces do not span the original space.*

To apply *Föppl* correctly, one needs to add the supremum on the space of function pairs orthogonal to both of the above subspaces. It is easily seen that this is a one-dimensional space which is spanned by the function pair $(r, -x)$, where

$$r(x) = \frac{1}{2}x \left(3 - \frac{x^2}{H^2}\right).$$

This adds the term

$$\frac{H^2}{5 \left(3D_0 + H^2 S\right)}$$

to the bound for $1/\lambda$. With this addition, the bound should be correct.

Example 2-10: Determine the critical load of the pin-ended laced column subjected to concentrated force N at its end (*Fig. 2-9a*). The column is built up of chord members of length a and cross sectional area A_0, and of brace members of length d_1 and d_2 and cross sectional area A_1 and A_2 such that the members are connected to each other by ideal joints. On the basis of the continuum analogy, the critical load of the laced column is calculated similar to that of the compressed bars having shearing deformation (sandwich bars with thin face). The difference between the continuum and the truss only is that for the continuum the shear rigidity is defined for the cross section of the bar, and the angular deformation (shearing strain) is defined for an infinitesimal length, but for the truss both the shear rigidity and the angular deformation is defined on a finite length a.

Let us suppose first that all the bracing members are infinitely rigid ($EA_1 = EA_2 = \infty$). Then the column has only bending deformation (due to the elongation of the chord mem-

bers), and the critical load of the column – with a good approximation – is equal to the
Euler load:

$$N_E = \frac{\pi^2 E I_0}{H^2}, \tag{2-63}$$

where $I_0 = A_0 b^2 / 2$.

Then let us suppose that the chord members are infinitely rigid ($EA_0 = \infty$), but the
bracing members are not. In this case the column can get to a critical state by shearing
deformation due to elongation of the bracing members. Let the critical load for shearing
deformation be denoted by N_S. According to the *Föppl* summation, the reciprocal of the
critical load N_{cr} of the column is

$$\frac{1}{N_{cr}} \approx \frac{1}{N_E} + \frac{1}{N_S}. \tag{2-64}$$

N_S itself can be determined again by the *Föppl* summation. Apart from its ends the column can be considered to be composed of equal rhombic cells braced by diagonal bars.
The shearing deformation is due to the elongation of these diagonal bars. Since the column can be composed from rhombi in two different ways (with diagonals with angle of
inclination (a) Φ_1 and (b) Φ_2), the shearing deformation is composed from two independent components. The critical loads associated to these two components can be determined
separately.

Consider first the case where all the bars, except the bracing members with inclination
of φ_1, are infinitely rigid ($EA_0 = \infty$, $EA_2 = \infty$). Now, in the critical state there is a
sway of one of the cells. Lengthening e_1 of the bracing bar in the cell causes a lateral displacement δ_1 (*Fig. 2-9b*). So, the axis of the deformed column – apart from the deformed
cell – has an angle δ_1/H of inclination to the original axis.

In accordance with *Fig. 2-9b*

$$\delta_1 = a\gamma_1$$

and

$$e_1 = \delta_1 \cos\varphi_1 = a\gamma_1 \cos\varphi_1.$$

Let us denote the load on this partially rigidized column by N_1. Its critical value can be
determined, for instance, by the energy method. The total potential is

$$\Pi = \Pi_{int} + \Pi_{ext},$$

where

$$\Pi_{int} = \frac{1}{2} \frac{EA_1}{d_1} e_1^2 = \frac{1}{2} \frac{EA_1}{d_1} a^2 \gamma_1^2 \cos^2\varphi_1,$$

$$\Pi_{ext} = -N_1 a \frac{\gamma_1^2}{2}.$$

From the condition $\partial\Pi / \partial\gamma_1 = 0$ it follows that

$$N_{1cr} = EA_1 \frac{a}{b} \cos^3\varphi_1. \tag{2-65}$$

Consider now the case where all the bars, except the bracing members with inclination
of φ_2, are infinitely rigid ($EA_0 = \infty, EA_1 = \infty$). The critical load in this case can be
determined from the lengthening of a bracing bar of inclination φ_2 in one of the cells of

the other kind (*Fig. 2-9c*), similar to the previous case. Its value is

$$N_{2cr} = EA_2 \frac{a}{b} \cos^3 \varphi_2.$$
(2-66)

Then the critical load of the column having shearing deformations only is obtained by the *Föppl* summation:

$$\frac{1}{N_S} = \frac{1}{N_{1cr}} + \frac{1}{N_{2cr}},$$
(2-67)

where N_{1cr} is given by (2-65) and N_{2cr} by (2-66). Knowing (2-63) and (2-67) we obtain the approximate value of the critical load of the laced column by the expression (2-64).

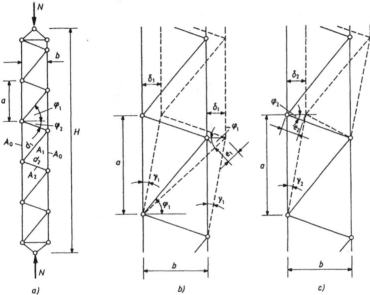

Fig. 2-9 Buckling of a laced column. *a*) Layout of the column. Shearing deformation from the elongation of the inclined member of length, *b*) d_1 and *c*) d_2.

2.3.3 The Kollár conjecture

KOLLÁR [1971] has published an approximate formula without proof for the determination of the least critical load parameter of compressed continuous bars on elastic supports (*Fig. 2-10*). Let λ_1 denote the least critical load parameter of the bar for the case where the spring rigidity of the elastic supports is infinitely large, that is, the springs are replaced by rigid supports. Let λ_2 denote the least critical load parameter of the bar for the case where the elastic rigidity of the supports are continuously distributed along the length of the bar, that is, the springs are replaced by an elastic foundation. If the least critical load parameter of the original structure is denoted by λ_0, then *Kollár* states that:

$$\frac{1}{\lambda_0} \le \frac{1}{\lambda_1} + \frac{1}{\lambda_2}.$$ (2-68)

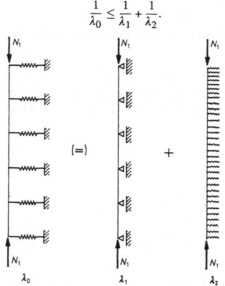

Fig. 2-10 Replacement of the elastic supports

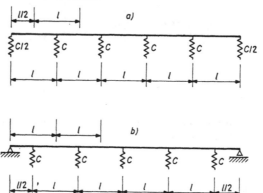

Fig. 2-11 Layout of supports resulting in uniform elastic foundation. *a*) with equal spacing, *b*) with equal spring rigidities

The correctness of formula (2-68) with general validity has not been proved or refuted so far. In this method it is not clear how to distribute the elasticity of the supports, and what kind of distribution function should be considered. Although we cannot prove it, we think that the *Kollár* approximation provides good results for those problems where the distribution of the elasticity of the supports are quite well defined. Such cases are where a bar is supported with equidistant springs of equal rigidity, but where the rigidity of the end springs is the half of the others. In such a case rigidity C of an internal spring should be distributed along the length l, but rigidity $C/2$ of the end spring along the length $l/2$ (*Fig. 2-11a*). For springs of equal rigidity the arrangement in *Fig. 2-11b* is suitable. In both cases a uniform elastic foundation is obtained.

We can say something similar about *Vierendel* columns (columns with batten plates) if the equally spaced horizontal beams have the same rigidity, but the uppermost and the bottom beams have the half of their rigidity (*Fig. 2-12*).

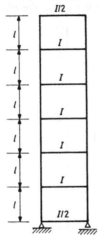

Fig. 2-12 Beam rigidities resulting in uniformly distributed elastic restraint for rotation

In the following the approximation by (2-68) will be illustrated by some examples. Let us define the *Kollár* critical load parameter λ_K by the expression

$$\frac{1}{\lambda_K} = \frac{1}{\lambda_1} + \frac{1}{\lambda_2}. \tag{2-69}$$

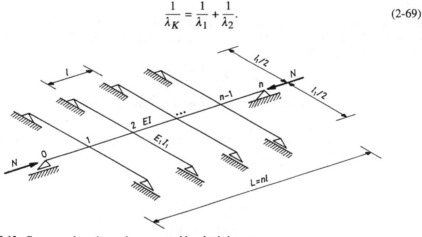

Fig. 2-13 Compressed continuous bar supported by elastic beams

Example 2-11: Let us consider buckling of a continuous beam on $n+1$ supports (*Fig. 2-13*) in which the middle $n-1$ supports are elastic beams. Let the compressive force be denoted by N, the length of the continuous beam by L and its bending rigidity by EI. Let the length of the equally spaced cross beams be denoted by l_1 and their bending rigidity by $E_1 I_1$. The cross beams provide elastic supports for the bar with spring rigidity

$$C = \frac{48 E_1 I_1}{l_1^3}.$$

Then we obtain the model shown in *Fig. 2-14a*. For the sake of simplicity let the reference value N_1 of the compressive force N be the unity. Therefore $\lambda = N$.

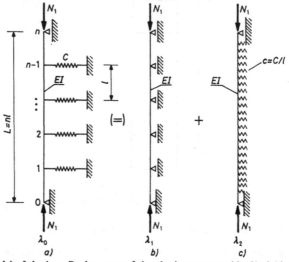

Fig. 2-14 *a)* Model of the bar. Replacement of the elastic supports with: *b)* rigid supports, *c)* elastic foundation

The task is to determine the least critical load parameter λ_0 of the bar on elastic supports seen in *Fig. 2-14a*. The exact solution is known from BLEICH [1952]. This critical load parameter is given as follows:

$$\lambda_0 = \left(\frac{\beta_0}{\pi}\right)^2 \frac{\pi^2 EI}{l^2}. \tag{2-70}$$

Here $\beta_0 = \min_{1 \leq i \leq n-1} \beta_i$, where β_i is the solution of the following equation for fixed i:

$$\beta^2 \frac{\left(1 - \cos \frac{i\pi}{n}\right) a - b}{1 - \dfrac{b}{1 - \cos \frac{i\pi}{n}}} - 24v = 0 \quad (i = 1, 2, \ldots, n - 1) . \tag{2-71}$$

Symbols a, b, v in (2-71) are:

$$a = \frac{\beta}{\beta - \sin\beta},$$

$$b = \frac{\beta (1 - \cos\beta)}{\beta - \sin\beta},$$

$$\nu = \frac{E_1 I_1 l^3}{E I l_1^3}.$$

For $n = \infty$ equation (2-71) is reduced to the form

$$\frac{\beta^3 (1 - \cos\beta)}{\left(\sqrt{\beta} + \sqrt{\sin\beta}\right)^2} - 24\nu = 0. \tag{2-72}$$

Let us determine now the *Kollár* critical load parameter λ_K of the bar. For that we need the least critical load parameter λ_1 and λ_2 of the bar on rigid supports (*Fig. 2-14b*) and on elastic foundation (*Fig. 2-14c*), respectively.

λ_1 is easy to determine as it is the *Euler* critical load parameter:

$$\lambda_1 = \frac{\pi^2 EI}{l^2} = \lambda_E. \tag{2-73}$$

The least critical load parameter of the bar on uniform elastic foundation (with modulus $c = C/l$ of the elastic foundation) can be determined from TIMOSHENKO and GERE [1961]:

$$\lambda_2 = \min_{m = 1,2,\ldots} \frac{\pi^2 EI}{l^2} \left[\left(\frac{m}{n}\right)^2 + \left(\frac{n}{m}\right)^2 \frac{48}{\pi^4}\nu \right]. \tag{2-74}$$

Then due to (2-69), λ_K can be determined by λ_1 and λ_2.

Let $E_1 I_1 = EI$ and $l_1 = 4l$. So, $\nu = 1/64$. Let us determine the load parameters and the error

$$\Delta = \frac{\lambda_0 - \lambda_K}{\lambda_0}$$

of the *Kollár* approximation for different values of n. We obtained the following results.

$n = 2$

$\lambda_1 = \lambda_E,$
$\lambda_2 = 0.2807980\lambda_E,$
$\lambda_K = 0.2192367\lambda_E,$
$\lambda_0 = 0.2807354\lambda_E,$
$\Delta = 21.9\%.$

$n = 4$

$\lambda_1 = \lambda_E,$
$\lambda_2 = 0.1856918\lambda_E,$
$\lambda_K = 0.1566105\lambda_E,$
$\lambda_0 = 0.1855357\lambda_E,$
$\Delta = 15.6\%.$

$n = \infty$

$\lambda_1 = \lambda_E,$
$\lambda_2 = 0.1754934\lambda_E,$
$\lambda_K = 0.1492934\lambda_E,$
$\lambda_0 = 0.1753758\lambda_E,$
$\Delta = 14.9\%.$

It can be seen that the *Kollár* formula (2-69) is on the safe side in all of the cases.

Example 2-12: Consider a two-legged n-storey plane frame with rigid restraints at the ends of its columns, each column subjected to a compressive force N (*Fig. 2-15*). Let the height of the frame be L, the distance between columns be l_1. Let the levels be equally spaced with distance l. Let the bending rigidity of each of the beams of the frame be $E_1 I_1$ and that of the columns of the frame be EI, suppose that the columns are infinitely rigid for axial compression. The beams provide elastic end restraints for the beams with spring rigidity

$$K = \frac{6E_1 I_1}{l_1}.$$

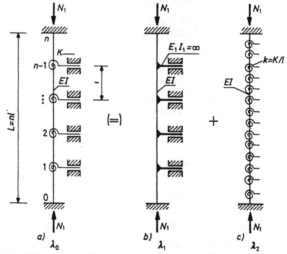

Fig. 2-15 Two-legged plane frame

In such a way the half of the frame can be modelled as a compressed bar with elastic restraint at every level (*Fig. 2-16a*). Let the reference value of the load be the unity.

Fig. 2-16 *a*) Model of the column of the frame. Replacement of the pointwise elastic rotational support with: *b*) pointwise rigid rotational support, *c*) continuous elastic restraint for rotation

The task is to determine the least critical load parameter λ_0 of the bar with pointwise rotational elastic restraints shown in *Fig. 2-16a*. The exact solution can be found in KÁRMÁN and BIOT [1940]. This critical load parameter can be obtained again from the expression (2-70) with the difference that now β_0 is the minimum of the positive solutions β of the equation

$$\cos\beta + 3\mu\frac{\sin\beta}{\beta} = \cos\frac{2\pi}{n}. \tag{2-75}$$

Here the meaning of parameter μ is:

$$\mu = \frac{E_1 I_1 l}{E I l_1}.$$

Then let us determine the *Kollár* critical load parameter λ_K of the bar. This can be done by the least critical load parameter λ_1 and λ_2 of the sway bar with pointwise rigid restraints for rotation (*Fig. 2-16b*) and of the bar on elastic foundation for rotation (*Fig. 2-16c*), respectively. λ_1 is obtained as the *Euler* load of a sway column of length l with built-in ends, so it is given by (2-73). The least critical load parameter λ_2 of the bar on elastic foundation for rotation – with modulus $k = K/l$ of continuous rotational elastic support – is the least eigenvalue of the eigenvalue problem

$$EIu'''' - ku'' + \lambda u'' = 0,$$

$$u\,(0) = u'\,(0) = u\,(l) = u'\,(l) = 0$$

and can be given in the form

$$\lambda_2 = \left[\left(\frac{2}{n}\right)^2 + \frac{6\mu}{\pi^2}\right]\frac{\pi^2 EI}{l^2}. \tag{2-76}$$

Hence, as (2-73) and (2-76) are known, it is easy to obtain λ_K due to (2-69).

Let $E_1 I_1 = EI$, $l_1 = l$. So, $\mu = 1$. Let us determine the parameters for different values of n and the error of the *Kollár* approximation. Our calculations have resulted in the following:

$n = 2$
$\qquad\qquad\lambda_1 = \lambda_E,$
$\qquad\qquad\lambda_2 = 1.6079271\lambda_E,$
$\qquad\qquad\lambda_K = 0.6165537\lambda_E,$
$\qquad\qquad\lambda_0 = \lambda_E,$
$\qquad\qquad\Delta = 38.3\%.$

$n = 4$
$\qquad\qquad\lambda_1 = \lambda_E,$
$\qquad\qquad\lambda_2 = 0.8579271\lambda_E,$
$\qquad\qquad\lambda_K = 0.4617658\lambda_E,$
$\qquad\qquad\lambda_0 = 0.6109857\lambda_E,$
$\qquad\qquad\Delta = 24.4\%.$

$n = 20$
$\qquad\qquad\lambda_1 = \lambda_E,$
$\qquad\qquad\lambda_2 = 0.6179271\lambda_E,$
$\qquad\qquad\lambda_K = 0.3819252\lambda_E,$
$\qquad\qquad\lambda_0 = 0.4046892\lambda_E,$
$\qquad\qquad\Delta = 5.6\%.$

$$n = \infty \qquad\qquad \lambda_1 = \lambda_E,$$
$$\lambda_2 = 0.6079271\lambda_E,$$
$$\lambda_K = 0.3780813\lambda_E,$$
$$\lambda_0 = 0.3958091\lambda_E,$$
$$\Delta = 4.5\%.$$

The *Kollár* formula (2-69) in this example also is conservative for any value of n. It is worth mentioning that, if elastic supports are replaced with elastic foundation only, and the modifying effect of the rigid supports are not taken into account, then in both Examples 2-11 and 2-12 the approximation is not on the safe side ($\lambda_2 > \lambda_0$).

Example 2-13: Let the task now be the investigation of torsional buckling of a compressed braced plane bar with open chord sections (*Fig. 2-17*). The braced bar can have the following buckling modes [KOLLÁR, 1971]:

(*a*) buckling in lateral direction (*Fig. 2-18a*),

(*b*) pure torsional buckling (*Fig. 2-18b*),

(*c*) twist of the chords with bending of the bracing members (*Fig. 2-18c*).

The critical load corresponding to the mode (*a*) and (*b*) can be determined quite easily. For mode (*c*), however, there is not a known exact solution in a closed form. Therefore, an approximatation of the critical load can be composed of parts by the *Kollár* summation. First, the critical load of the chord should be determined for torsional buckling under the assumption that at the nodes the chord is rigidly restrained against twist. This critical load is $P_{cr,rigid}$. Then we determine the critical load of the chords considered as a bar on a continuous elastic torsional support (elastic restraints provided by the bracing members are continuously distributed along the length of the chords). This critical load is $P_{cr,elastic}$. The critical load $P_{cr,twist}$ for torsional buckling of the chord with elastic torsional restraints at the nodes can be obtained from these two critical loads:

$$\frac{1}{P_{cr,twist}} \approx \frac{1}{P_{cr,rigid}} + \frac{1}{P_{cr,elastic}}.$$

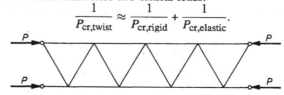

Fig. 2-17 Plane braced bar

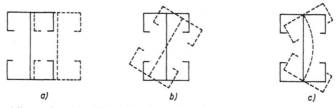

Fig. 2-18 Buckling modes. *a*) Buckling in lateral direction, *b*) pure torsional buckling, *c*) twist of the chords with bending of the bracing members

2.3.4 The Melan theorem

It is known that the natural frequencies of a transversely vibrating bar change under a static axial force. The problem of the free vibration of a bar subjected also to a static axial force is very old. We do not know exactly who was the first to treat this problem. MELAN [1917] published an approximate solution (2-77) of this problem for the case of a concentrated mass at the mid-span of a simply supported bar (but not in this form):

$$\frac{\omega_k^2}{\omega_0^2} = 1 - \frac{N}{N_E}, \tag{2-77}$$

where N_E is the *Euler* critical load of the bar, ω_0 is the least natural circular frequency of the bar free from axial force, ω_k is the least natural circular frequency of the bar subjected to axial static force N. Here N is negative if it is a tensile force. Later it was shown that (2-77) is valid with $\leq$, instead of $=$, and equality holds for the case only when the vibration mode and the buckling mode are the same.

In the following we present a general theorem for this problem, and in the applications the *compressive* force will be considered positive.

Theorem 2-4: Let A, B, C be symmetric linear operators in a *Hilbert* space H such that A is positive definite, B and C are positive and $D_A \subset D_B \cap D_C \subset H$. Suppose that (Bu, u) and (Cu, u) is completely continuous with respect to (Au, u), $u \in D_A$. Let λ_0 and μ_0 be a pair of associated eigenvalues of the two-parameter eigenvalue problem

$$(A - \lambda B - \mu C) u = 0, \quad u \in D_A, \tag{2-78}$$

such that for a fixed λ_0, μ_0 is the minimum of the eigenvalues μ. Let λ_1 and μ_1 be the least eigenvalue of the eigenvalue problem

$$(A - \lambda B) u = 0, \quad u \in D_A \tag{2-79}$$

and

$$(A - \mu C) u = 0, \quad u \in D_A \tag{2-80}$$

respectively. If $\lambda_0 \leq \lambda_1$ and $\mu_0 \geq 0$, then

$$\frac{1}{\lambda_0} \leq \frac{1}{\lambda_1} + \frac{\mu_0}{\lambda_0} \frac{1}{\mu_1}. \tag{2-81}$$

Remark: Inequality (2-81) in fact provides a lower bound in a domain, and an upper bound in another domain disjoint from the previous one. The two bounds can be obtained by multiplying (2-81) by λ_0:

$$1 \leq \frac{\lambda_0}{\lambda_1} + \frac{\mu_0}{\mu_1} \quad (0 \leq \lambda_0 < \lambda_1), \tag{2-82}$$

$$1 \geq \frac{\lambda_0}{\lambda_1} + \frac{\mu_0}{\mu_1} \quad (\lambda_0 < 0). \tag{2-83}$$

Example 2-14. Consider the vibration problem of the axially loaded bar mentioned at the beginning of Section 2.3.4. If EI is the bending rigidity of the cross section of the bar, m is the mass of the bar on a unit length, N is an axial force on the bar (considered positive if it causes compression), and ω is the circular frequency of the vibration, then the equation of transverse vibration of the bar, after eliminating the time variable, is:

$$\left(EIu''\right)'' + \left(Nu'\right)' - \omega^2 mu = 0. \tag{2-84}$$

Under usual boundary conditions (for instance, simple supports), Eq. (2-84) determines a symmetric problem, that is, using notation

$$\lambda N_1 = N,$$

operators A, B, C defined by the relationships

$$Au = \left(EIu''\right)'',$$

$$Bu = -\left(N_1 u'\right)',$$

$$Cu = mu$$

under the given boundary conditions are symmetric, and meet conditions of the Theorem 2-4. If notation $\mu = \omega^2$ is introduced, then statement (2-81) holds, where λ_0 is the given axial force parameter, μ_0 is the square of the least natural circular frequency of the bar for λ_0, λ_1 is the least critical load parameter of the nonvibrating bar ($\mu = 0$), and μ_1 is the square of the least natural circular frequency of the bar free from axial force ($\lambda = 0$). Inequality (2-81), or the pair of inequalities (2-82) and (2-83) are suitable to give bounds of the vibration and stability parameters of the bar.

   Bounds for the least natural frequency of a bar subject to given axial force. From (2-82) and (2-83) it immediately follows that

$$\mu_0 \geq \mu_1 \left(1 - \frac{\lambda_0}{\lambda_1}\right), \quad 0 \leq \lambda_0 < \lambda_1,$$

$$\mu_0 \leq \mu_1 \left(1 - \frac{\lambda_0}{\lambda_1}\right), \quad \lambda_0 < 0.$$

It can be seen that, for the square of the least natural frequency of a bar with axial force parameter λ_0, a lower bound is obtained if the bar is in compression, and an upper bound if a tensile force is in the bar (*Fig. 2-19*).

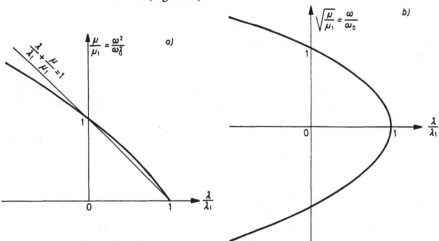

Fig. 2-19 *a*) Square of the natural frequency versus the load parameter, and its *Dunkerley* approximation, *b*) natural frequency versus load parameter

Bound for the least critical load parameter of a bar. An approximate value of the critical load parameter of the bar free from vibration is sought. Expressing λ_1 from (2-81) we obtain

$$\lambda_1 \leq \frac{\lambda_0}{1 - \dfrac{\mu_0}{\mu_1}} \tag{2-85}$$

independent of the sign of the force in the bar.

The *Melan* formula (2-81) approximates the least critical load parameter of a structure always on the unsafe side. In most practical cases, however, the error of the approximation (2-85) is not significant. In this situation there is the possibility of determining an approximate value of the least critical load parameter of a structure by measuring its vibrations. Formula (2-85) is a special case of a more general relationship, which can be obtained in the following way. For two different axial force parameters λ_{01} and λ_{02}, we measure the squares μ_{01} and μ_{02} of the least natural circular frequencies of the structure. These values determine two points in the co-ordinate system λ, μ, and the two points determine a straight line. Points of intersections of this straight line and the co-ordinate axes provide approximate values of λ_1 and μ_1

$$\lambda_1 \leq \lambda_{01} + \frac{\lambda_{02} - \lambda_{01}}{1 - \dfrac{\mu_{02}}{\mu_{01}}}, \tag{2-86}$$

$$\mu_1 \leq \mu_{01} + \frac{\mu_{01} - \mu_{02}}{1 - \dfrac{\lambda_{02}}{\lambda_{01}}}, \quad \lambda_{01}\lambda_{02} \geq 0, \tag{2-87}$$

$$\mu_1 \geq \mu_{01} + \frac{\mu_{01} - \mu_{02}}{1 - \dfrac{\lambda_{02}}{\lambda_{01}}}, \quad \lambda_{01}\lambda_{02} \leq 0. \tag{2-88}$$

By means of formula (2-86), the least critical load (and from it the bending rigidity) of structures which actually cannot be subjected to compressive forces, can be determined with good approximation. Such structures are, for instance, cables. Interestingly, in such a way, a characteristic compression parameter (the least critical load parameter) of a structure can be determined by a tensional experiment.

Stability domain. The stability domain is a set of points in the load parameter space where the structure is in a pre-critical, stable state of equilibrium. Looking at the stability domain of a transversely vibrating bar subjected to an axial force (*Fig. 2-19b*), it is easy to discover a close analogy with the stability domain in *Fig. 2-1b*. Only the least positive critical load parameters are taken into account in the *Dunkerley* theorem. If the load is considered as a multi-parameter system of forces, then well-posed problems can be defined where some of the parameters are negative. For example, for a two-parameter system the buckling problem makes sense if it is considered not only on the positive quarter of the parameter plane but on the remaining three-quarters also. In this way the complete stability domain can be produced, and bounds can be given for the different segments of the boundary curve (boundary hypersurface) of the stability domain, although if some of the parameters are negative the *Dunkerley* theorem is not valid. There exist several different stability domains for different buckling problems. Some of them are shown in *Figs. 2-20, 2-21, 2-22, 2-23*. *Figs. 2-20* and *2-21* are characteristic of problems where loss of stability occurs, for every subsystem, only for positive load parameter. *Figs. 2-22* and *2-23* are characteristic of problems where loss of stability can occur for both positive and negative values of the load parameters. Lateral buckling is such a problem, for instance.

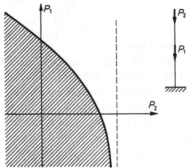

Fig. 2-20 Interaction between two forces in buckling; stability domain

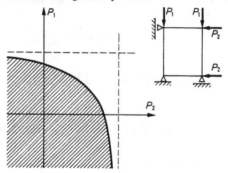

Fig. 2-21 Interaction between two forces in buckling; stability domain

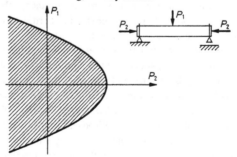

Fig. 2-22 Interaction between forces causing lateral buckling and *Euler* buckling; stability domain

Although complete stability domains are used in practice very rarely, it is important to know them, since they provide insight into some unusual interaction relationships like that in *Fig. 2-24*, for lateral buckling of a continuous beam on three supports, subjected to transverse forces at the middle of the spans. *Fig. 2-24* is in fact similar to a part of *Fig. 2-23*, and the unusual knee-point on the boundary is explained by the fact that the lens-like stability domain in *Fig. 2-23* is an intersection of two domains with smooth boundaries. Intersection of the two boundaries produces the vertex of the interaction curve.

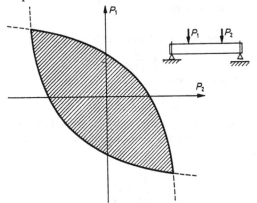

Fig. 2-23 Interaction between two forces causing lateral buckling; stability domain

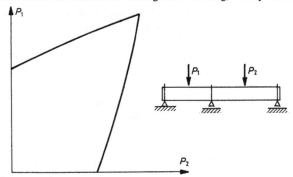

Fig. 2-24 Interaction between two forces causing lateral buckling

2.3.5 The Rankine formula

The *Rankine* formula in fact does not belong to the *Dunkerley* type formulae because it does not relate to elastic structures but elastic-plastic structures. The reason why the *Rankine* formula is mentioned here is that its algebraic form in special cases is similar to that of the *Dunkerley* formula.

The original form of RANKINE's [1863] formula of a compressed elastic-plastic bar is the following:

$$P_F \approx \frac{P_P}{1 + a\,(l/r)^2},\qquad(2\text{-}89)$$

where the notation below is used
P_F failure load,

P_P plastic collapse load ($P_P = A\sigma_y$ where A is the area of the cross section and σ_y is the yield stress),

l the length of the bar,

r radius of gyration of the cross section,

a a constant.

If the bar is pin-ended, and constant a is selected as

$$a = \frac{\sigma_y}{\pi^2 E},$$

then the *Rankine* formula takes the form

$$\frac{1}{P_F} \approx \frac{1}{P_P} + \frac{1}{P_C}. \tag{2-90}$$

The *Rankine* formula (2-90) can be applied not only for individual bars but bar structures subjected to a system of loads. In this case, the formula is called *Rankine-Merchant* formula, and it has the form

$$\frac{1}{\lambda_F} \approx \frac{1}{\lambda_P} + \frac{1}{\lambda_C} \tag{2-91}$$

where $\lambda_F, \lambda_P, \lambda_C$ are the failure load factor of the elastic-plastic structure, the rigid-plastic collapse load factor of the structure, the least elastic critical load factor of the structure, respectively. The so-called *Rankine-Merchant* load factor λ_R is defined by the relationship

$$\frac{1}{\lambda_R} = \frac{1}{\lambda_P} + \frac{1}{\lambda_C}.$$

The *Rankine-Merchant* load factor λ_R for certain problems gives a lower bound of the true failure load factor λ_F, but for other problems an upper bound. The *Rankine-Merchant* load factor is in fact an empirical result, therefore it is not possible to find useful conditions under which the *Rankine-Merchant* load factor is a lower bound of the failure load factor with absolute certainty [HORNE, 1995].

Experiments show that, for certain structures, the *Rankine-Merchant* formula (2-91) is too conservative. Thus, it was reasonable to fit the formula to the experimental results. This intention has led to the nonlinear interaction formulae, many of them have been applied in different design codes. One of them is similar to the *Strigl* formula (2-35):

$$\frac{\lambda_F}{\lambda_P} + \frac{\lambda_F}{\lambda_C} - \frac{\lambda_F^2}{\lambda_P \lambda_C} \approx 1.$$

But another one is also well known:

$$\left(\frac{\lambda_F}{\lambda_P}\right)^m + \left(\frac{\lambda_F}{\lambda_C}\right)^n \approx 1, \quad m \geq 1, \quad n \geq 1.$$

All these formulae are also empirical; to our knowledge there are no conditions under which these formulae provide lower bounds of the true failure load.

2.4 CONCLUSIONS

From point of view of practical structural design, we summarize the statements of this chapter as follows.

1. The method based on the subdivision of rigidity into parts, due to the *Southwell* theorem, gives a conservative approximation of the least elastic critical load parameter only if the buckling problem cannot be defined for the opposite direction of the load, that is the problem has positive eigenvalues only.

2. The method based on the subdivision of the load system into parts, due to the *Dunkerley* theorem, in the case of all the structures occurring in practice, gives a conservative approximation of the least positive elastic critical load parameter of the structure.

3. The method based on partial rigidizing of a structure, due to the *Föppl-Papkovich* theorem, gives a conservative approximation of the least elastic critical load parameter only if the deformations occurring at different partial rigidizing of the structure are independent, and the buckling problem cannot be defined for the opposite direction of the load, that is, the problem has positive eigenvalues only.

4. The *Kollár* method based on a transformation of the elastic properties of the investigated structure, according to experience so far, gives a conservative approximation of the least elastic critical load parameter of the structure. Correctness of this method, however, has not been proved yet. A proof of the *Kollár* conjecture is still needed.

5. For a transversely vibrating structure subjected to a static axial force, in every case, the *Melan* formula gives a nonconservative estimate of the least elastic critical load parameter of the structure.

6. Linear and nonlinear variants of the *Rankine* formula can give both conservative and nonconservative approximations of the failure load parameter of an elastic-plastic structure. The *Rankine* formula in fact is an empirical relationship which, as an inequality, cannot be proved theoretically with general validity.

7. The summation theorems and formulae are valid only for bifurcation critical loads in the linear theory of stability. Critical loads representing limit points cannot be included.

8. The summation theorems are valid only for conservative forces. This follows from the fact that the equation $(A - \lambda B)u = 0$ is in fact an equilibrium equation expressed by displacement u. Forming the scalar product of this equation with u we obtain

$$(Au, u) - \lambda(Bu, u) = 0$$

that yields the *Rayleigh* quotient, if λ is expressed from it. The expression $(Au, u) - \lambda(Bu, u)$ is twice the total potential energy of the system. Thus, it is tacitly supposed that there exists a potential, that is, the forces are conservative.

9. An important term in the theory of elastic stability is orthogonality with respect to a symmetric linear operator A, that is, A-orthogonality. [$u, v \in D_A$ are A-orthogonal, if $(Au, v) = 0$ (Section 2.3.2).] It is easy to show that eigenfunctions (buckling forms) corresponding to different eigenvalues of the problem $(A - \lambda B)u = 0$ are A-orthogonal. The approximation due to the *Southwell* and *Dunkerley* theorems is the more exact, the closer the buckling forms of the subproblems are to each other. The formulae are exact, if the buckling forms of the subproblems are identical. The error of the approximation is a maximum if the buckling forms of the subproblems are A-orthogonal. In this case the buckling modes corresponding to the subproblems are not combined in the original problem. In connection with the *Dunkerley* theorem, this fact is illustrated in *Fig. 2-25* where u_0, u_1, u_2 are buckling shape functions (eigenfunctions) corresponding to critical load parameters (eigenvalues) λ_0, λ_1, λ_2.

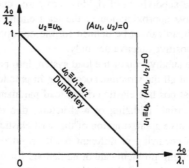

Fig. 2-25 Effect of buckling forms upon the exactness of the *Dunkerley approximation*

Historical remarks on the *Southwell* and *Dunkerley* type theorems and formulae, and additional references can be found in the Hungarian edition of this book [KOLLÁR, 1991].

3

Interaction of different buckling modes in the post-buckling range

Lajos KOLLÁR

3.1 DESCRIPTION OF THE PHENOMENON

Complex structures, consisting of various elements, may lose their stability, as a rule, in several ways: one of their elements may buckle (local buckling), or the whole structure may lose its stability (global buckling). These two modes of buckling interact in such a way that, on the one hand, the buckling of one of the elements reduces the stiffness of the whole structure (since this consists of the 'sum' of the rigidities of the elements). On the other hand, due to the buckling deformation of the whole structure, some elements undergo overloading, so that these will buckle under a lower load intensity than as parts of the undeformed structure.

Global and local buckling shapes are in most cases orthogonal to each other, so that, on the basis of what has been said in Chapter 2, we could think that they do not influence each other, and the bifurcation critical load can be computed for both buckling modes separately, disregarding the other mode. This is true, however, only in the frame of the linear theory, operating with infinitely small deformations. In the nonlinear theory, necessary for taking into account the large deformations in the post-buckling range, this interaction becomes significant. In this chapter we intend to investigate this phenomenon.

Due to the interaction in the post-buckling range, the (initial) post-buckling load-bearing capacity of the structure changes: in most cases it becomes decreasing even if the post-buckling load-bearing capacity of the elements or of the whole structure is constant or increasing. The post-buckling load-bearing capacity decreases most, as a rule, if the critical loads causing local and global buckling are equal to each other. In this case the imperfection sensitivity of the structure is also the greatest. This is a remarkable fact, since design engineers generally try to equalize the critical loads of global and local buckling, because by so doing they want to 'optimize' the design. This problem thus deserves a detailed analysis.

The different buckling modes do not always interact in such a way as to result in a sharper decreasing post-buckling load-bearing capacity. In [REIS and ROORDA, 1979] the mathematical criteria can be found on the basis of which we can decide whether the interaction of the different buckling modes is detrimental or not. In the following, however, we shall show only examples in which this interaction causes (or may cause) the post-buckling load-bearing capacity to decrease. In these cases even if the curves describing the post-buckling load-bearing capacity have an ascending character, having no descending section, the asymptotic values to which they tend are inferior to those which would be valid without interaction.

In our investigations we assume a perfectly elastic material. We will consider the post-buckling load-bearing capacity of a buckling bar as constant, and that of a buckling plate as increasing according to the diagram of *Fig. 1.1*.

So far, in our description of the phenomena we assumed that local and global buckling always interact. There are, however, cases in which the post-buckling load-bearing capacity becomes decreasing due to the interaction of buckling modes of 'equal rank'. We shall show an example for this in Sect. 3.5.

3.2 THE POST-BUCKLING LOAD-BEARING CAPACITY OF A BRACED COLUMN

Let us consider the braced column of *Fig. 3-1*, supported at both ends by hinges; its critical load is denoted by P_E. The bar force causing buckling of one chord bar be S_{cr}, so that – assuming central compression of the whole column – the load causing local buckling is $P_l = 2S_{cr}$. We assume that the inclined bars joining the supports do not buckle.

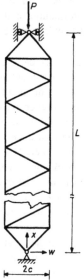

Fig. 3-1 The braced column

We want to investigate first the centrally compressed, *perfectly straight* column.

Let us first assume that $P_E < P_1$, i.e. the whole column buckles prior to the chord bar. The buckling shape be

$$w = W \sin \frac{\pi}{L} x, \tag{3-1}$$

where W is the horizontal displacement of the middle cross section, i.e. the amplitude of the buckling shape.

In the buckled state, in the middle chord bar on the compressed side of the column a bar force

$$S = \frac{P_E}{2} + \frac{P_E W}{2c} = \frac{P_E}{2}\left(1 + \frac{W}{c}\right) \tag{3-2}$$

arises. Let us denote by W_{cr} the value of W at which S becomes equal to S_{cr}. Thus, writing S_{cr} instead of S, and W_{cr} instead of W, we can express W_{cr} from (3-2):

$$\frac{W_{cr}}{c} = \frac{2S_{cr}}{P_E} - 1. \tag{3-3}$$

If $W > W_{cr}$, then (3-2) assumes the form:

$$S_{cr} = \frac{P}{2}\left(1 + \frac{W}{c}\right), \tag{3-4}$$

and from this we can determine the value of P which keeps the equilibrium with the constant S_{cr} at increasing W:

$$\frac{P}{P_E} = \frac{2S_{cr}/P_E}{1 + \frac{W}{c}} = \frac{P_1/P_E}{1 + \frac{W}{c}}. \tag{3-5}$$

That is, P has to decrease with increasing W.

On the other hand, if we start from the assumption that $P_E > P_1$, then the chord bar buckles first. So P can increase only up to P_1. After that, due to the fact that the buckled chord bar does not develop any additional stiffness, the whole column also begins to buckle, and the relation between P and S is given by (3-4), from which we obtain the relation (3-5), valid now from $W = 0$ on.

After buckling of the chord bar the buckling shape of the column will deviate from the sine curve. We will, however, neglect this deviation.

As we can see, the whole column becomes curved either at P_E or at a smaller load value. Let P_b denote the value of the load at which the column begins to become curved. The variation of P_b with the ratio P_E/P_1, together with the character of the initial post-buckling behaviour, is shown in *Fig. 3-2a*, and the load P carried by the column (i.e. the post-buckling load-bearing capacity) as a function of the amplitude W of the curved form is depicted in *Fig. 3-2b*. (The two little diagrams in *Fig. 3-2a* also show the post-buckling load-bearing capacity of the structure as a function of the amplitude W of the curved form in the domains $P_E < P_1$ and $P_E > P_1$). The diagrams clearly show that after buckling of the chord bar the load-bearing capacity begins, sooner or later, to decrease. The slope of the corresponding curve is the steepest when the absolute value of dP/dW is the greatest. From (3-5) we obtain

$$\frac{dP}{dW} = -\frac{P_1}{\left(1 + \frac{W}{c}\right)^2}\frac{1}{c}, \tag{3-6}$$

and the absolute value of dP/dW reaches its maximum when P_1 assumes its highest value at $W = 0$. *Fig. 3-2b* shows that this occurs in the case of $P_1 = P_E$, i.e. when the column has been designed 'optimally' in such a way that the column and the chord bar buckle at the same load. Thus the 'optimum' design according to the simple engineering sense resulted in the steepest falling post-buckling load-bearing capacity, i.e. in the greatest imperfection sensitivity.

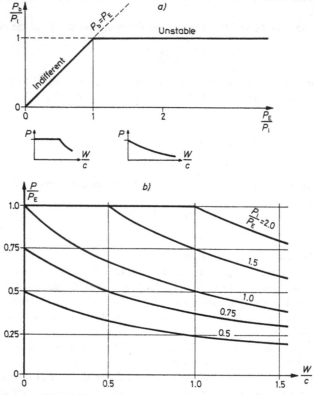

Fig. 3-2 The post-buckling behaviour of the braced column. *a*) The domains of the various initial post-buckling behaviours, *b*) the post-buckling behaviour

We can visually describe the post-buckling behaviour of the column also in the following way. Up to the buckling of the chord bar the internal ('resisting') bending moment of the column increases at the same pace as the external bending moment of the load. Since the chord bar is not able to take any additional compressive force after buckling, the column is also not able to take any additional bending moment, but the internal bending moment 'accumulated' until buckling of the chord bar has to equilibrate the moment of the external force which must decrease since its eccentricity increases. We can put this also in the following way. The critical load of the structure, computed with its bending rigidity valid after buckling of one of its element (in our case: zero rigidity), will be the value to which the external load, equilibrated by the structure, has to tend. This value can be called 'asymptotic critical load'. Since, in the case of the braced column, this value is equal to zero, the curves of *Fig. 3-2b* also tend to zero.

Let us investigate now the *initially curved* column. Assume that this initial curve w_0 has the shape of the buckling mode, i.e.

$$w_0 = W_0 \sin \frac{\pi}{L} x. \tag{3-7}$$

Due to the compressive force P the middle amplitude W_0 of this curve increases to

$$W = \frac{W_0}{1 - \dfrac{P}{P_E}}, \tag{3-8}$$

according to Eq. (1-14). From this P can be expressed as a function of W:

$$\frac{P}{P_E} = 1 - \frac{\left(\dfrac{W_0}{c}\right)}{\left(\dfrac{W}{c}\right)}. \tag{3-9}$$

The relation (3-9) remains valid until the bar force S reaches the value S_{cr}. Since S is defined by (3-2), in this latter W has to be replaced by (3-8).

When S reaches S_{cr}, P assumes its maximum value P_{max}, which the column is able to carry. Thus, after some rearrangement, Eq. (3-2) assumes the form:

$$\frac{P_1}{P_E} = \frac{P_{max}}{P_E} \left[1 + \frac{W_0}{c\left(1 - \dfrac{P_{max}}{P_E}\right)} \right]. \tag{3-10}$$

Solving this equation of the second degree for P_{max}, we obtain P_{max} as a function of W_0. The result is to be seen in *Fig. 3-3*. We can see that P_{max} decreases with increasing W_0 steepest if $P_1/P_E = 1$: here it starts with a vertical tangent. This is in accordance with what has been said in Section 1.2: the perfect structure with the steepest falling post-buckling load-bearing capability (the curve belonging to P_1/P_E in *Fig. 3-2b*) exhibits at the same time the greatest imperfection sensitivity.

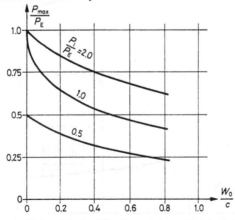

Fig. 3-3 The imperfection sensitivity of the braced column to the eccentricity due to the initial curvature

After reaching P_{max} the chord bar buckles, and the relation between P and W is again described by Eq. (3-5), as in the case of central compression.

Hence the function $P(W)$ consists of two parts: up to $P \le P_{max}$ the relation (3-9), and after that (3-5) is valid. At $P = P_{max}$ there is a kink. In *Fig. 3-4* we have plotted the diagrams $P(W)$ for two values of W_0/c and for three ratios P_1/P_E.

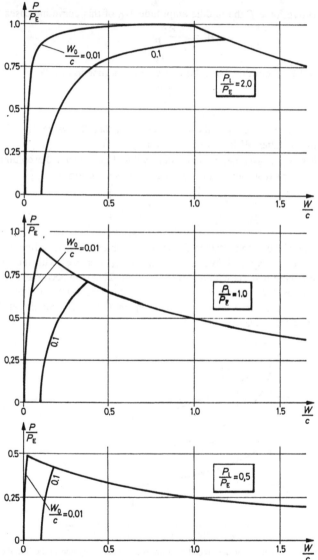

Fig. 3-4 The behaviour of the eccentrically loaded braced column

We still have to mention the fact that in the domain $P_l/P_E < 1$, where the curves of the post-critical load-bearing capacity start with inclined tangents (*Fig. 3-2b*), the load P_{max}, causing the divergence of equilibrium of the eccentrically compressed column, decreases with increasing initial eccentricity W_0 not according to the law $\sqrt{W_0}$, given in Chapter 1, i.e. it does not start with a vertical tangent, but with an inclined one (*Fig. 3-3*). The explanation is the following. We have here to do with two different structures. Before reaching the maximum load a more rigid structure deforms, whose deformation is governed by Eq. (3-8), i.e. by the critical force P_E which is twice the load P_l causing buckling of the chord bar of the centrally compressed column. The deformation before reaching the maximum load is thus much smaller than it would be if it increased according to the law $W = W_0/(1 - P/P_l)$, corresponding to the actual critical load. After reaching the load causing buckling of the chord bar, however, the rigidity of the structure suddenly drops. Consequently, in the domain $P < P_{max}$ a structure deforms which is different from that which buckles after reaching $P = P_{max}$. The phenomenon is the same as the behaviour of the structure made of perfectly plastic material and having a two-chord cross section (see *Fig. 1-12*): as there the plastification of one flange, here the buckling of one chord bar renders the load-bearing capacity of one part of the structure constant.

In the cases $P_l/P_E \geq 1$ the situation is different not only in the respect that the structure deforms corresponding to the actual critical force $P_b = P_E$, according to Eq. (3-8), much longer, but also here we arrive at P_{max} when, reaching the critical force of the chord bar, the rigidity of the structure suddenly drops.

From the foregoing it can be seen that the buckling of the chord bar strongly reduces the post-buckling load-bearing capacity of the structure. This can be explained visually, according to what has been said in Section 3.1, if we consider that the bending stiffness of the whole column, which determines the post-buckling load-bearing capacity, consists of the tensile stiffnesses of the chord bars. The buckling of the chord bar means that the two end points of the bar can approach each other also under constant load, so that the bending rigidity of the whole column drops. Consequently, the local buckling reduces the post-buckling load-bearing capacity of the whole structure in all cases when the rigidity of the whole structure consists of the stiffnesses of the locally buckling elements, and this is true with most structures.

Let us now investigate the influence of *the initial curvature of the chord bars*. For the sake of simplicity we assume that all chord bars have the same initial curvature (*Fig. 3-5a*), but the axis of the whole column is straight.

The initially curved shape ζ_0 of one chord bar should be identical with its own buckling shape, i.e.

$$\zeta_0 = a_0 \sin \frac{\pi}{l} x . \tag{3-11}$$

This increases, due to the compressive force $S = P/2$ acting in the chord bar, according to (1-14), to

$$\zeta = \frac{a_0}{1 - \dfrac{S}{S_{cr}}} \sin \frac{\pi}{l} x = a \sin \frac{\pi}{l} x, \tag{3-12}$$

where

$$a = \frac{a_0}{1 - \dfrac{S}{S_{cr}}} = \frac{a_0}{1 - \dfrac{P}{P_1}} \qquad (3\text{-}12b)$$

is the increased amplitude of curvature.

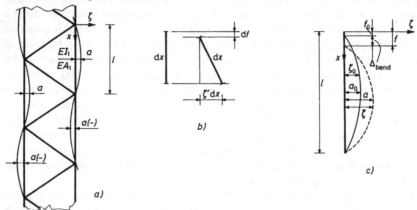

Fig. 3-5 Investigation of the initial curvature of the chord bars. *a*) The assumed initial curvature of the chord bars, *b*) the apparent shortening due to the inclination of the elementary section of the bar axis, *c*) the apparent shortening due to the curvature of the bar axis

The approach of the bar end points due to the curvature ζ can be computed – since we have to deal with small displacements – in the following way.

Due to the inclination $d\zeta/dx \equiv \zeta'$ of an elementary section of the length dx its two end points approach each other in the direction of x by the value df (*Fig. 3-5b*):

$$df = dx - \sqrt{(dx)^2 - (\zeta' dx)^2} \approx dx - dx\left[1 - \frac{1}{2}\left(\zeta'\right)^2\right] = dx\frac{1}{2}\left(\zeta'\right)^2, \qquad (3\text{-}13)$$

so that after the curvature ζ the approach Δ_{bend} of the two end points of the bar with the original curvature ζ_0, becomes (*Fig. 3-5c*):

$$\Delta_{bend} = f - f_0 = \frac{1}{2}\int_0^l \left(\zeta'\right)^2 dx - \frac{1}{2}\int_0^l \left(\zeta_0'\right)^2 dx = \frac{\pi^2}{4l}\left(a^2 - a_0^2\right) =$$

$$\frac{\pi^2 a_0^2}{4l}\left[\left(\frac{1}{1 - \dfrac{P}{P_1}}\right)^2 - 1\right] = \frac{\pi^2 a_0^2}{4l}\frac{2\dfrac{P}{P_1} - \dfrac{P^2}{P_1^2}}{1 - 2\dfrac{P}{P_1} + \dfrac{P^2}{P_1^2}}. \qquad (3\text{-}14)$$

To this the shortening caused by the compressive force has to be added:

$$\Delta_{compr} = \frac{Sl}{EA_1} = \frac{Pl}{2EA_1}, \qquad (3\text{-}15)$$

The total shortening due to P is thus:

$$\Delta = \Delta_{bend} + \Delta_{compr} \qquad (3\text{-}16)$$

and, hence, the compressive stiffness T_1 of one bar becomes:

$$T_1 = \frac{P/2}{\Delta/l} = \cfrac{1}{\cfrac{\pi^2 a_0^2}{2l^2}\cfrac{\dfrac{2}{P_1} - \dfrac{P}{P_1^2}}{1 - 2\dfrac{P}{P_1} + \dfrac{P^2}{P_1^2}} + \cfrac{1}{EA_1}}. \tag{3-17}$$

The bending stiffness of the whole column is composed of two compressive stiffnesses T_1:

$$EI(P) = 2c^2 T_1, \tag{3-18}$$

and the maximum force $P = P_b$ which can be carried by the column in its straight state will be the critical force computed with the bending stiffness given by Eq. (3-18):

$$P_b = \frac{\pi^2 EI(P)}{L^2} = \frac{2c^2\pi^2}{L^2} \cfrac{1}{\cfrac{\pi^2 a_0^2}{2l^2}\cfrac{\dfrac{2}{P_1} - \dfrac{P_b}{P_1^2}}{1 - 2\dfrac{P_b}{P_1} + \dfrac{P_b^2}{P_1^2}} + \cfrac{1}{EA_1}} \tag{3-19}$$

Since the critical force of one chord bar is

$$S_{cr} = \frac{P_1}{2} = \frac{\pi^2 EI_1}{l^2} \tag{3-20}$$

and the critical force of the whole column, computed with straight chord bars, is

$$P_E = \frac{\pi^2 EI_0}{L^2} = \frac{\pi^2 E2c^2 A_1}{L^2}, \tag{3-21}$$

we can transcribe Eq. (3-19) as follows, introducing also the radius of gyration $i_1 = \sqrt{I_1/A_1}$ of the cross section of the chord bar:

$$\frac{1}{2}\frac{a_0^2}{i_1^2}\frac{P_b}{P_1}\left(2 - \frac{P_b}{P_1}\right) = 2\frac{P_E}{P_1}\left(1 - \frac{P_b}{P_1}\right)^2 - 2\frac{P_b}{P_1}\left(1 - \frac{P_b}{P_1}\right)^2. \tag{3-22}$$

In knowledge of the ratios a_0/i_1 and P_E/P_1, we can determine P_b from the equation of the third degree (3-22). P_b is the bifurcation critical load of the braced column with straight axis (but with curved chord bars), which is plotted against P_E/P_1 for various initial curvatures a_0/i_1 in *Fig. 3-6*.

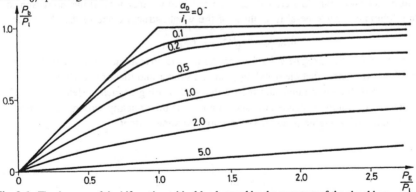

Fig. 3-6 The decrease of the bifurcation critical load caused by the curvature of the chord bars

After reaching the load P_b the whole column begins to buckle. During deformation the bar force on the compressed side would increase, if the load P acting on the column maintained its value P_b = constant [see Eq. (3-4)]. The increasing bar force, however, causes a decreasing compressive stiffness in the bar [cf. Eq. (3-17), where we must now substitute S, computed from Eq. (3-4), for $P/2$]. S cannot reach S_{cr}, because in this case the stiffness of the chord bar would become zero, and this can be avoided only if P decreases, during buckling of the whole column, to values lower than P_b, i.e. if the post-buckling load-bearing capacity of the column decreases after reaching the load value $P = P_b$.

The complete investigation of the braced column should also include the cases when only the chord bars on one side of the column have an initial curvature, and, furthermore, when both the chord bar and the column are initially curved. However, since our aim is only the explanation of the phenomenon, and not the complete, detailed analysis, we refer to the literature [THOMPSON and HUNT, 1973], where similar investigations are to be found.

The investigations shown so far are, however, sufficient to draw some conclusions.

The buckling of the chord bars influences the post-buckling behaviour of the column in the same way as the plastification of the chords of the two-chord section shown in Section 1.4 (compare the curves of *Fig. 1-11b* with those of *Fig. 3-4*): the load-bearing capacity of the chord becomes constant in both cases. So we can say also of the post-buckling behaviour of the column that – if the whole column buckles first – the buckling of the chord bar changes the structure, and also its initial post-buckling behaviour (cf. the curves of *Fig. 3-2b* pertaining to $P_E < P_l$).

There is a difference between the two cases only during unloading: the braced column returns to its original position unequivocally along the equilibrium path, while the plastified structure returns along different paths, depending on when its unloading begins. It is interesting to compare our results with those presented in Chapter 2 on braced columns. There it was stated that since the buckling modes of the chord bars and the whole column are orthogonal to each other, they do not interact, and thus it has no sense to apply the *Föppl-Papkovics* theorem

$$\frac{1}{N_{cr,complex}} \le \frac{1}{N_{cr,global}} + \frac{1}{N_{cr,local}}$$

since it furnishes an unduly low critical load. To put it in another way, it has no reason to compute a 'complex slenderness' from those of the whole structure and of the chord bar:

$$\lambda_{complex} = \sqrt{\lambda_{global}^2 + \lambda_{local}^2}$$

which follows from the *Föppl-Papkovich* theorem. The investigations presented in the foregoing, however, show that the buckling of the chord bar and that of the whole structure interact in the post-buckling range and reduce the load-bearing capacity. Hence, although the formula for the complex slenderness shown above is theoretically wrong, nevertheless it takes into this unfavourable phenomenon account to some extent.

3.3 THE POST-BUCKLING LOAD-BEARING CAPACITY OF THE RIBBED PLATE

In the following, based on [WALKER, 1975], we shall treat an example which shows that the interaction of global and local buckling may result in a decreasing post-buckling load-bearing capacity even if the structure has during global buckling a constant post-buckling load-bearing capacity, and during local buckling an increasing one.

Let us consider a ribbed plate depicted in *Fig. 3-7*, loaded by uniformly distributed forces over its cross section. The static model of the whole structure be a hinged-hinged bar. The cross-sectional area and moment of inertia of a section of the width b be A_0 and I_0 respectively.

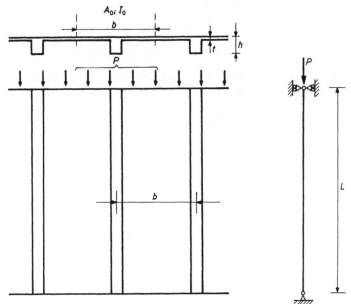

Fig. 3-7 The ribbed plate

For the sake of simplicity let us assume that the ribs do not become twisted during buckling and, consequently, they rigidly clamp the plate in the transverse direction. The fields of the plate be sufficiently long in order to allow the buckling dimples to freely develop (independently of the actual distance of the two supporting hinges).

Let us denote the critical load of the whole structure, as a bar with a T-shaped cross section, by P_E:

$$P_E = \frac{\pi^2 E I_0}{L^2}.$$ (3-23)

Its post-critical load-bearing capacity will be considered constant, according to *Fig. 3-8b*.

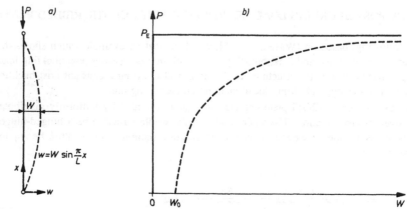

Fig. 3-8 The model of the whole structure a), and its post-critical behaviour b)

The critical stress of a plate field between two adjacent ribs is – in the case of a sufficiently large ratio L/b – according to [TIMOSHENKO and GERE, 1961], taking $\nu = 0.25$ for *Poisson's* ratio:

$$\sigma_{cr} = 7\frac{\pi^2 E}{12(1 - \nu^2)} \left(\frac{t}{b}\right)^2 = 6.14E \left(\frac{t}{b}\right)^2. \qquad (3\text{-}24)$$

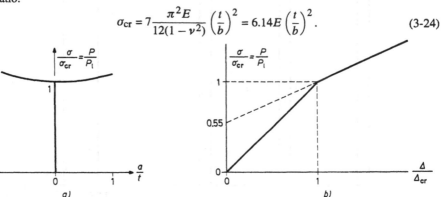

Fig. 3-9 The behaviour of the plate. a) The post-buckling load-bearing capacity, b) the relation between the total shortening of the plate and the arising stress

The post-buckling behaviour of the plate is shown in *Fig. 3-9a*, as a function of the maximum displacement a, perpendicular to the plate. The post-critical load-bearing curve of the originally perfectly plane plate can be described, according to what has been said in Section 1.1.1, with sufficient accuracy by the expression

$$\frac{\sigma}{\sigma_{cr}} = 1 + c_1 \left(\frac{a}{t}\right)^2. \qquad (3\text{-}25)$$

Since after buckling the stresses taken by the plate cease to be uniformly distributed, in the foregoing equation σ means the average stress acting on the plate. The total compressive force N acting on the plate field is

$$N = bt\sigma. \qquad (3\text{-}26a)$$

The force acting on the whole cross section will be denoted by P:

$$P = A_0\sigma. \tag{3-26b}$$

The value of the constant c_1 can be taken from WILLIAMS and WALKER [1975], *Table 4.* Neglecting the term ψ^3 in their equation (1)

$$\frac{a}{t} = 1.846\sqrt{\frac{\sigma}{\sigma_{cr}} - 1},$$

is arrived at, and comparing this with (3-25), the value

$$c_1 \approx 0.30$$

will be obtained.

Since the plate is the flange of the beam with a T section, we need its compressive stiffness. To this purpose we have to determine the average shortening Δ of the distance between the two edges $x = $ constant of the plate field of the dimensions lb (*Fig. 3-10a*), compressed in the x direction.

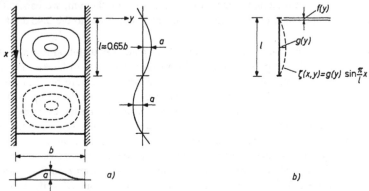

Fig. 3-10 The (apparent) shortening in the x direction of a buckled plate field. *a*) The buckled plate field, *b*) the shortening of the distance between the end points of the fibres running in the x direction

Knowing the maximum, middle displacement a of the buckled plate field, perpendicularly to the plane of the plate, we can compute, by making use of the simple geometric relations (3-13) and (3-14), the shortening of the distance between the end points of the fibers, running in the x direction, of the plate field lb, caused by the fact that they become inextensionally curved. Since this will be a function of y, we have to integrate it over the length b and divide it by b.

We know the ratio of the side lengths of a buckled plate field and the function describing the buckled shape ζ [TIMOSHENKO and GERE, 1961]:

$$l/b \approx 0.65; \tag{3-27}$$

and

$$\zeta = a \left(\frac{1}{2} - \frac{1}{2}\cos\frac{2\pi}{b}y\right)\sin\frac{\pi}{l}x. \tag{3-28}$$

Performing the calculation we obtain for the apparent specific shortening f_1 caused by the curvature: $f_1 = 2.20\frac{a^2}{b^2}$. We have to add this to the physical compression $\varepsilon = \sigma/E$ in order to obtain the total specific shortening Δ:

$$\Delta = \frac{\sigma}{E} + f_1. \tag{3-29}$$

Introducing here f_1, dividing both sides by the compression caused by σ_{cr}: $\quad \Delta_{cr} = \sigma_{cr}/E$, substituting (3-25) for $(a/t)^2$ and using also (3-24) we obtain:

$$\frac{\Delta}{\Delta_{cr}} = \frac{\sigma}{\sigma_{cr}} + 1.21 \left(\frac{\sigma}{\sigma_{cr}} - 1 \right) = 2.21 \frac{\sigma}{\sigma_{cr}} - 1.21. \tag{3-30}$$

from which we can express σ:

$$\frac{\sigma}{\sigma_{cr}} = 0.55 + 0.45 \frac{\Delta}{\Delta_{cr}}. \tag{3-31}$$

The compression of the plate before buckling is given by the relation

$$\frac{\sigma}{\sigma_{cr}} = \frac{\Delta}{\Delta_{cr}}. \tag{3-32}$$

The expressions (3-31) and (3-32) are depicted in *Fig. 3-9b*. Since the compressive stiffness is defined as the direction tangent of the stress-strain diagram, we can state that in the range $\sigma > \sigma_{cr}$ the compressive stiffness of the plate decreases to the 0.45-fold of the original stiffness.

We can now investigate the behaviour of the ribbed plate. Let us introduce the notation P_1, denoting the (central) compressive force which causes the plate to buckle, i.e. which causes stress σ_{cr} in the plate. We first assume that the whole structure buckles first as a bar at the load P_E, see Eq. (3-23), i.e. that $P_E < P_1$. If the plate is on the compressed side, then, due to the bending, the stress in the plate increases faster than the average stress in the structure, but the load carried by the structure remains constant, since it buckles like a bar. When the stress in the plate reaches its critical value σ_{cr}, the plate buckles and its compressive rigidity drops to the 0.45-fold of its original value. We can visualize this as shown in *Fig. 3-11*: the plate behaves as if its width decreased from b to $0.45b$ and this part remained unbuckled. This modified T section thus has a centroid C^* shifted from that of the original plate C and, consequently, the originally central compressive force becomes eccentric. The moment of inertia of the cross section also decreases from the original value I to I^*, so that the critical load of the structure with buckled plate becomes, instead of P_E,

$$P^* = \frac{\pi^2 E I^*}{L^2} \tag{3-33}$$

which we call the 'reduced' critical load. Taking all this into account, the load which can be carried by the structure has to decrease to P^*.

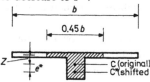

Fig. 3-11 The original cross section and that effective after buckling of the plate

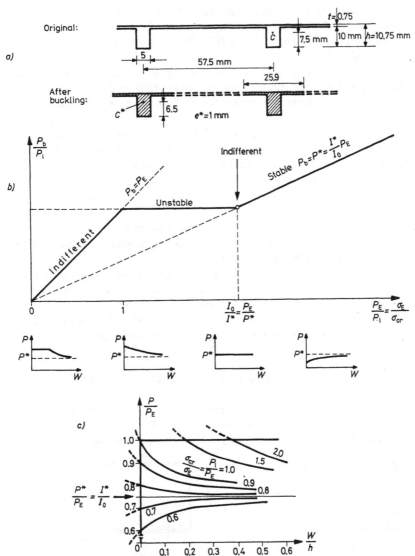

Fig. 3-12 Numerical example for the behaviour of the ribbed plate. *a*) The original cross section and that after plate buckling, *b*) the domains of the various kinds of post-buckling behaviour, *c*) the post-buckling behaviour of the ribbed plate

Second, we assume that the plate buckles first, i.e. $P_E > P_l$. Since the structure with buckled plate cannot carry a load higher than P^*, the load must decrease from the value P_l to P^* if $P_l > P^*$, and increase to P^* if $P_l < P^*$. This value P^* can also be called the 'asymptotic critical load' of the structure. In *Fig. 3-12* we show the results of the numerical example of WALKER [1975]. The cross section can be seen in *Fig. 3-12a*. The variation of the bifurcation load P_b in dependence of the ratio P_E/P_1 is shown in *Fig. 3-12b*, the curves describing the post-buckling behaviour are depicted in *Fig. 3-12c*. Since the ratio P^*/P_E was 0.755, the curves belonging to $P_1 < 0.755 P_E$ are ascending, those belonging to $P_1 > 0.755 P_E$ are descending.

The figure clearly shows that the curve $P(W)$ belonging to $P_1 = P_E$ starts with the steepest initial tangent. Hence the interaction of the two buckling modes causes also here the greatest imperfection sensitivity if the critical forces of the two modes are equal to each other.

Comparing *Fig. 3-12* with *Fig. 3-3* describing the behaviour of the braced column (where we assumed the post-buckling load-bearing capacity of the buckled chord bar to be constant) we can state that that the increasing post-buckling load-bearing capacity of the plate caused the following differences:

- the curve belonging to $P_1 = P_E$ starts with an inclined tangent, instead of a vertical one;
- the load-bearing curves tend to a finite value instead of to zero;
- in the cases when $P_1 < P^*$ the structure has an increasing post-buckling load-bearing capacity (but it tends to a value lower than P_E).

In the literature we can find a solution for the case when the whole structure has an initial curvature [WALKER, 1975]. TULK and WALKER [1976] carried out model tests which confirmed the computation method of *Walker* shown in the foregoing. The torsional buckling of the ribs have been included by FOK, RHODES and WALKER [1976] furthermore by FOK, WALKER and RHODES [1977] into the investigation of the interaction.

3.4 THE POST-BUCKLING LOAD-BEARING BEHAVIOUR OF THE BOX BAR

The compressed bar with a closed hollow section (*Fig. 3-13a*) is also a good example of those complex structures in which local and global bucklings interact and markedly influence the post-buckling load-bearing capacity. Here the post-buckling load-bearing

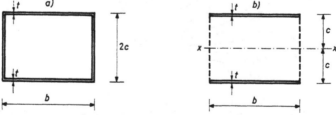

Fig. 3-13 The box bar and its simplified model *a*) The hollow cross section, *b*) *van der Neut's* (two-flange) cross section

capacity of the locally buckling elements is increasing, nevertheless that of the whole structure is decreasing, or, more precisely, the maximum load the structure can carry is always less than that of a structure with unbuckled plates, even if the post-buckling load-bearing curves show a steadily ascending character.

In the following we treat this problem on the basis of the pioneer work of VAN DER NEUT [1969], omitting the derivations.

VAN DER NEUT investigated the behaviour of an idealized hollow cross section (*Fig. 3-13b*), consisting of two flange plates of the width b and of a thickness which is sufficiently small, so that it is permissible to assume a constant normal stress all over the thickness; and of the two connecting 'webs'. The flange plates are simply supported along their two edges parallel to the bar axis. The webs do not take any part of the compressive force, they only transmit shear from one flange to the other, and provide the support of the flange plates (We can imagine them as bracings). This idealized cross section has the advantage that it shows the basic features of the phenomenon correctly, but eliminates all those side effects which would unduly complicate the investigation, but would hardly modify the result.

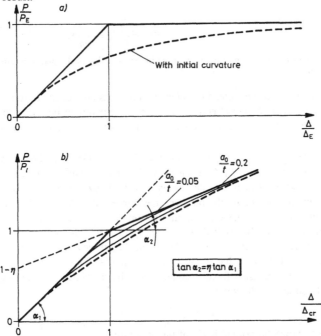

Fig. 3-14 The load-bearing capacity of the elements of the box bar plotted against the longitudinal shortening. *a*) The load-bearing capacity of the bar, *b*) the load-bearing capacity of the plate

VAN DER NEUT started from the assumptions used in the preceding sections: the material is perfectly elastic; the post-buckling load-bearing capacity of the whole bar is constant (see *Fig. 3-8b*), while that of the plates is increasing (see *Fig. 3-9a*). Plotting the load-bearing capacity of the bar against the longitudinal shortening (Δ/Δ_E) we obtain the diagram of *Fig. 3-14a* (Δ_E is the shortening caused by the critical force P_E). Compared with Section 3.3 the only difference is that the two edges of the plates are not clamped, but simply supported, and, consequently, the $\sigma/\sigma_{cr} = P/P_1$ diagram plotted against the longitudinal shortening (Δ/Δ_{cr}) assumes the shape shown in *Fig. 3-14b*. Thus the original compressive rigidity $T_0 = \tan\alpha_1$ of the plate decreases after buckling to its η-fold:

$$T = \eta T_0, \tag{3-34}$$

where $T = \tan\alpha_2$ (*Fig. 3-14b*). In the case of the assumed simple supports $\eta = 0.41$.

The notations P_E, P_1, P_b and P_{max}, introduced earlier, are valid also here.

The rigidity of the geometrically perfect, centrally compressed bar is shown in *Fig. 3-15a*.

Fig. 3-15 The rigidity of the two-flange cross section. *a*) Before buckling of the plates, *b*) after buckling of one of the plates

If $P_E < P_1$, then the whole bar buckles first. Hence, in this case the original, total compressive rigidities of the plates remain effective, and, when the compressive force reaches the value $P_b = P_E$, the equilibrium state of the bar becomes indifferent (neutral), according to *Fig. 3-14a*.

If $P_E > P_1$, then when reaching $P_b = P_1$, the plates buckle, and their compressive rigidities drop to the value $T = \eta T_0$. Consequently, the bending stiffness of the whole bar drops to its η-fold, and its critical load decreases to ηP_E. If $\eta P_E < P_1$, i.e. $P_E/P_1 < 1/\eta$, the whole bar begins to buckle. As the bar becomes curved, the plate on the tension side undergoes unloading, becomes plane again, and its compression stiffness increases to T_0. The bending stiffness of the bar will thus be the $2\eta/(1 + \eta)$-fold of the original value EI_0, cf. *Fig. 3-15b*. The phenomenon is similar to the *Engesser-Kármán* theory of plastic buckling. (We disregard the fact that – due to the unequal stiffnesses of the plates – the geometrically central compressive force becomes statically eccentric.) According to what has been said in Section 3.2, the value

$$P_\eta = \frac{2\eta}{1 + \eta} P_E \tag{3-35}$$

can be called the 'asymptotic critical load' of the structure.

If $P_1 > P_\eta$ (or $1 < \frac{P_E}{P_1} < \frac{1+\eta}{2\eta}$), then when P reaches the value P_1, the equilibrium state of the column becomes unstable, since its asymptotic critical load P_η is lower than P_1. If $P_1 = P_\eta$, then the equilibrium becomes indifferent (neutral), and in the domain $\frac{1+\eta}{2\eta} < \frac{P_E}{P_1} < \frac{1}{\eta}$ the post-critical equilibrium state will be stable, since we can increase the load from P_1 up to P_η, causing buckling.

In the domain where $\eta P_E > P_1$, or $P_E/P_1 > 1/\eta$, the column does not start to buckle at $P = P_1$, because we can increase the load up to ηP_E, computed with the compressive stiffnesses ηT_0 of the buckled plates. When we reach ηP_E and the column begins to bow out, the stiffness of the plate on the tension side does not increase to T_0, but remains ηT_0, because ηP_E is by a finite measure greater than P_l, and thus the plate becomes plane only after a finite elongation during unloading. Until then it exhibits only the stiffness ηT_0, valid in the buckled state. (This means a deviation from the *Engesser-Kármán* theory, namely, even after the slightest unloading, the original value E is valid). When the load reaches ηP_E, the equilibrium state becomes indifferent, according to *Fig. 3-14a.*

The variation of P_b, at which the column begins to bow out, is plotted against the ratio P_E/P_l in *Fig. 3-16a*, together with the indication of the initial post-buckling states of equilibrium. The further behaviour of the structure in the domains mentioned before is shown in the *Figs 3-16b* to f, as functions of the amplitude W of the curved shape.

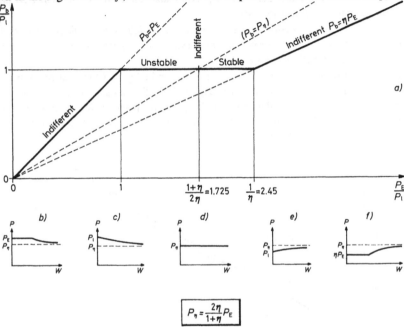

$$P_\eta = \frac{2\eta}{1+\eta} P_E$$

Fig. 3-16 The domains of the different initial post-buckling behaviours of the box bar

According to *Fig. 3-16b*, the equilibrium state of the buckling bar remains indifferent as long as the force in the plate on the compressed side does not reach the critical load of the plate. If this value is reached, the stiffness of the plate drops to ηT_0 and thus the load must also gradually decrease to the asymptotic critical value P_η, because otherwise it cannot be equilibrated by the now slower increasing force in the flange plate while the eccentricity is increasing.

Fig. 3-16c shows a similar phenomenon, but here the drop of the stiffness occurs already at $W = 0$, due to the buckling of the plates at $P = P_l$, so that the load-bearing capacity begins to decrease from $P = 0$ on.

Fig. 3-16d shows the indifferent post-critical behaviour which comes about if $P_l = P_\eta$. That is, in this case, the bar begins to bow out together with the buckling of the plates;

thus one of the plates becomes plane again at once, and the bar buckles with the bending stiffness shown in *Fig. 3-15b*.

Fig. 3-16e presents the phenomenon of *Fig. 3-16c* with 'opposite sign': due to the bowing out of the bar, beginning after buckling of the plates, the plate on the tension side regains its original stiffness T_0, and thus the load-bearing capacity of the structure increases up to the asymptotic critical force P_η.

Finally, in *Fig.3-16f* we see that the load-bearing capacity of the buckling structure remains constant as long as the plate on the tension side does not straighten and its stiffness does not increase up to the original value T_0. After that, the load can be increased until reaching the asymptotic critical load P_η.

VAN DER NEUT also investigated the case of *the initially curved flange plates*. He assumed that both plates have the same initial curvature (waviness) and that this shape is the same as that of the buckling mode.

The behaviour of the initially curved plate is shown in *Fig. 3-14b* by dashed lines. As can be seen, the $P(\Delta)$ diagram of the curved plate is in the domain $\Delta < \Delta_{cr}$ flatter, but in the domain $\Delta > \Delta_{cr}$ steeper than that of the perfectly plane plate. Accordingly, the compressive stiffness of the plate $T = \tan \alpha$ decreases due to the initial waviness when $P < P_1$, but increases when $P > P_1$.

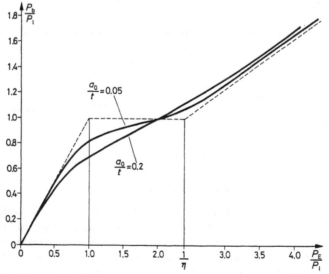

Fig. 3-17 The influence of the initial waviness of the plate on the load-bearing capacity of the box bar

Performing the stability analysis of the whole column with these modified T values, we arrive at the curves of *Fig. 3-17*, which show the decrease of the bifurcation load P_b of the column with perfectly plane plates when $P_E < 2P_1$, and its increase when $P_E > 2P_1$, in accordance with the domains $\Delta < \Delta_{cr}$ and $\Delta > \Delta_{cr}$ of *Fig. 3-14b*, respectively. The boundary point between them, $P_E/P_1 = 2$, does not coincide with the point $(1+\eta)/(2\eta)$ of *Fig. 3-16a*, because the stable, indifferent and unstable sections of *Fig. 3-16a* refer to the initial post-buckling behaviour of the whole bar and, consequently, they reflect the influence of the initial curvature of the axis of the whole column, and not of the initial waviness of the plates.

VAN DER NEUT also investigated the influence of the *initial crookedness of the bar axis*. He arrived essentially at the result that as long as $P_E < \frac{1+\eta}{2\eta}P_1$, i.e. $P_\eta < P_1$ (see *Fig. 3-16a*), that is, as long as the post-buckling load-bearing capacity of the perfect bar sooner or later becomes decreasing, the load-bearing capacity of the bar is described by curves similar to those of *Fig. 3-4*, but instead of having a kink in the peak point they are smooth. However, since the descending branches of the diagrams of *Figs. 3-16b* and *c* do not tend to zero – as contrasted to those of *Fig. 3-2b* – but to the finite value of the asymptotic critical load P_η, the load-bearing curves have, at eccentricities greater than a certain limit value W_{OL}, no peak points, but they are steadily ascending (*Fig. 3-18*).

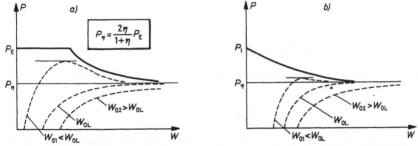

Fig. 3-18 The load-bearing capacity of the initially curved bar. *a*) In the case of $P_E < P_1$, *b*) in the case of $P_E > P_1$.

In the range $P_E > \frac{1+\eta}{2\eta}P_1$, i.e. if $P_\eta > P_1$, where the load-bearing capacity is essentially increasing, the load-bearing curve of the initially curved bar is, of course, monotonically ascending and has no peak point.

VAN DER NEUT has determined the P_{max}/P_b curve of the initially curved bar for two values of P_E/P_1, taking into account also the influence of various values of flange plate waviness (*Fig. 3-19*). This waviness practically does not influence the ratio P_{max}/P_b as long as $a_0/t \le 0.1$, but P_b itself strongly depends on a_0/t (see *Fig. 3-17*). The curves of *Fig. 3-19* show that the initial crookedness of the bar axis reduces P_{max} as compared to P_b only slightly. The end points of the curves on the right correspond to the limit eccentricity values W_{OL}.

THOMPSON and LEWIS [1972] further developed the theory of *Van der Neut* and stated that for optimum design it is expedient if P_E is somewhat less than P_1. SVENSSON and CROLL [1975] included also plasticity into the investigation of box bars.

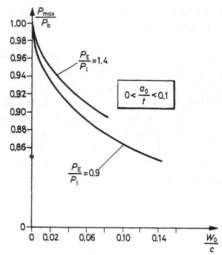

Fig. 3-19 The influence of the initial curvature of the bar axis and of the waviness of the plate on the load-bearing capacity of the bar

3.5 THE INTERACTION OF THE BUCKLING MODES OF CYLINDRICAL SHELLS

In this section we present an example in which not local and global buckling modes, but those of of 'equal rank' interact.

We shall show this phenomenon on the cylindrical shell. It is well known (see e.g. [TIMOSHENKO and GERE, 1961], or [KOLLÁR and DULÁCSKA, 1984]) that the axially compressed cylinder (also when subjected to internal pressure too) buckles, according to the linear theory, in small, local waves. Hence we can consider the shell shallow over such a buckle and we can describe its behaviour by the comparatively simple equations of the shallow-shell theory. We assume the shell to be sufficiently long, so that the end conditions do not influence the buckling shape.

In this section we follow the derivations of HUTCHINSON [1965].

3.5.1 Nonlinear shell equations

When a shell buckles, its points undergo displacements u and v in the shell surface, and w perpendicularly to it, parallel to the co-ordinates x, y and z (*Fig. 3-20*). As a rule, w is much larger than u and v. Nonlinear shell theory means that, in addition to the linear terms of the displacements appearing in the equilibrium and compatibility equations of the linear theory [FLÜGGE, 1973], also the second powers of the displacement w and its derivatives are included, together with their product with the stress function ϕ and its derivatives. This is called *Donnell's* theory. Hence let us start from the nonlinear equilibrium and compatibility equations of shallow shells:

$$\frac{Et^3}{12\left(1-v^2\right)}\Delta^2 w - \left[(z^{\cdot\cdot}+w^{\cdot\cdot})\phi'' - 2\left(z'^{\cdot}+w'^{\cdot}\right)\phi'^{\cdot} + \left(z''+w''\right)\phi^{\cdot\cdot}\right] -$$

$$- \left(n_{x,0}w'' + 2n_{xy,0}w'^{\cdot} + n_{y,0}w^{\cdot\cdot}\right) = 0. \tag{3-36}$$

$$\frac{1}{Et}\Delta^2\phi + \left(z^{\cdot\cdot}w'' - 2z'^{\cdot}w'^{\cdot} + z''w^{\cdot\cdot}\right) + w''w^{\cdot\cdot} - \left(w'^{\cdot}\right)^2 = 0. \tag{3-37}$$

where x,y are co-ordinates on the plane with respect to which the shell can be considered as shallow,

$$' = \partial/\partial x;$$
$$\cdot = \partial/\partial y;$$
$$\Delta = \partial^2/\partial x^2 + \partial^2/\partial y^2;$$

noindent
$z(x,y)$ is the shape function of the shell,
E and v are *Young*'s modulus and *Poisson*'s ratio,
t thickness of the shell,
p_z load acting in the z direction,
$\phi(x,y)$ stress function, whose second derivatives yield the membrane forces arising during buckling:

$$n_x = \phi^{\cdot\cdot} \tag{3-38a}$$
$$n_{xy} = -\phi'^{\cdot}, \tag{3-38b}$$
$$n_y = \phi''. \tag{3-38c}$$

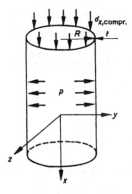

Fig. 3-20 The cylinder subjected to axial compression and internal pressure

Due to the shallowness of the shell, the membrane forces can be taken equal to their projections on the plane xy.

In the case of the cylinder being investigated,

$$z'' = z'^. = 0; \quad z'' = -\frac{1}{R}. \tag{3-39}$$

During buckling the following membrane forces arise in the shell:

Normal force from axial compression: $-t\sigma_{x,\text{compr.}}$

Tensile forces from internal pressure:

$$n_{x,p} = \frac{pR}{2},$$

$$n_{y,p} = pR,$$

that is:

$$n_{x,0} = -t\sigma_{x,\text{compr}} + \frac{pR}{2}, \tag{3-40a}$$

$$n_{y,0} = pR. \tag{3-40b}$$

Introducing all these into (3-36) and (3-37), we arrive at the following equations:

$$\frac{Et^3}{12\left(1-v^2\right)}\Delta^2 w + \frac{1}{R}\phi'' - \left(w''\phi'' - 2w'^.\phi'^. + w''\phi''\right) +$$

$$+ \left(t\sigma_{x,\text{compr}} - \frac{pR}{2}\right)w'' - pRw'' = 0, \tag{3-41}$$

$$\frac{1}{Et}\Delta^2\phi - \frac{1}{R}w'' + w''w'' - (w'^.)^2 = 0. \tag{3-42}$$

3.5.2 The eigenfunctions of the cylindrical shell

To investigate the post-critical behaviour we first have to establish the eigenfunctions of the shell, i.e. the buckling shape(s) that are valid in the frame of the linear theory. To this purpose we omit the nonlinear terms in Eqs (3-41) and (3-42), and eliminate from them the stress function ϕ. This can be done by applying the operator Δ^2 to (3-41), the operator $-\frac{1}{R}\frac{\partial^2}{\partial x^2}$ to (3-42), and adding them. We thus arrive at the differential equation:

$$-\frac{Et^3}{12(1-v^2)}\Delta^2 w + \left(t\sigma_{x,\text{compr}} - \frac{pR}{2}\right)\Delta^2 w'' -$$

$$-pR\Delta^2 w'' + \frac{Et}{R^2}w'''' = 0. \tag{3-43}$$

Let us introduce the notations:

$$\sigma_{x,\text{cr}}^{\text{lin}} = \frac{Et}{\sqrt{3\left(1-v^2\right)}R}, \tag{3-44}$$

for the critical stress of the axial compression according to the linear theory;

$$q_0 = \sqrt[4]{12\left(1-v^2\right)R^2/t^2}, \tag{3-45}$$

a geometric parameter;

$$\bar{p} = \frac{pR^2}{Et^2}\sqrt{3(1-\nu^2)} = \frac{pR}{\sigma_{x,cr}^{lin}t} \tag{3-46}$$

the parameter of the internal pressure; and

$$\lambda = \frac{\sigma_{x,compr}q_0^2}{2E} - \frac{\bar{p}}{2}, \tag{3-47}$$

the load parameter of the axial pressure.

At $\lambda = 1$ the actual axial compressive stress is: $-\sigma_{x,0} = \sigma_{x,compr} - \frac{pR}{2t} = \sigma_{x,cr}^{lin}$.

Let us assume the solution of (3-43) in the form of a 'chessboard-like' buckling mode:

$$w = W \cos\frac{\alpha q_0 x}{R} \cos\frac{\beta q_0 y}{R}, \tag{3-48}$$

where α and β are for the time being unknown quantities, determining the lengths of the buckling half-waves, l_x and l_y, in the directions x and y respectively:

$$l_x = \frac{\pi R}{\alpha q_0}, \tag{3-49a}$$

$$l_y = \frac{\pi R}{\beta q_0}. \tag{3-49b}$$

Introducing (3-48) into (3-43), and simplifying by w, we obtain the following equation for the load parameter λ:

$$\lambda = \frac{1}{2}\left[\frac{(\alpha^2+\beta^2)^2}{\alpha^2} + \frac{\alpha^2}{(\alpha^2+\beta^2)^2}\right] + \bar{p}\frac{\beta^2}{\alpha^2}. \tag{3-50}$$

The amplitude W is undetermined; the dimensionless quantities α and β are to be determined from the condition that they render λ a minimum.

If no internal pressure acts on the shell ($\bar{p} = 0$), then, making the first derivative of λ with respect to $(\alpha^2+\beta^2)^2/\alpha^2$ equal to 0, we obtain the equation

$$\alpha^2 + \beta^2 = \alpha \tag{3-51}$$

for the two half wave lengths. This means an infinite number of pairs of values α and β, i.e. an infinity of eigenfunctions, among which also the axisymmetric

$$w_1 = W_1 \cos\frac{q_0 x}{R}, \tag{3-52a}$$

and the asymmetric (chessboard-type)

$$w_2 = W_2 \cos\frac{q_0 x}{2R} \cos\frac{q_0 y}{2R}, \tag{3-52b}$$

buckling shapes appear.

All buckling shapes yield the same critical load

$$\lambda_{cr}^{(\bar{p}\,=\,0)} = 1, \tag{3-53}$$

i.e.

$$\sigma_{x,cr}^{compr} = \sigma_{x,cr}^{lin}, \tag{3-54}$$

see also in [KOLLÁR and DULÁCSKA, 1984].

If internal pressure also acts ($\bar{p} > 0$), then the critical load parameter will be again

$$\lambda_{cr}^{(\bar{p})} = 1,$$

but now only one single buckling shape pertains to it, that is, the axisymmetric one, corresponding to (3-52a). All other buckling shapes require greater loads: thus e.g. to the shape (3-52b) $\lambda_{cr}^{(\bar{p})} = 1 + \bar{p}$, belongs. This can be interpreted visually in the following way: the internal pressure 'irons out' those buckling shapes which contain several half waves in hoop direction, but does not influence the axisymmetric buckling mode.

3.5.3 The post-buckling behaviour of the shell

To investigate the post-critical behaviour of the shell we have to use Eqs (3-41) and (3-42).

HUTCHINSON [1965] took the two eigenfunctions of the cylindrical shell without internal pressure, (3-52a) and (3-52b), as the buckling shape. In order to obtain the result as a function of a positive ξ_1, we give a negative sign to (3-52a):

$$w = -\xi_1 t \cos\frac{q_0 x}{R} + \xi_2 t \cos\frac{q_0 x}{2R}\cos\frac{q_0 y}{2R}. \tag{3-55}$$

The first term represents an axisymmetric, the second one an asymmetric (chessboard-like) buckling shape. Introducing (3-55) into the compatibility equation (3-42), the quadratic expressions of the trigonometric functions can be linearized by the well-known relations

$$\cos\alpha\,\cos\beta = \frac{1}{2}\left[\cos(\alpha - \beta) + \cos(\alpha + \beta)\right], \tag{3-56a}$$

$$\sin\alpha\,\sin\beta = \frac{1}{2}\left[\cos(\alpha - \beta) - \cos(\alpha + \beta)\right]. \tag{3-56b}$$

Of Eq. (3-42), made linear in the sine and cosine functions, the stress function ϕ can be determined, having the shape

$$\phi = \sum_i \sum_j \phi_{ij} \cos\frac{i q_0 x}{2R}\cos\frac{j q_0 y}{2R} \qquad i = 0,1,2,3; \quad j = 0,1,2 \tag{3-57}$$

The detailed derivation, for $p = 0$, is to be found in the literature: [BÜRGERMEISTER, STEUP and KRETSCHMAR, 1963], [PFLÜGER, 1964].

By so doing we exactly fulfill the compatibility equation (3-42), but the equilibrium equation (3-41) cannot be fulfilled, since the assumed two-term expression for w [Eq. (3-55)] does not coincide with the exact post-buckling shape. Hence, by appropriate choice of the undetermined constants ξ_1 and ξ_2 we try to fulfil Eq. (3-41) with the least possible error. To this purpose *Galerkin*'s method suits best [PFLÜGER, 1964], which approximates the exact solution as the energy method, but has the advantage that it is not necessary to write the strain energy and external work, which is rather complicated in the case of shells, because in our case, according to the differential equation of the second degree, the energy expressions would have to be written down up to the fourth power terms.

Galerkin's method requires that we use the assumed w-function (3-55), containing two free parameters, in the form

$$w = w_1 + w_2 = \xi_1\eta_1(x) + \xi_2\eta_2(x,y), \tag{3-58}$$

and – denoting, for the sake of simplicity, the differential equation to be fulfilled by $DE(w,\phi)$ – we construct the following two integral expressions:

$$\int_{-\frac{\pi R}{q_0}}^{+\frac{\pi R}{q_0}} \int_{-\frac{\pi R}{q_0}}^{+\frac{\pi R}{q_0}} DE(w,\phi)\eta_i\,dx\,dy = 0, \quad i = 1,2. \tag{3-59}$$

The limits of integration are given by the two end points of the (larger) buckling half wave corresponding to w_2.

It should be remarked that in *Galerkin*'s method, besides the integral expression (3-59), the difference of some combinations of the values of the functions and their derivatives, assumed at the two limits of integration, should appear. In our case, however, this is equal to zero, since the assumed buckling shape function is symmetrical to the middle point between the limits of integration.

In (3-42), since it is of the second degree, the parameters ξ appear on the second power, so that ϕ (3-57) also contains second powers of ξ. In Eq. (3-41) we still have to multiply ϕ by w. So, finally, (3-59) yields the following two homogeneous equations of the third degree for the buckling amplitudes ξ_1, ξ_2:

$$-\xi_1 (1 - \lambda) + \frac{3}{32} c \xi_2^2 - \left\{ \frac{13}{200} c^2 \xi_1 \xi_2^2 \right\} = 0, \tag{3-60}$$

$$\xi_2 (1 + \bar{p} - \lambda) - \frac{3}{2} c \xi_1 \xi_2 + \left\{ \frac{13}{25} c^2 \xi_1^2 \xi_2 + \frac{1}{16} c^2 \xi_2^3 \right\} = 0, \tag{3-61}$$

where

$$c = \sqrt{3 \left(1 - \nu^2 \right)}. \tag{3-62}$$

The terms appearing in the curly brackets are of the third degree in ξ. If we neglect these terms, then our investigation will be sufficiently accurate only for 'small' (but finite) buckling deformations. The accuracy is, however, generally sufficient to determine the snap-through load with a good approximation. This is the essential feature of *Koiter*'s method, used for studying the post-critical behaviour [KOITER, 1945], [KOITER, 1963a], [KOITER, 1963b], [BUDIANSKY and HUTCHINSON, 1964], [KOLLÁR and DULÁCSKA, 1984], see also Chapter 1. After neglecting the terms of the third degree we arrive at two equations of the second degree which are sufficiently simple in order to study the post-critical behaviour of the shell visually:

$$-\xi_1 (1 - \lambda) + \frac{3}{32} c \xi_2^2 = 0; \tag{3-63}$$

$$\xi_2 (1 + \bar{p} - \lambda) - \frac{3}{2} c \xi_1 \xi_2 = 0. \tag{3-64}$$

As long, as $\lambda < 1$, there is no deformation. Accordingly, the equation system, homogeneous in ξ-s, has the solution $\xi_i = 0$.

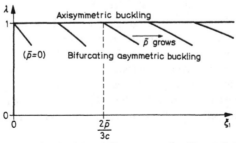

Fig. 3-21 The post-critical behaviour of the axially compressed and internally pressurized cylindrical shell

When λ reaches unity, ξ_1 can assume, according to (3-63), any arbitrary value: the shell buckles in an axisymmetric shape, with a constant post-buckling load-bearing capacity (and with an undetermined amplitude), independently of the magnitude of the internal pressure (*Fig. 3-21*). During this process $\xi_2 = 0$.

When, with $\lambda = 1$, the amplitude of axisymmetric buckling reaches the value

$$\xi_1 = \frac{2\bar{p}}{3c}, \tag{3-65}$$

then in (3-64) the sum of the coefficients of ξ_2 becomes zero:

$$1 + \bar{p} - \lambda - \frac{3}{2}c\xi_1 = 0. \tag{3-66}$$

Consequently, ξ_2 could from now on differ from zero, if Eq. (3-63) allowed it. In (3-63), however, λ must now decrease, in order to prevent the first term from becoming equal to zero, because the second term is – due to the nonzero value of ξ_2 – also not equal to zero.

Accordingly, (3-63) and (3-64) assume the form:

$$-\xi_1(1 - \lambda) + \frac{3}{32}c\xi_2^2 = 0; \tag{3-67}$$

$$\lambda = 1 + \bar{p} - \frac{3}{2}c\xi_1. \tag{3-68}$$

Introducing (3-68) into (3-67) we obtain

$$\xi_1\left(\frac{3}{2}c\xi_1 - \bar{p}\right) = \frac{3}{32}c\xi_2^2. \tag{3-69}$$

Due to $|\xi_2| > 0$ the right-hand side of Eq. (3-69) is greater than 0, and thus on the left-hand side ξ_1 must be greater than $2\bar{p}/(3c)$, that is ξ_1 must further increase as compared to the value given by (3-65). Consequently, as required by (3-68), λ must decrease. Thus, at $\xi_1 = 2\bar{p}/(3c)$, from the axisymmetric buckling an asymmetric (chessboard-like) one bifurcates which exhibits a decreasing load-bearing capacity (*Fig. 3-21*).

The gradient of this decrease can be measured by the steepness of the function $\lambda(\xi_1)$, i.e. by its derivative at $\xi_2 = 0$, i.e. at the beginning of the asymmetric buckling. To this purpose it is expedient to use the equation of the third degree (3-61), in which we can neglect ξ_2^3, since at the point investigated ξ_2 tends to zero. The coefficients of ξ_2 thus give

$$\lambda = 1 + \bar{p} - \frac{3}{2}c\xi_1 + \frac{13}{25}c^2\xi_1^2, \tag{3-70}$$

from which the direction tangent

$$\frac{d\lambda}{d\xi_1} = -\frac{3}{2}c + \frac{26}{25}c^2\xi_1 \tag{3-71}$$

is obtained. The absolute value of this is the greatest, i.e. the curve decreases the steepest, if ξ_1 is the smallest possible. Since we confined our investigations to positive ξ_1, this happens when $\xi_1 = 0$, to which, according to (3-65), $\bar{p} = 0$ belongs. Thus, in this case no internal pressure acts; the two types of buckling (axisymmetric and asymmetric) start simultaneously, and this yields the sharpest decreasing post-buckling capacity.

Similarly to the other examples treated earlier, this sharpest decreasing post-buckling load-bearing capacity of the perfect shell also represents the greatest imperfection sensitivity.

In *Fig. 3-21* we depicted the initial direction tangents of the decreasing branches of the load-bearing curves for different values of $\bar{p}$. According to what has been said in connection with *Fig. 1-8*, to the load-bearing diagrams $P(w)$ of the perfect structures, decreasing with an inclined initial tangent, a $P_{max}(w_0)$ curve of the type $A - C\sqrt{w_0}$, starting with a vertical tangent, of the imperfect structure belongs. Hence we can expect that P_{max} drops rather steeply in our case too, see also Chapter 4.

HUTCHINSON [1965] included three terms of the buckling shape function in his investigations, and he also studied the cylindrical shell with initial waviness. A similar investigation of the post-buckling behaviour of the *spherical shell* is to be found in [HUTCHINSON, 1967].

4

Stability of elastic structures with the aid of the catastrophe theory

Zsolt Gáspár

4.1 STATEMENT OF THE PROBLEM

It was shown in Chapter 1 that structures can loose their stability in different ways: the equilibrium paths may have limit or bifurcation points. The bifurcation might be symmetric or asymmetric. The secondary equilibrium path might be stable, unstable or indifferent. Analysing the different types of the loss of stability the following questions arise:

- Are there also other types?
- If there are some, how many and which types?
- Which types do arise very often and which rarely?
- How stable are these results? Does changing our mathematical model change the type of the loss of stability or not?

In this chapter we try to answer these questions. The elastic structure is supposed to be a finite-degree-of-freedom structure (or a discretized conservative system) and the total potential energy function $V(q)$ will be examined, where q is a vector of n generalized co-ordinates. A stationary value of the total potential energy with respect to the generalized co-ordinates is the necessary and sufficient condition for the equilibrium of the system, i.e.

$$V_i(q) = 0, \tag{4-1}$$

where the subscript on V denotes partial differentiation with respect to the generalized co-ordinate. This nonlinear equation system might have some solutions.

The load is supposed to depend on a scalar parameter Λ. To every value of Λ a different potential energy function belongs. We also might analyse not only our original structure (the so called 'perfect' structure) but also imperfect structures. The imperfections are supposed to be depended on generalized co-ordinates ($\alpha_i, \ i = 1, 2, ..., m$). So not a total potential energy function but a family of functions $V(q, \Lambda, \alpha)$ have to be analysed.

Considering that the structure loses its stability at a singular point of its potential energy function and that the elementary catastrophe theory deals with the classification and examination of singular points of families of functions, we will use the results of the catastrophe theory. (Here catastrophe means that smooth alteration in a situation causes sudden changes.)

The beginning of the elementary catastrophe theory is connected to the appearance of the book of THOM [1972], although only ZEEMAN [1977] proved *Thom*'s theorem, and very important results were published also earlier. The catastrophe theory is published in a most popular way for researchers of other sciences (e.g. POSTON and STEWART [1978], GILMORE [1981], THOMPSON [1982]).

4.2 DEFINITIONS

A *function* or *map* (f) consisting of m scalar functions with n variables will be denoted by

$$f : R^n \rightarrow R^m.$$

It is a *smooth function* if all derivatives of the function exist in every point of it. Usually the functions will be analysed *locally*, i.e. in a neighbourhood of the examined point. *Diffeomorphism* is a smooth and locally reversible map

$$R^n \rightarrow R^n. \tag{4-2}$$

Two smooth functions

$$f, g : R^n \rightarrow R \tag{4-3}$$

are said to be *equivalent* around 0 if there is a local diffeomorphism

$$y : R^n \rightarrow R^n \tag{4-4}$$

around 0 and a constant γ such that, around 0,

$$g(x) = f(y(x)) + \gamma. \tag{4-5}$$

The *gradient vector* of a function

$$f : R^n \rightarrow R \tag{4-6}$$

consists of its first derivatives:

$$Df = \begin{bmatrix} \frac{\partial f}{\partial x_1} \\ \vdots \\ \frac{\partial f}{\partial x_n} \end{bmatrix}. \tag{4-7}$$

The *Hessian matrix* of a function $f : R^n \rightarrow R$ consists of its second derivatives:

$$D^2 f \equiv Hf = \begin{bmatrix} \frac{\partial^2 f}{\partial x_1^2} & \cdots & \frac{\partial^2 f}{\partial x_1 \partial x_n} \\ \vdots & & \vdots \\ \frac{\partial^2 f}{\partial x_n \partial x_1} & \cdots & \frac{\partial^2 f}{\partial x_n^2} \end{bmatrix}. \tag{4-8}$$

A function of the form

$$x \rightarrow -x_1^2 - \ldots - x_l^2 + x_{l+1}^2 + \ldots + x_n^2 \tag{4-9}$$

is called a *Morse l-saddle*. When $l = n$ we have a maximum, when $l = 0$ a minimum.

Let $f : R^n \rightarrow R$ be a smooth function. A point $u \in R^n$ is a *critical point* of f if $Df|_u = 0$. We say that f has a non-degenerate critical point at u if $Df|_u = 0$ and the *Hessian* matrix $Hf|_u$ is nonsingular.

Morse Lemma: Let u be a non-degenerate critical point of a smooth function $f : R^n \to R$. Then there is a local co-ordinate system $(y_1 \cdots y_n)$ in a neighbourhood U of u, with $y_i(u) = 0$ for all i, such that $f - f(u)$ is a Morse l-saddle in U.

The function $f : R^n \to R$ is *structurally stable* if, for all sufficiently small smooth functions $p : R^n \to R$, f and $f + p$ have the same numbers and same types of critical points; or, in other words, if f and $f + p$ are equivalent after a suitable translation of the origin. (We note, that the expression 'structurally stable' does not concern to the analysed elastic structure.) A single function is typically structurally stable, but we want to deal with r-parameter families of functions $R^n \to R$,

$$f_{u_1, \cdots, u_r}(x_1 \cdots, x_n) \tag{4-10}$$

or, in an equivalent way:

$$f : R^n \times R^r \to R$$

Our aim is to classify these, up to suitable co-ordinate changes, in such a way that the critical point structure is not affected qualitatively by the co-ordinate changes involved.

Two smooth r-parameter *families of functions* $f, g : R^n \times R^r \to R$ are said to be equivalent if there exist a diffeomorphism $e : R^r \to R^r$, a smooth map $y: R^n \times R^r \to R^n$ (such that for each $s \in R^r$ the map $y_s : R^n \to R^n$, $y_s(x) = y(x, s)$ is a diffeomorphism) and a smooth map $\gamma : R^r \to R$, all defined in a neighbourhood of 0, such that

$$g(x, s) = f(y_s(x), e(s)) + \gamma(s)$$

for all $(x, s) \in R^n \times R^r$ in that neighbourhood.

Splitting Lemma: Let $f : R^n \times R^r \to R$ be smooth. Denote a point in $R^n \times R^r$ by (x, c). Suppose that the rank of the *Hessian* matrix $\mathbf{H}$ at $(x, c) = 0$ is $n - m$. Then f is equivalent to a family of the form

$$\tilde{f}(y_1(x, c), \cdots, y_m(x, c), c) \pm y_{m+1}^2 \pm \cdots \pm y_n^2.$$

We call $y_1, \cdots, y_m$ the *active* or *essential variables*, $y_{m+1}^2, \cdots, y_n^2$ the *passive* or *inessential variables*.

If $f : R^n \times R^r \to R$ is equivalent to any family $f + p : R^n \times R^r \to R$, where p is a sufficiently small family $R^n \times R^r \to R$, then f is *structurally stable*. An *r-unfolding* of a function $f : R^n \to R$ is a function $F : R^{n+r} \to R$, such that $F(x_1, \cdots, x_n, 0, \cdots, 0) = f(x_1, \cdots, x_n)$. F is an *universal unfolding* of f if all other unfoldings of f can be induced from F, and r is as small as possible.

4.3 THOM'S THEOREM

This theorem classifies the typical singularities of families of functions with less than 6 parameters.

Thom's theorem: Typically an r-parameter family of smooth functions $R^n \to R$, for any n and for $r \leq 5$, is structurally stable, and is equivalent around any point to one of the following forms:

1. u_1
2. $u_1^2 + \ldots + u_i^2 - u_{i+1}^2 - \ldots - u_n^2$ $(0 \leq i \leq n)$
3. $u_1^3 + t_1 u_1 + (M)$

4. $\pm(u_1^4 + t_2 u_1^2 + t_1 u_1) + (M)$

5. $u_1^5 + t_3 u_1^3 + t_2 u_1^2 + t_1 u_1 + (M)$

6. $\pm(u_1^6 + t_4 u_1^4 + t_3 u_1^3 + t_2 u_1^2 + t_1 u_1) + (M)$

7. $u_1^7 + t_5 u_1^5 + t_4 u_1^4 + t_3 u_1^3 + t_2 u_1^2 + t_1 u_1 + (M)$

8. $u_1^2 u_2 - u_2^3 + t_3 u_1^2 + t_2 u_2 + t_1 u_1 + (N)$

9. $u_1^2 u_2 + u_2^3 + t_3 u_1^2 + t_2 u_2 + t_1 u_1 + (N)$

10. $\pm(u_1^2 u_2 + u_2^4 + t_4 u_2^2 + t_3 u_1^2 + t_2 u_2 + t_1 u_1) + (N)$

11. $u_1^2 u_2 - u_2^5 + t_5 u_2^3 + t_4 u_2^2 + t_3 u_1^2 + t_2 u_2 + t_1 u_1 + (N)$

12. $u_1^2 u_2 + u_2^5 + t_5 u_2^3 + t_4 u_2^2 + t_3 u_1^2 + t_2 u_2 + t_1 u_1 + (N)$

13. $\pm(u_1^3 + u_2^4 + t_5 u_1 u_2^2 + t_4 u_2^2 + t_3 u_1 u_2 + t_2 u_2 + t_1 u_1) + (N)$

   where $(u_1, \ldots, u_n) \in r^n$, $(t_1, \ldots, t_r) \in R^r$ and

   $(M) = u_2^2 + \ldots + u_i^2 - u_{i+1}^2 - \ldots - u_n^2$ $(1 \leq i \leq n)$

   $(N) = u_3^2 + \ldots + u_i^2 - u_{i+1}^2 - \ldots - u_n^2$ $(2 \leq i \leq n)$.

The first function has no critical point, the second function is a Morse $(n - i)$-saddle hence it has a nondegenerate critical point. These two types are not catastrophe forms: they do not change with t. The other functions are catastrophes. They can be put in two groups. Functions 3 to 7 have only one active variable, they are called *cuspoid catastrophes*. Functions 8 to 13 have two active variables, they are called *umbilic catastrophes*.

   Each catastrophe has a pet name and a symbol in the systematic classification of ARNOL'D [1972]. These are:

Cuspoid catastrophes:

   3. the fold (A_2)

   4. the cusp (A_3)

   5. the swallowtail (A_4)

   6. the butterfly (A_5)

   7. the wigwam (A_6)

Umbilic catastrophes:

   8. the elliptic umbilic (D_4^-)

   9. the hyperbolic umbilic (D_4^+)

   10. the parabolic umbilic (D_5)

   11. the second elliptic umbilic (D_6^-)

   12. the second hyperbolic umbilic (D_6^+)

   13. the symbolic umbilic (E_6).

   Four functions contain two signs. In the case of a positive sign they are called standard forms, those with a negative sign are referred to as dual forms.

   The first statement of the theorem is that the family of functions is structurally stable, i.e. if a process depending on maximum 5 parameters is examined under a certain parameter change, then the process will be similar under a similar parameter change.

   The second part of the theorem classifies the points of the family of functions. Each form consists of a function and a universal unfolding. These forms are canonical. (Later we will use other canonical forms.)

   A smooth function $f : R^n \to R$ is k-determinate at 0 if $f + g$ for any g of order $k + 1$ is locally equivalent by a smooth change of co-ordinates to the Taylor series of f up to degree k. Determinacy of f is the lowest k for which f is k-determinate. The determinacy of the catastrophes is as follows.

Determinacy	Types
3	A_2, D_4^+, D_4^-
4	A_3, D_5, E_6
5	A_4, D_6^+, D_6^-
6	A_5
7	A_6

In the following two sections the most important catastrophes will be examined (only the active part of the function). To simplify notation, we use x and y as active variables rather than u_1, u_2; and a, b, c as control variables (parameters) in place of t_1, t_2, t_3.

4.4 THE CUSPOID CATASTROPHES

4.4.1 The fold catastrophe

The following canonical form will be analysed

$$f(x,a) = \frac{1}{3}x^3 + ax,$$ (4-11)

where the first coefficient has been introduced to simplify the subsequent calculations. Equation (4-11) can be drawn in the three-dimensional space (*Fig. 4-1a*), where we have shown three members of the family.

At a stationary point (critical point in catastrophe theory) the gradient of f is zero:

$$\frac{\partial f}{\partial x} = x^2 + a = 0.$$ (4-12)

In *Fig. 4-1a* the stationary points are shown, their types are marked as follows:

$\oplus$ maximum
$\ominus$ minimum
$\bigcirc$ saddle
$\textcircled{\scriptsize{4-2}}$ catastrophe point.

In applications the value of f is usually not important, so it is enough to use a two-dimensional picture (in the space x,a). The so-called *equilibrium surface* (*Fig. 4-1b*) is given by Eq. (4-12). There is a catastrophe when the *Hessian* matrix (now: one scalar) is singular:

$$\frac{\partial^2 f}{\partial x^2} = 2x = 0.$$ (4-13)

Eliminating x from Eqs (4-12) and (4-13) one has

$$a = 0,$$ (4-14)

so this is the only parameter at which the function (4-11) has a catastrophe point.

In applications sometimes only this parameter is important, and the dangerous parameters can be shown in a one-dimensional picture on the line a (*Fig. 4-1c*). The set given by

Eq. (4-14) is called a bifurcation set, because two different kinds of function appear when the parameter a varies in both directions (if $a > 0$ then there is no stationary point, and if $a < 0$ then there are a maximum and a minimum).

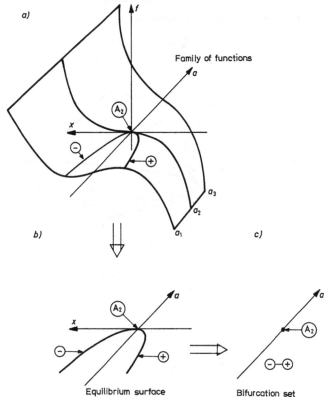

Fig. 4-1 The fold catastrophe

4.4.2 The cusp catastrophe

A canonical form of the standard cusp catastrophe (which is marked by A_3^+) is

$$f(x,a,b) = \frac{1}{4}x^4 + \frac{1}{2}ax^2 + bx, \qquad (4\text{-}15)$$

and at a dual cusp (which is marked by A_3^-)

$$f(x,a,b) = -\left(\frac{1}{4}x^4 + \frac{1}{2}ax^2 + bx\right) \qquad (4\text{-}16)$$

To show the family of functions (4-15) four dimensions ought to be used. The equilibrium surface is given by equation

$$\frac{\partial f}{\partial x} = x^3 + ax + b = 0. \tag{4-17}$$

and is shown in *Fig. 4-2a.* At the catastrophe point

$$\frac{\partial^2 f}{\partial x^2} = 3x^2 + a = 0. \tag{4-18}$$

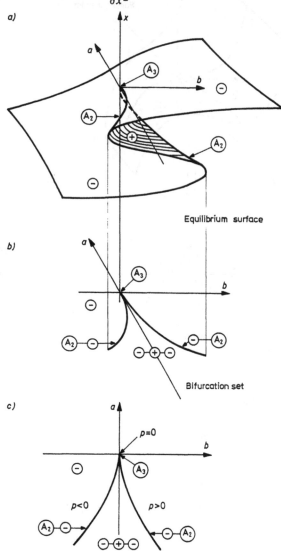

Fig. 4-2 The standard cusp catastrophe

Eliminating x from Eqs (4-17) and (4-18) one gets the equation of the bifurcation set (*Fig. 4-2b,c*):

$$4a^3 + 27b^2 = 0. \tag{4-19}$$

Eq. (4-19) can be written in a parametric form:

$$a = -3p^2$$
$$b = 2p^3. \tag{4-20}$$

which is sometimes much more useful for calculations.

In the case of the dual cusp (Eq. (4-16)) we have the same equations, only the maximum and minimum are inverted. Five points are marked in the parameter space on and around the bifurcation set of the dual cusp in *Fig. 4-3a*, and the functions belonging to these points are shown in *Fig. 4-3b*.

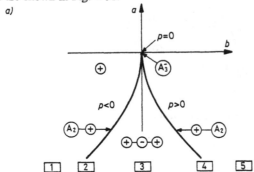

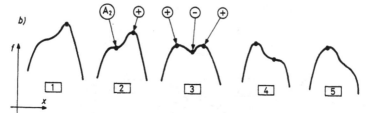

Fig. 4-3 The dual cusp catastrophe

4.4.3 The swallowtail catastrophe

A canonical form of the swallowtail catastrophe is

$$f(x,a,b,c) = \frac{1}{5}x^5 + \frac{a}{3}x^3 + \frac{b}{2}x^2 + cx \tag{4-21}$$

and the bifurcation set (*Fig. 4-4*) in a parametric form is the following:

$$a = 3q - 6r^2,$$
$$b = -6rq + 8r^3,$$
$$c = 3qr^2 - 3r^4.$$

(4-22)

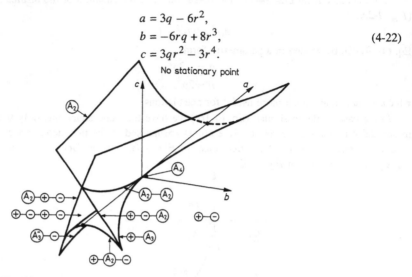

Fig. 4-4 The bifurcation set of the swallowtail catastrophe

4.4.4 The butterfly catastrophe

A canonical form of the butterfly catastrophe is

$$f(x,a,b,c,d) = \frac{1}{6}x^6 + \frac{a}{4}x^4 + \frac{b}{3}x^3 + \frac{c}{2}x^2 + dx$$

(4-23)

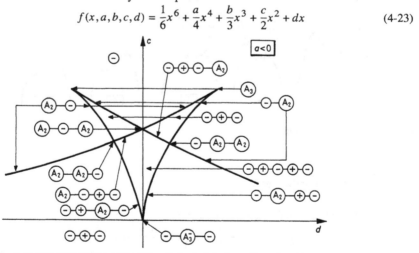

Fig. 4-5 A characteristic section of the bifurcation set of the butterfly catastrophe

and the bifurcation set in a parametric form is the following:

$$a = 4r - 10s^2,$$
$$b = 3q - 12rs + 20s^3,$$
$$c = -6qs^2 + 12rs^2 - 15s^4, \tag{4-24}$$
$$d = 3qs^2 - 4rs^3 + 4s^5.$$

This ought to be shown in the four-dimensional space, so only a characteristic section of it is shown in *Fig. 4-5*.

4.5 THE UMBILIC CATASTROPHES

Two-variable cubic forms can be classified by their root structure:

Root structure	Umbilic point
three distinct lines	elliptic
a single real line	hyperbolic
three lines, two of which coincide	parabolic
three coincident lines	symbolic

These can be illustrated by an umbilic bracelet (see POSTON and STEWART [1978]).

4.5.1 The elliptic umbilic

The following canonical form will be analysed:

$$f(x,y,a,b,c) = y^3 - 3x^2y + a(x^2 + y^2) + bx + cy, \tag{4-25}$$

The equilibrium surface is determined by the equations

$$\frac{\partial f}{\partial x} = -6xy + 2ax + b = 0, \tag{4-26}$$

$$\frac{\partial f}{\partial y} = 3y^2 - 3x^2 + 2ay + c = 0. \tag{4-27}$$

The elements of the *Hessian* are:

$$H_{11} = \frac{\partial^2 f}{\partial x^2} = -6y + 2a,$$

$$H_{12} = H_{21} = \frac{\partial^2 f}{\partial x \partial y} = -6x,$$

$$H_{22} = \frac{\partial^2 f}{\partial y^2} = 6y + 2a,$$

so the matrix is singular if

$$(6y + 2a)(-6y + 2a) - 36x^2 = 0. \tag{4-28}$$

Eliminating x and y from Eqs (4-26) to (4-28) one obtains the equation of the bifurcation set, but the following parametric form is more useful:

$$a = a,$$

$$b = \frac{a^2}{3}(\sin 2\Theta - 2\cos \Theta), \qquad (4\text{-}29)$$

$$c = \frac{a^2}{3}(\cos 2\Theta - 2\sin \Theta).$$

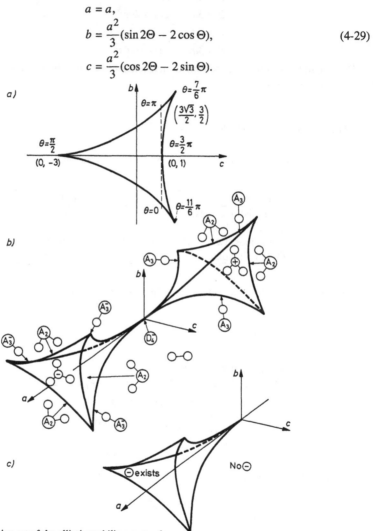

Fig. 4-6 Bifurcation set of the elliptic umbilic catastrophe

Apart from the parabolically increasing factor $a^2/3$ the cross sections for a constant a have the same Θ-dependence. The cross section for $a = \sqrt{3}$ is a three-cusped hypocycloid, as shown in *Fig. 4-6a*. The whole bifurcation set and the types of the stationary points are drawn in *Fig. 4-6b*. The parameter space is divided into three parts by the bifurcation set, but only one of them contains a minimum. Usually only the border of this part (given by the restriction $a \geq 0$ is important, and it is shown in *Fig. 4-6c*.

4.5.2 The hyperbolic umbilic

The following canonical form will be analysed:

$$f(x,y,a,b,c) = x^3 + y^3 + axy + bx + cy,$$ (4-30)

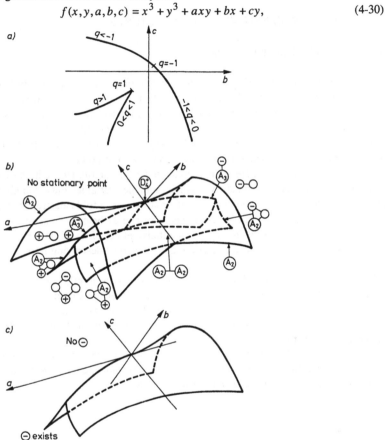

Fig. 4-7 Bifurcation set of the hyperbolic umbilic catastrophe

The bifurcation set can be written in the following parametric form:

$$a = 6p,$$
$$b = -3p^2(2q^{-1} + q^2),$$ (4-31)
$$c = -3p^2(2q + q^{-2})$$

but we have to add to it the negative b-axis and the negative c-axis. The cross sections for constant a are similar, one of them is shown in *Fig. 4-7a*. The whole bifurcation set and the types of the stationary points are drawn in *Fig. 4-7b*. The parameter space is divided into four parts by the bifurcation set, and two of them contain a minimum. So only one half of the bifurcation set has a different number of minimum points on its both sides. This half satisfies the condition $pq > 0$ and is shown in *Fig. 4-7c*.

4.6 IMPERFECTION-SENSITIVITY OF STRUCTURES

4.6.1 What kinds of catastrophes arise?

The behaviour of an engineering structure depends only on a single control parameter, the time (the variation of the forces and imperfections also occur in time). We know from Thom's theorem that with a single control parameter only one type of structurally stable singularity can arise, namely the fold catastrophe. Indeed, a real structure (with its inevitable manufacturing imperfections) will always lose its stability at a fold, in which an equilibrium path merely reaches a maximum load at a limit point.

The structure and the load are often assumed to be symmetric. Hence in this case a Taylor expansion of the potential function in some (suitable) variables must contain vanishing coefficients of odd powers. In the case of perfect structures, a fold catastrophe cannot arise, but perfect symmetry on the drawing board can give rise to the cusp catastrophe. To model the structure successfully, allowing for the qualitative effect of all random perturbations of the system and its environment, we must include symmetry-destroying imperfection in the analysis.

Obviously a designer would like his system to carry loads as efficiently as possible, and this often calls for the simultaneity of failure loads so that the structure loses its stability with respect to two buckling modes simultaneously. In these cases there are two active variables, and umbilic catastrophes arise, typically the elliptic and hyperbolic umbilic (hence these need the fewest control parameters).

What is more, if both buckling modes are symmetric, then the perfect system must exhibit one of the double-cusp catastrophes, which are not included in Thom's theorem (because they need more than five control parameters).

4.6.2 The method

To start the stability analysis of a structure, first of all we have to make a discrete model. The deformation of the structure is approached by a linear combination of some basic functions. The scalar factors of these functions are the generalized co-ordinates (q). The total potential energy function is written as a function of these co-ordinates and the load parameter ($V(q, \Lambda)$). The critical load parameter (Λ^{cr}) of the perfect structure and the co-ordinates of the critical state (q^{cr}) can be computed by solving the equation system

$$DV = 0,$$

$$\det(HV) = 0, \tag{4-32}$$

where DV denotes the gradient and HV the *Hessian* matrix of the energy function.

In the next step we consider the potential energy function of the imperfect structure ($V(q, \Lambda, \alpha,)$). According to Section 4.6.1, we usually know the type of the arising catastrophe. The determinacy of this catastrophe shows which is the highest power we need in the Taylor expansion of the energy function to describe the singularity correctly. The Taylor expansion is made around the point ($V(q^{cr}, \Lambda^{cr}, 0,)$). (If later we realize that there is a higher type of catastrophe, we have to continue the Taylor expansion according to the new determinacy.)

The Splitting Lemma declares that the active and passive variables can be separated. A process for this separation is given in page 160 of POSTON and STEWART [1978]. To describe the singularity it is enough to analyse only the active part of the energy function. The new variables of this active part are the following:

- u and perhaps v the co-ordinates of the active state ($q - q^{cr}$ is transformed)
- $\lambda = \Lambda - \Lambda^{cr}$
- the new imperfection parameters (ε) which are the coefficients of the terms being zero in the perfect system.

The active part and the analysed canonical forms of the catastrophes do not have the same form. So we have to find a suitable transformation which, when put into the canonical form, gives the form of our active energy function.

Substituting this transformation into the equation of the equilibrium surface of the catastrophe, we obtain the *equilibrium surface* of our structure. Fixing the imperfection parameters, we have the *equilibrium path* of the structure (these are sections of the equilibrium surface).

Substituting the earlier transformation into the equations of the bifurcation set of the catastrophe, we arrive at the *imperfection-sensitivity surface* of the structure, which shows the critical load parameter against the imperfection parameters.

Let us introduce the concept of the *critical imperfection territory* (GÁSPÁR [1982]). This territory covers all imperfections resulting in a value of the critical load smaller than a prescribed value. These territories have two advantages with regard to imperfection-sensitivity surfaces:

- Imperfection-sensitivity surfaces are usually governed by multi-valued functions, so it is not easy to determine the smallest critical load for a prescribed imperfection. Critical imperfection territories directly show whether the imperfect structure does carry a prescribed load without losing its stability or not.
- Critical imperfection territory can be represented in a space of fewer dimensions.

The following steps are needed to determine the critical imperfection territories:

- To determine the envelopes of levels of imperfection-sensitivity surfaces.
- To show which parts of the levels and which parts of the envelopes give the boundary of the critical imperfection territory.

In the case of some catastrophes, one can make a subclassification considering whether

- it is a standard or a dual form,
- a general coefficient has a special value or not,
- the type of the boundary changes or not.

4.6.3 The fold catastrophe

The active part of the potential energy function exhibits a fold catastrophe at its critical point if the first few terms of its Taylor series have the form

$$V' = c_1 u^3 + \varepsilon_1 u^2 + \varepsilon_2 u - \lambda(c_2 u^2 + c_3 u), \tag{4-33}$$

where u is the only active generalized co-ordinate, $c_1 > 0$ (if $c_1 < 0$, then we change the direction of axis u), ε_1 and ε_2 are functions of the imperfections α, and $\varepsilon_1(0) = \varepsilon_2(0) = 0$. For a perfect structure typically $c_3 \neq 0$. If $c_3 = 0$, then this coefficient will be denoted by ε_3 and typically $c_2 \neq 0$. The case $c_2 = c_3 = 0$ will not be considered.

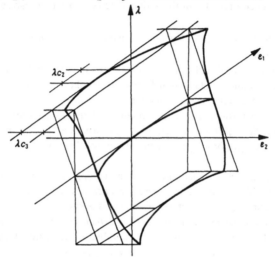

Fig. 4-8 The imperfection-sensitivity surface

First applying the linear transformation

$$u = \frac{v}{(3c_1)^{1/3}} - \frac{\varepsilon_1 - \lambda c_2}{3c_1} \tag{4-34}$$

we eliminate the second-order terms from Eq. (4-33):

$$V(v,\lambda,\varepsilon_1,\varepsilon_2) = \frac{1}{3}v^3 + \left(\frac{\varepsilon_2 - \lambda c_3}{(3c_1)^{1/3}} - \frac{(\varepsilon_1 - \lambda c_2)^2}{(3c_1)^{4/3}}\right)v. \tag{4-35}$$

Function (4-35) can be induced from the canonical form (4-11), because the transformations

$$x = v,$$

$$a = \frac{\varepsilon_2 - \lambda c_3}{(3c_1)^{1/3}} - \frac{(\varepsilon_1 - \lambda c_2)^2}{(3c_1)^{4/3}} \tag{4-36}$$

give

$$f(x,a) = V(v,\lambda,\varepsilon_1,\varepsilon_2).$$

Substituting the transformation (4-36) into the equation (4-14) of the bifurcation set, one obtains the equation of the imperfection-sensitivity surface:

$$\varepsilon_2 = \frac{(\varepsilon_1 - \lambda c_2)^2}{3c_1} + \lambda c_3. \tag{4-37}$$

A typical form of it is shown in *Fig. 4-8*.

The envelope of the level lines (λ = const.) of this surface can be written in the form:

$$\varepsilon_2 = \frac{c_3}{c_2}\varepsilon_1 - \frac{3c_1 c_3^2}{4c_2^2}. \tag{4-38}$$

(If $c_2 = 0$, there is no envelope.) The envelope touches the level lines at the points

$$\varepsilon_1^e = c_2\lambda + \frac{3c_1 c_3}{2c_2},$$

$$\varepsilon_2^e = c_3\lambda + \frac{3c_1 c_3^2}{4c_2^2}. \tag{4-39}$$

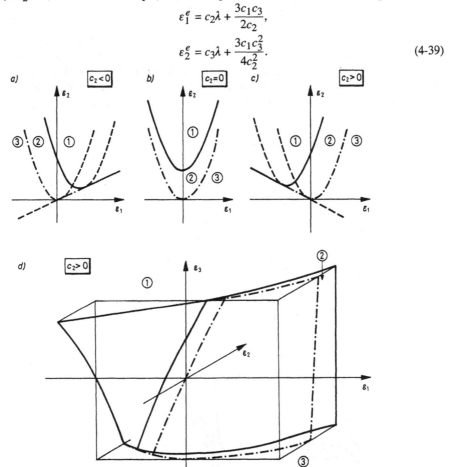

Fig. 4-9 Critical imperfection territories

Figs 4-9a-c show the typical shape of the critical imperfection territories for the case $c_3 < 0$. Here and in some other figures the territories are denoted by numbers as follows:

1. Critical imperfection territory belonging to a given load parameter $\lambda_0 < 0$.
2. Imperfections resulting in critical loads that are smaller than the critical load of the perfect system, but greater than λ_0.
3. Imperfections that result in no loss of stability if the load is smaller than Λ^{cr}.

The distinction between territories 1 and 2 is of no immediate theoretical importance, but helps us to visualize the rules governing the change of the critical imperfection territory.

Fig. 4-9d shows the critical imperfection territory for the special case when $c_3 = 0$ in a perfect structure.

According to Eqs (4-12) and (4-36) the equilibrium paths are given by

$$v = \left(\frac{\lambda c_3 - \varepsilon_2}{(3c_1)^{1/3}} - \frac{(\varepsilon_1 - \lambda c_2)^2}{(3c_1)^{4/3}} \right)^{1/2} \tag{4-40}$$

which can have three different types (*Fig. 4-10a-c*), depending on the values of its constants. The distance between the two curves in *Fig. 4-10a* increases when the value of $|c_2|$ decreases, and the upper curve disappears when $c_2 = 0$. According to *Fig. 4-1b* the equilibrium state is stable (denoted by a solid line) if $v > 0$ while it is unstable (broken line) if $v < 0$. Using transformation (4-34) the path can be written in dependence of the original active state variables:

$$u = \left(\frac{\lambda c_3 - \varepsilon_2}{3c_1} + \left(\frac{\varepsilon_1 - \lambda c_2}{3c_1} \right)^2 \right)^{1/2} - \frac{\varepsilon_1 - \lambda c_2}{3c_1}. \tag{4-41}$$

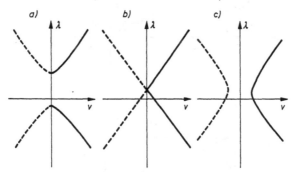

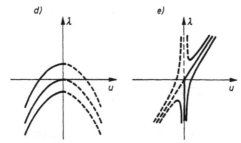

Fig. 4-10 Equilibrium paths

This gives also the well-known pictures of the limit point (*Fig. 4-10d*) if $\varepsilon_1 = c_2 = 0$, and the asymmetric point of bifurcation (*Fig. 4-10e*).

Example 4-1: Limit point

Let us consider the shallow arch analysed in Chap. 1, but now we do not suppose it to be shallow. For simplicity let us put $c = 1$ and $l = 1$. The generalized co-ordinate (is positive as shown in *Fig. 4-11*. The perfect unloaded structure is in equilibrium if $\varphi = \alpha$.

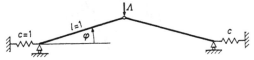

Fig. 4-11 Model for the limit point

The total potential energy function of the perfect structure is:
$$V = (\cos\varphi - \cos\alpha)^2 + \Lambda\sin\varphi.$$
The structure is in a critical equilibrium state if $\varphi = \beta$ and $\Lambda = P$, where
$$\beta = \arccos(\cos^{1/3}\alpha)$$
$$P = 2\sin^3\beta.$$
The unloaded imperfect structure is in equilibrium when $\varphi = \alpha + \varepsilon$. Introducing the new variables
$$u = \varphi - \beta,$$
$$\lambda = P - \Lambda$$
the total potential energy function becomes
$$V = [\cos(\beta + u) - \cos(\alpha + \varepsilon)]^2 + (P + \lambda)\sin(\beta + u).$$
Its truncated Taylor series is (first and second degree terms are calculated in the critical state $u = 0$, while the third degree term at the critical point of the perfect structure where $u = \lambda = \varepsilon = 0$)
$$V = 0.5\sin(2\beta) \cdot u^3 + [\cos(\alpha + \varepsilon) - \cos\alpha]\cos\beta \cdot u^2 +$$
$$+2[\cos(\alpha + \varepsilon) - \cos\alpha]\sin\beta \cdot u - \lambda(0.5\sin\beta \cdot u^2 - \cos\beta \cdot u).$$
So, according to Eq. (4-33):
$$c_1 = 0.5\sin(2\beta), \quad c_2 = 0.5\sin\beta, \quad c_3 = -\cos\beta,$$
$$\varepsilon_1 = [\cos(\alpha + \varepsilon) - \cos\alpha]\cos\beta, \quad \varepsilon_2 = 2[\cos(\alpha + \varepsilon) - \cos\alpha]\sin\beta.$$
Set $\alpha = \pi/4$, and let us determine the largest ε where the critical load of the imperfect structure is not less then the 95% of the critical load of the perfect structure. At this angle α, $\beta = 0.4714763, P = 0.1874033$, so we have to determine the critical imperfection territory belonging to $\lambda = -0,00937016$.

Hence $c_1 = 0.4046480, c_2 = 0.2271010, c_3 = -0.8908988$, and the boundary of the critical imperfection territory is given by the level line (Eq. (4-37)):
$$\varepsilon_2 = 0.8237612(\varepsilon_1 + 0.0021279736)^2 + 0.0083478668 \tag{4-42}$$
and the envelope (Eq. (4-38)):
$$\varepsilon_2 = -3.9229206\varepsilon_1 - 4.6704378 \tag{4-43}$$
This will be achieved when $\varepsilon = -0.0132225$.

Catastrophe theory

Example 4-2: Asymmetric point of bifurcation
The structure showed in Section 1.1.3 will be analysed here, but according to THOMPSON and HUNT [1973] two imperfections are also considered. A state of the loaded imperfect structure is shown in *Fig. 4-12.*

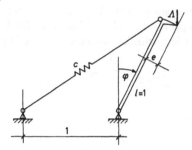

Fig. 4-12 Model for the asymmetric point of bifurcation

The structure is perfect if $e = 0$, and the state $\varphi = 0$ is in equilibrium for the unloaded case. The unloaded imperfect structure is in equilibrium if $\varphi = \alpha$. Its total potential energy function is

$$V(\varphi, \Lambda, \alpha, e) = \frac{c}{2}\left((2 + 2\sin\varphi)^{1/2} - (2 + 2\sin\alpha)^{1/2}\right)^2 + \Lambda(\cos\varphi - e\sin\varphi).$$

At the critical point $\varphi = 0$ and $\Lambda = c/2$. Its truncated Taylor series is, up to the third degree terms:

$$V = -\frac{3}{24}\varphi^3 + \frac{c}{4}((1 + \sin\alpha)^{1/2} - 1)\varphi^2 + c\left(1 - (1 - \sin\alpha)^{1/2} - \frac{e}{2}\right)\varphi - \lambda\left(\frac{1}{2}\varphi^2 + e\varphi\right),$$

i.e.

$$u = -\varphi, \qquad c_1 = c/8, \qquad c_2 = 1/2, \qquad c_3 \equiv \varepsilon_3 = -e,$$
$$\varepsilon_1 = c\left[(1 + \sin\alpha)^{1/2} - 1\right]/4, \qquad \varepsilon_2 = c\left[e/2 + (1 - \sin\alpha)^{1/2} - 1\right].$$

$c_3 = 0$ for the perfect structure, so there is indeed an asymmetric point of bifurcation.

4.6.4 The cusp catastrophe

The active part of the potential energy function has a cusp catastrophe at its critical point if the first few terms of its Taylor series have the form

$$V' = c_1 u^4 + \varepsilon_1' u^3 + \varepsilon_2' u^2 + \varepsilon_3' u - \lambda(c_2 u^3 + c_3 u^2 + c_4 u), \tag{4-44}$$

where $c_1 \neq 0$. There is a standard cusp if $c_1 > 0$ and a dual cusp if $c_1 < 0$. There is usually a cusp catastrophe if the structure and the load are both symmetric. Then, only even degree terms in u are present in the case of perfect systems, so we suppose that $c_3 \neq 0$ in the case of perfect system, but $c_2 = \varepsilon_4'(\alpha)$, $c_4 = \varepsilon_5'(\alpha)$ and $\varepsilon_4'(0) = \varepsilon_5'(0) = 0$.

To make the analysis manageable, we neglect the effect of the imperfection ε_4' (this is usually much smaller than the effect of ε_5').

Applying the linear transformation

$$u = \frac{v}{|4c_1|^{1/4}} - \frac{\varepsilon_1'}{4c_1} \tag{4-45}$$

we eliminate the third order terms from Eq. (4-44):

$$V(v,\lambda,\varepsilon_1,\varepsilon_2,\varepsilon_3) = \alpha \left[\frac{1}{4}v^4 + \varepsilon_1 v + \varepsilon_2 v - \lambda \left(cv^2 + \varepsilon_3 v \right) \right] \tag{4-46}$$

where

$$\varepsilon_1 = \frac{\alpha \varepsilon_2'}{2|c_1|^{1/2}} - \frac{3(\varepsilon_1')^2}{16|c_1|^{3/2}}, \quad \alpha = \text{sign } c_1,$$

$$\varepsilon_2 = \frac{\alpha \varepsilon_3'}{|4c_1|^{1/4}} - \frac{2\varepsilon_1'\varepsilon_2'}{|4c_1|^{5/4}} + \frac{2\alpha(\varepsilon_1')^3}{|4c_1|^{9/4}},$$

$$\varepsilon_3 = \frac{\alpha \varepsilon_5'}{|4c_1|^{1/4}} - \frac{2c_3\varepsilon_1'}{|4c_1|^{5/4}}, \quad c = \frac{\alpha c_3}{|2c_1|^{1/2}}.$$

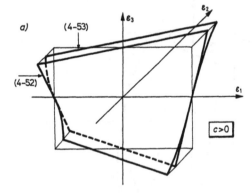

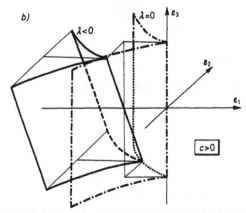

Fig. 4-13 The envelope and the surface of the singular points *a*) and the level surfaces *b*)

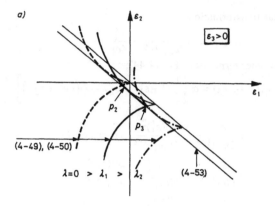

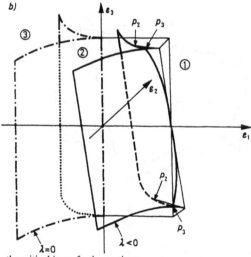

Fig. 4-14 Dual cusp; the critical imperfection territory

Function (4-46) can be induced from the canonical form (4-15) or (4-16), because the transformations

$$x = v,$$

$$a = 2(\varepsilon_1 - \lambda c), \tag{4-47}$$

$$b = \varepsilon_2 - \lambda\varepsilon_3 \tag{4-48}$$

yield

$$f(x,a,b) = V(v,\lambda,\varepsilon_1,\varepsilon_2,\varepsilon_3).$$

Substituting the transformations (4-47), (4-48) into Eq. (4-20) of the bifurcation set, one obtains the equation of the imperfection-sensitivity surface:

$$\varepsilon_1 = c\lambda - 3p^2/2, \tag{4-49}$$

$$\varepsilon_2 = \lambda \varepsilon_3 + 2p^3 \tag{4-50}$$

The envelope of the level lines (λ = const., ε_3 = const.) of this surface also has to satisfy the equation:

$$c \cdot 6p^2 + 3p \cdot \varepsilon_3 = 0 \tag{4-51}$$

This is true for two values of p : $p_1 = 0$ and $p_2 = -\varepsilon_3/(2c)$. Substituting these values into Eqs (4-49) and (4-50), λ can be eliminated and we arrive at the equations of two surfaces:

$$\varepsilon_2 = \varepsilon_1 \varepsilon_3/c, \tag{4-52}$$

$$\varepsilon_2 = \varepsilon_1 \varepsilon_3/c + \varepsilon_3^3 \left(8c^3\right). \tag{4-53}$$

The first surface belongs to the singular points of the level curves, so it is not a part of the envelope, while (4-53) defines the envelope. Both surfaces are shown in *Fig. 4-13a*, and two level surfaces in *Fig. 4-13b*.

There are three different cases to analyse:
– the dual cusp,
– the standard cusp, when the corresponding point is outside of the cusp for a perfect system if $\lambda < 0$,
– the standard cusp, starting from outside of the cusp.

Unstable-symmetric point of bifurcation

In the case of dual cusp there is only one minimum point and it is inside of the cusp (*Fig. 4-3*). For perfect systems $b = 0$, so c must be negative to have negative a if $\lambda < 0$.

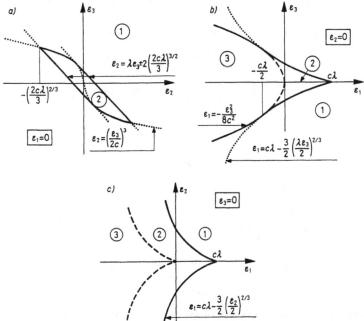

Fig. 4-15 Dual cusp; sections of the critical imperfection territory

Catastrophe theory

In the case of a fixed $\varepsilon_3 > 0$ *Fig. 4-14a* shows the variation of the sections of the level surfaces and the envelope. The point of interaction belongs to parameter $p_3 = \varepsilon_3/(4c)$. *Fig. 4-14b* shows the three territories of the imperfection space.

Fig. 4-15 shows the sections of the critical imperfection territory by the planes $\varepsilon_1 = 0$, $\varepsilon_2 = 0$ and $\varepsilon_3 = 0$.

Eqs (4-47) and (4-48) show that we have a line in the parameter space if we fix the imperfections. Some cases are shown in *Fig. 4-16a*. The corresponding equilibrium paths are shown in *Fig. 4-16b-e*. The well-known equilibrium paths (*Fig. 4-16f*) do not show the possibility of the type shown in *Fig. 4-16e*. In the literature, usually only the effect of ε_2 is considered in imperfection-sensitivity (*Fig. 4-16g*).

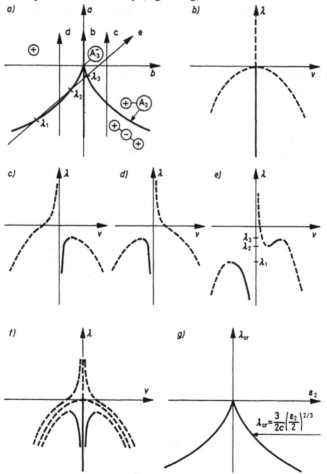

Fig. 4-16 Dual cusp; parameter space a), equilibrium paths $b) - f$), imperfection-sensitivity g)

Stable-symmetric point of bifurcation

Fig. 4-2 shows that in the case of standard cusp there are minimum points in both territories of the parameter space, so one can start from any point. There is a stable-symmetric point of bifurcation if $c > 0$, i.e. we start from the territory containing one minimum point. *Figs 4-17* and *4-18* are the analogies of *Figs 4-14* and *4-16.*

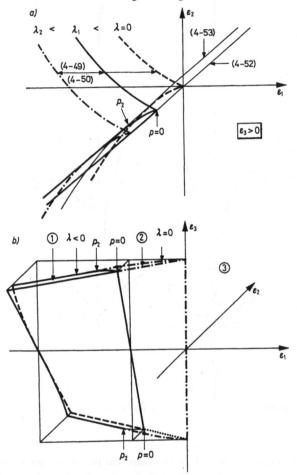

Fig. 4-17   Standard cusp; $c > 0$; the critical imperfection territory

Catastrophe theory

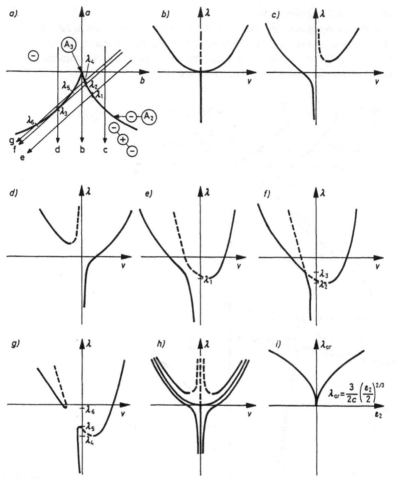

Fig. 4-18 Standard cusp; $c > 0$; parameter space a), equilibrium paths b) $-$ h), imperfection-sensitivity i)

The 'upside down' case

Now we analyse the standard cusp when $c < 0$, i.e. there are two minimum points if $\lambda < 0$. We suppose that the minimum point characterizing the original state of the structure coincides with the maximum point on the branch p<0 (the other branch can be analysed similarly). The analogies of *Figs 4-14* and *4-15* are shown in *Figs 4-19* and *4-20*.

The equilibrium paths are similar to those shown in *Fig. 4-18*, only the direction of the λ-axis is reserved.

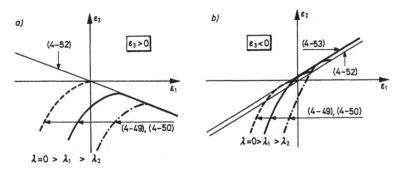

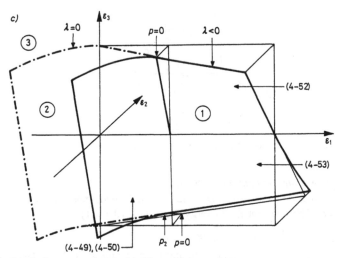

Fig. 4-19 Standard cusp; $c < 0$; the critical imperfection territory

Example 4-3: Unstable-symmetric point of bifurcation

Let us consider the structure shown in *Fig. 1-2a*. The unloaded imperfect structure is in equilibrium if $\varphi = \alpha$. The total potential energy function is (if $l = 1$):

$$V(\varphi, P, \alpha) = \frac{c}{2}(\sin\varphi - \sin\alpha)^2 + P\cos\varphi.$$

The perfect structure is in equilibrium if $\varphi = 0$, and the critical load is $P = c$. The first few terms of the Taylor series at the critical point have the form

$$V = -\frac{c}{8}\varphi^4 + \frac{c}{6}\sin\alpha \cdot \varphi^3 - c\sin\alpha \cdot \varphi - \frac{\lambda}{2}\varphi^2.$$

According to Eq. (4-44), $c_1 = -c/8 < 0$, so we really have a dual cusp.

We note that GÁSPÁR and DOMOKOS [1991] analysed a simple structure producing another type of dual cusp catastrophe when the ratio of two stiffness constants was chosen as a special value.

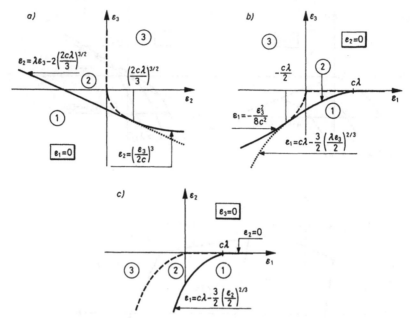

Fig. 4-20 Standard cusp; $c < 0$; sections of the critical imperfection territory

Example 4-4: Stable-symmetric point of bifurcation

Let us consider the structure shown in *Fig. 1-1a*. The unloaded imperfect structure is in equilibrium if $\varphi = \alpha$. The total potential energy function is (if $l = 1$):

$$V(\varphi, P, \alpha) = \frac{1}{2}c(\varphi - \alpha)^2 + P\cos\varphi.$$

The perfect structure is in equilibrium if $\varphi = 0$, and the critical load is $P = c$. The first few terms of the Taylor series at the critical point are:

$$V = \frac{c}{24}\varphi^4 - c\alpha\varphi - \frac{1}{2}\lambda\varphi^2.$$

According to Eq. (4-44) $c_1 = c/24 > 0$, and thus, applying transformation (4-45), the c coefficient will be positive, so we have a stable-symmetric point of bifurcation.

Example 4-5: The 'upside down' case

Let us consider a buckled bar loaded by a compressive force which is somewhat larger than the critical load. Let the structure undergo a change of temperature nonuniform along the cross section. If we decrease the force, a snap may occur.

4.6.5 Higher order cuspoid catastrophes

In this sections only two examples are shown. They produce higher order cuspoid catastrophes.

Example 4-6: Butterfly catastrophe
Let us consider the structure shown in *Fig. 4-21*. It is the combination of the structures analysed in Examples 4-3 and 4-4. Be $k_1 = 3$ and $k_2 = 6$. We introduce an imperfection parameter ε_1, which shows that the first spring is longer by ε_1 than it would be in the vertical position of the bar.

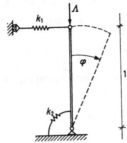

Fig. 4-21 The model for the butterfly catastrophe

The total potential function of the imperfect structure is:

$$V = 0.5k_1(\sin\varphi - \varepsilon_1)^2 + 0.5k_2\varphi^2 + \Lambda\cos\varphi. \tag{4-54}$$

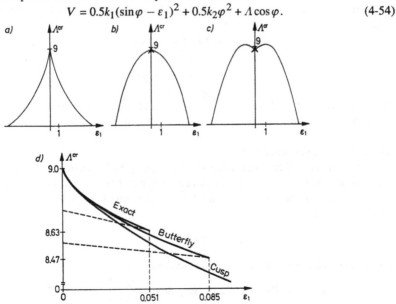

Fig. 4-22 Imperfection sensitivity curves: cusp *a*), butterfly *b*), exact *c*) and the enlarged parts *d*)

The critical load parameter of the perfect structure is $\Lambda = k_1 + k_2$. The coefficient of the fourth order term at the critical point is equal to -0.125, so this is an unstable-symmetric point of bifurcation. Neglecting the cubic terms, the imperfection-sensitivity curve (*Fig. 4-22a*) can be written in parametric form:

$$\lambda = -2^{1/2} \cdot 3p^2,$$

$$\varepsilon_1 = 2^{3/4} p^3 / 3.$$

Let us notice that if $k_1 = 2$ and $k_2 = 6$ then a butterfly catastrophe occurs. We introduce two more imperfections: ε_2 is the error in length of the second spring, ε_3 is the rigidity error of the first spring. The total potential function of the imperfect structure is:

$$V = 0.5(2 + \varepsilon_3)(\sin\varphi - \varepsilon_1)^2 + 3(\varphi - \varepsilon_2)^2 + (8 + \lambda)\cos\varphi.$$

To analyse the original structure we choose $\varepsilon_2 = 0$ and $\varepsilon_3 = 1$, so the first few terms of the Taylor series of the energy function are

$$V = \frac{1}{30}\varphi^6 - \frac{\varepsilon_1}{40}\varphi^5 + \frac{\lambda - 4}{24}\varphi^4 + \frac{\varepsilon_1}{2}\varphi^3 + \frac{1 - \lambda}{2}\varphi^2 - 3\varepsilon_1\varphi,$$

which (omitting the fifth order term) can be induced from Eq. (4-21) by the transformations:

$$x = 5^{-1/6}\varphi,$$

$$a = 5^{2/3}(\lambda - 4)/6,$$

$$b = 1.5 \cdot 5^{1/2}\varepsilon_1,$$

$$c = 5^{1/3}(1 - \lambda),$$

$$d = -3 \cdot 5^{1/6}\varepsilon_1.$$

Substituting these transformations into Eqs (4-22) we obtain the equations of the imperfection-sensitivity curve (*Fig. 4-22b*):

$$\lambda = \frac{-3.6p^6 + 16p^4 - 18p^2 + 12}{p^4 + 12},$$

$$\varepsilon_1 = \frac{0.2p^7 - 9.6p^5 + 12p^3}{3p^4 + 36}.$$

Although the value of ε_3 is large, these curves yield much better approximations of the exact curves (*Fig. 4-22c*).

Example 4-7: Cuspoid catastrophes

Let us consider the highly degenerated structure analysed in Section 1.1.4. The perfect structure is shown in *Fig. 4-23a*. We introduce r imperfection parameters. The eccentricity of the load is ε_1 and the force (S) in the horizontal spring is not a linear function of its elongation (Δ), but

$$S = k\Delta + \sum_{i=2}^{r} \varepsilon_i \Delta^i.$$

The total potential function of the imperfect structure is:

$$V = (k - P)(1 - \cos\varphi) - P\varepsilon_1 \sin\varphi + \sum_{i=2}^{r} \frac{1}{i + 1}\varepsilon_i(\sin\varphi)^{i+1}.$$

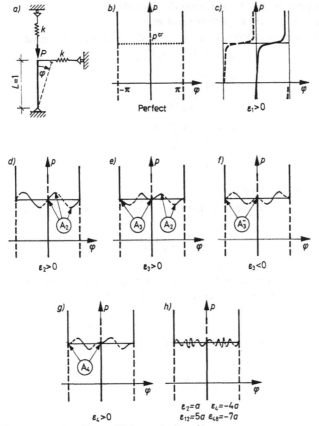

Fig. 4-23 The structure *a*) and its equilibrium paths *b*) − *h*)

There is a horizontal secondary equilibrium path for the perfect structure (*Fig. 4-23b*). The structure does not loose its stability if only ε_1 is different from zero (*Fig. 4-23c*). If only one $\varepsilon_j (j > 1)$ is different from zero, then a cuspoid catastrophe (A_j) occurs at the original critical point (*Fig. 4-23d-g*). If $j > 6$, then there will be catastrophes which do not appear in Thom's theorem. Very waved secondary paths can be produced by appropriate imperfections (*Fig. 4-23h*).

4.6.6 The umbilic catastrophes

As in HUNT [1978], the following form will be examined:

$$\hat{V} = c_1 \hat{v}^3 + c_2 \hat{u}^2 \hat{v} - \hat{\lambda} \left(c_3 \hat{u}^2 + c_4 \hat{v}^2 \right) + \hat{\sigma} \hat{u}^2 + \hat{\varepsilon}_1 \hat{u} + \hat{\varepsilon}_2 \hat{v}, \qquad (4\text{-}55)$$

where $\hat{u}$, $\hat{v}$ stand for the active generalized state co-ordinates; $c_1 c_2 \neq 0$, $c_3 > 0$, $c_4 > 0$ are constants; $\hat{\lambda}$ is deviation of the load parameter from the critical value; $\hat{\sigma}$, $\hat{\varepsilon}_1$, and $\hat{\varepsilon}_2$

Catastrophe theory

are functions of imperfections. The parameter $\hat{\sigma}$ shows the difference of both critical loads (GÁSPÁR [1985]), so it is called a splitting parameter.

For simplicity, function (4-55) will be normalized. Using the transformations

$$\hat{u} = c_1^{1/3} c_3^{-1/2} c_4^{1/2} u,$$

$$\hat{v} = c_1^{-1/3} v,$$

$$\hat{\lambda} = c_1^{2/3} c_4^{-1} \lambda,$$

$$\hat{\sigma} = c_1^{2/3} c_3 \dot{c}_4^{-1} \sigma, \tag{4-56}$$

$$\hat{\varepsilon}_1 = c_1^{1/3} c_3^{1/2} c_4^{-1/2} \varepsilon_1,$$

$$\hat{\varepsilon}_2 = c_1^{1/3} \varepsilon_2$$

function (4-55) assumes the form

$$V(u,v,\lambda,\sigma,\varepsilon_1,\varepsilon_2) = v^3 + \alpha A u^2 v - \lambda \left(u^2 + v^2\right) + \sigma u^2 + \varepsilon_1 u + \varepsilon_2 v \tag{4-57}$$

where

$$A = \left| c_1^{-1} c_2 c_3^{-1} c_4 \right|,$$

$$\alpha = \text{sign}(c_1 c_2).$$

If $\alpha = -1$ (i.e. $c_1 c_2 < 0$), then function (4-55) has an elliptic umbilic point at $\hat{\lambda} = \hat{\sigma} = \hat{\varepsilon}_1 = \hat{\varepsilon}_2 = 0$, and if $\alpha = 1$, then there is a hyperbolic umbilic point.

The elliptic umbilic

Function (4-57) can be induced from (4-25) by the transformations

$$x = \sqrt{A/3}u,$$

$$y = v + \frac{(3-A)\lambda + A\sigma}{6A},$$

$$a = \frac{A\sigma - (A+3)\lambda}{2A}, \tag{4-58}$$

$$b = \sqrt{A/3}\varepsilon_1, \tag{4-59}$$

$$c = \varepsilon_2 + \frac{[(3-A)\lambda + A\sigma][(A+1)\lambda - A\sigma]}{4A^2} \tag{4-60}$$

if $\alpha = -1$. Let us substitute Eqs (4-58), (4-59), (4-60) into Eqs (4-29) which determine the bifurcation set. Ordering the equations we obtain

$$\varepsilon_1 = B\sqrt{A/3}(\sin 2\Theta - 2\cos\Theta), \tag{4-61}$$

$$\varepsilon_2 = C + B(\cos 2\Theta - 2\sin\Theta) \tag{4-62}$$

where

$$B = \frac{[(A+3)\lambda - A\sigma]^2}{12A^2}, \tag{4-63}$$

$$C = \frac{[(A-3)\lambda - A\sigma][(A+1)\lambda - A\sigma]}{4A^2}. \tag{4-64}$$

Using parameter Θ, Eqs (4-61) and (4-62) define the imperfection-sensitivity surface in the four-dimension-space (ε_1, ε_2, σ, λ). This cannot be drawn unless one parameter is fixed at different values. Fixing σ we obtain the imperfection-sensitivity surface (*Fig. 4-24*) of a structure at which the critical points of the perfect structure are at a distance σ (near-coincidence of critical loads).

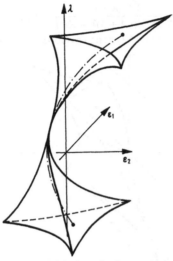

Fig. 4-24 Level surface (σ = const.) of the imperfection-sensitivity surface

Now let us fix λ rather than σ in Eqs (4-61) and (4-62). According to (4-58), the condition $a \leq 0$ can be written in the form:

$$\sigma > \sigma^{(1)} = \frac{A+3}{A}\lambda. \tag{4-65}$$

The parts satisfying this condition are shown in *Fig. 4-25* for a positive and a negative λ.

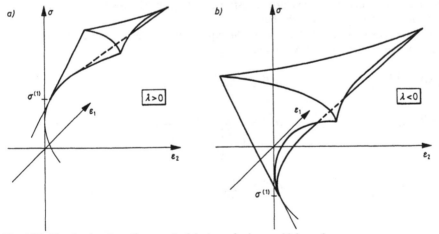

Fig. 4-25 Two level surfaces (λ = const.) of the imperfection-sensitivity surface

The envelope surface of these level surfaces can be given as the function of two parameters (λ, σ). First the value of Θ is computed:

$$\Theta = \arcsin\left(\frac{2A^2(\lambda - \sigma)}{(A+3)[(A+3)\lambda - A\sigma]} - 1\right),$$

then it is substituted into Eqs (4-61) and (4-62). The envelope is shown in *Fig. 4-26*.

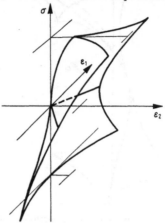

Fig. 4-26 The envelope surface

The critical imperfection territory belonging to a λ_0 is bounded partly by the level surfaces and partly by the envelope. It will be characterized by its sections $\sigma = $ const. The types of the sections are shown in *Fig. 4-27*. Here the boundary of the critical imperfection territory is drawn by full line, parts of the level line not belonging to the boundary by dotted lines, and parts of the envelope (satisfying $\lambda < \lambda_0$) by dashed lines. The critical imperfection territory lies outside the boundary.

Fig. 4-28 shows the validity intervals of the types of the sections for different intervals of λ and A. The boundaries of the types are:

$$\sigma^{(1)} = \frac{A+3}{A}\lambda, \qquad\qquad \sigma^{(2)} = \frac{2A+3}{A}\lambda,$$

$$\sigma^{(3)} = \lambda, \qquad\qquad \sigma^{(4)} = \frac{\lambda}{1 - \left[\dfrac{3}{(2A+3)}\right]^{1/2}},$$

$$\sigma^{(5)} = \frac{(A-3)(A+1)}{A(A-1)}\lambda, \qquad \sigma^{(6)} = \frac{2A+3}{2A}\lambda.$$

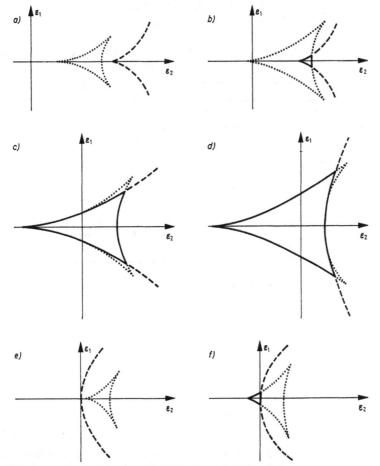

Fig. 4-27 Types of sections (σ = const.) of critical imperfection territories

The hyperbolic umbilic

Function (4-57) can be induced from (4-30) by the transformations

$$x = 2^{-1/3}\left(v + \sqrt{\frac{A}{3}}u + \frac{3\sigma - (3+A)\lambda}{6A}\right),$$

$$y = 2^{-1/3}\left(v - \sqrt{\frac{A}{3}}u + \frac{3\sigma - (3+A)\lambda}{6A}\right),$$

$$a = 2^{-1/3}\frac{(3-A)\lambda - 3\sigma}{A}, \tag{4-66}$$

Catastrophe theory

$$b = 2^{-2/3}\left(\varepsilon_2 + \sqrt{\frac{3}{A}}\varepsilon_1 + \frac{[(3+A)\lambda - 3\sigma][(1-A)\lambda - \sigma]}{4A^2}\right),\tag{4-67}$$

$$c = 2^{-2/3}\left(\varepsilon_2 - \sqrt{\frac{3}{A}}\varepsilon_1 + \frac{[(3+A)\lambda - 3\sigma][(1-A)\lambda - \sigma]}{4A^2}\right)\tag{4-68}$$

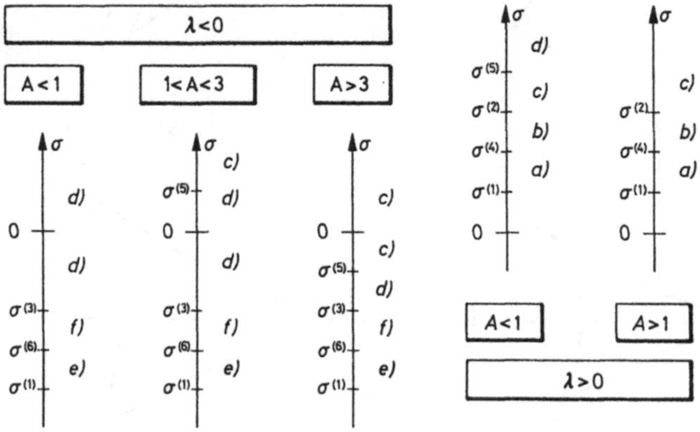

Fig. 4-28 Validity intervals of the types of sections (letters a to f refer to the curves in *Fig. 4-27*)

if $\alpha = 1$. Let us substitute Eqs (4-66), (4-67), and (4-68) into Eqs (4-29) which determine the bifurcation set. Ordering the equations we obtain

$$\varepsilon_1 = B\sqrt{A/3}(-q^2 + 2q - 2q^{-1} + q^{-2}),\tag{4-69}$$

$$\varepsilon_2 = C - B(q^2 + 2q + 2q^{-1} + q^{-2})\tag{4-70}$$

where

$$B = \frac{[(3-A)\lambda - 3\sigma]^2}{24A^2},\tag{4-71}$$

$$C = \frac{[(3+A)\lambda - 3\sigma][(A-1)\lambda + \sigma]}{4A^2}.\tag{4-72}$$

Using parameter q, Eqs (4-69) and (4-70) define the imperfection-sensitivity surface in the four-dimension-space (ε_1, ε_2, σ, λ). This cannot be drawn unless one parameter is fixed at different values. Fixing σ we obtain the imperfection-sensitivity surface (see e.g. *Figs 8* and *9* in HUNT [1978]) of a structure at which the critical points of the perfect structure are at a distance σ (near-coincidence of critical loads). Now let us fix λ rather than σ in Eqs (4-69) and (4-70). According to (4-31), (4-66) and the condition $pq \geq 0$, it can be seen that if

$$\sigma < \sigma^{(1)} = \frac{3-A}{3}\lambda$$

then only the part corresponding to $q > 0$, in other cases the part corresponding to $q < 0$ has to be considered. This surface is shown in *Fig. 4-29*.

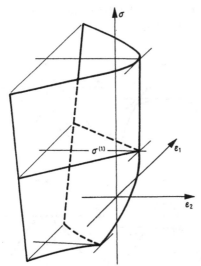

Fig. 4-29 A level surface (λ = const.) of the imperfection-sensitivity surface

The envelope surface of these level surfaces can be given as the function of two parameters (λ, σ). First the values of q_2 and q_3 are computed:

$$q_{2,3} = \frac{E - D \pm \sqrt{(D - E)^2 - 4D^2}}{2D} = \frac{A \pm \sqrt{6A - 9 + (9 - 3A)\sigma/\lambda}}{A \mp \sqrt{6A - 9 + (9 - 3A)\sigma/\lambda}},$$

then they are substituted into Eqs (4-69) and (4-70). The envelopes for different cases are shown in *Fig. 4-30.*

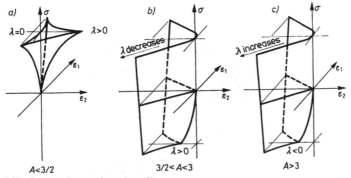

Fig. 4-30 The envelope surfaces ($\sigma > 0$)

The critical imperfection territory belonging to a λ_0 is bounded partly by the level surfaces and partly by the envelope. It will be characterized by its sections σ = const. The types of the sections are shown in *Fig. 4-31.* Here the boundary of the critical imperfection territory is drawn by full line, parts of the level line not belonging to the boundary by dotted lines,

Catastrophe theory

and parts of the envelope (satisfying $\lambda < \lambda_0$) by dashed lines. The critical imperfection territory lies on the right side of the boundary.

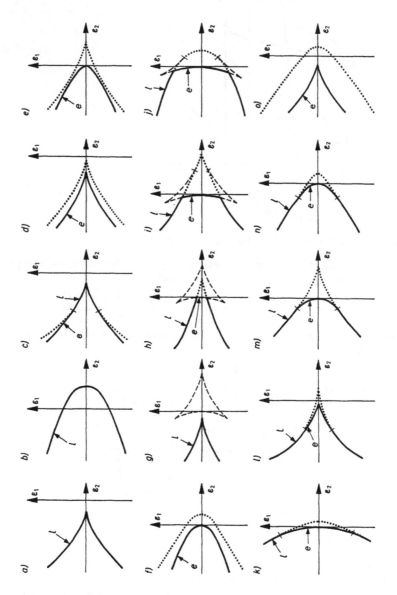

Fig. 4-31 Types of sections (σ = const.) of critical imperfection territories

Fig. 4-32 shows the validity intervals of the types of sections for different intervals of λ and A. The boundaries of the types are:

$$\sigma^{(1)} = \frac{(3 - A)\lambda}{3}, \qquad \sigma^{(2)} = \frac{(2A - 3)\lambda}{A - 3},$$

$$\sigma^{(3)} = \frac{(8A - 12)\lambda}{A - 3}, \qquad \sigma^{(5)} = \frac{(3 - 2A)\lambda}{3}.$$

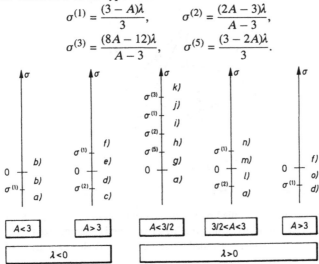

Fig. 4-32   Validity intervals of the types of sections (letters *a* to *o* refer to the curves in *Fig. 4-31*)

In the case of hyperbolic umbilic catastrophe a further classification can be made. For simplicity, $\sigma = 0$ is supposed. Details are shown in GÁSPÁR [1983*b*], the results are summarized below:

	Type	Comment
$A = 0$	symbolic umbilic	fourth degree terms are needed
$0 < A < 3/2$	monoclinal bifurcation	
$A = 3/2$	triple root	
$3/2 < A < 3$	homeoclinal bifurcation	no bifurcation if $\lambda < 0$
$A = 3$	spherical shell	another parameter is needed
$3 < A < \infty$	homeoclinal bifurcation	can be bifurcation at $\lambda < 0$
$A = \infty$	parabolic umbilic	fourth degree terms are needed

Example 4-8: Umbilic catastrophes

This simple model was first analysed by THOMPSON and GÁSPÁR [1977], then it was developed by others (HUNT, REAY and YOSHIMURA [1979], THOMPSON and HUNT [1984], GIONCU and IVAN [1984], HACKL [1990], PAJUNEN and GÁSPÁR [1996]). Here only some basic equations are given.

The physical model of our investigation is shown in *Fig. 4-33* and consists of a weightless rigid strut of unit length carrying a vertical dead load P. It is pinned at its bottom and supported by three linear springs inclined initially at $45°$. The first spring lies in the plane yz, while the others are placed symmetrically to the plane yz, the horizontal angle being denoted by β. The stiffnesses of the equal second and third springs are denoted by k_2, and the stiffness of the first springs is $k_1 = 1 - 2k_2$.

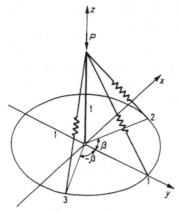

Fig. 4-33 Model for the umbilic catastrophes

A state of the structure is specified by the two co-ordinates (x,y) of the top of the strut. The perfect system rests vertically, while the imperfect one is characterized by x_0, y_0 (and the springs are stress free) under zero load. The truncated Taylor expansion of the total potential energy function is:

$$V = k_1 \left(\frac{1}{4} y^2 + \frac{1}{8} y^3 + \frac{5}{64} y^4 \right) +$$

$$+ k_2 \left[\frac{1}{2} \left(x^2 s^2 + y^2 c^2 \right) + \frac{1}{4} \left(3x^2 y s^2 c + y^3 c^3 \right) + + \frac{5}{32} \left(x^4 s^4 + 6 x^2 y^2 s^2 c^2 + y^4 c^4 \right) \right] +$$

$$+ P \left(-\frac{1}{2} x^2 - \frac{1}{2} y^2 - \frac{1}{8} x^4 - \frac{1}{8} y^4 - \frac{1}{4} x^2 y^2 \right) -$$

$$- \frac{1}{2} k_1 y_0 y - k_2 s^2 x_0 x - k_2 c^2 y_0 y, \qquad (4\text{-}73)$$

where $s = \sin\beta$ and $c = \cos\beta$.

The perfect structure has the trivial fundamental path $x = y = 0$. The Hessian matrix is diagonal along the fundamental path. One of these diagonal elements must vanish at critical loads, so there are two of them:

$$P^{\text{cr1}} = k_2 s^2,$$
$$P^{\text{cr2}} = k_1 / 2 + k_2 c^2.$$

If $k_2 = 1/4s^2$ then the two values will be equal (1/4), and we have an umbilic catastrophe. To avoid a negative stiffness, β is restricted to the interval $(45°, 135°)$. The cubic part of the potential energy function at the critical point is

$$\frac{1}{16} \frac{s^2 + c}{1 + c} y^3 + \frac{3}{16} c x^2 y,$$

so according to (4-55) and (4-57)

$$\alpha A = \frac{s^2 + c}{3c(1 + c)}$$

i.e. the structure has the following types of critical points for the different horizontal angles β:

$45° \le \beta < 64.26°$	hyperbolic umbilic, homeoclinal point of bifurcation
$\beta = 64.26°$	hyperbolic umbilic, triple root
$64.26° < \beta < 90°$	hyperbolic umbilic, monoclinal point of bifurcation
$\beta = 90°$	symbolic umbilic
$90° < \beta < 128.17°$	elliptic umbilic, anticlinal point of bifurcation
$\beta = 128.17°$	parabolic umbilic
$128.17° < \beta \le 135°$	hyperbolic umbilic, homeoclinal point of bifurcation.

4-9. Example: Higher catastrophes

GÁSPÁR [1977] investigated a similar model to that shown in *Fig. 4-33*, but there the strut is supported by n linear springs. It has been proven that there is an anticlinal point of bifurcation if $n = 3$, but if $n > 3$, then higher catastrophes arise which do not appear in Thom's theorem. They are the so-called double-cusp catastrophes (ZEEMAN [1976]).

GÁSPÁR and MLADENOV [1995] analysed a simple in-plane column loaded by a polar force. This structure had a highly degenerate critical point where a secondary equilibrium surface arises instead of equilibrium paths.

4.7 PROBABILITY OF THE INSTABILITY

The allowable imperfections can be given by a territory in the imperfection space. It is sure that the structure is able to carry a prescribed load if the corresponding critical imperfection territory and the allowable imperfection territory have no common points. If they have common points, then one can ask about the probability of instability. To elucidate this we have to know the probability density functions of the imperfections.

BOLOTIN [1965, 1982] gave general formulas, but he analysed only distinct critical points in the case of a single imperfection parameter. ELISHAKOFF [1983] suggests to use the Monte-Carlo method.

When determining the critical imperfection territories, in this chapter some transformations were made to achieve the canonical form of the potential energy functions. First we have to make these transformations with the density functions of the imperfections. The probability of the instability below a given load (λ) is

$$P(\lambda) = \int_A h_\varepsilon(\varepsilon) dA$$

where A is the critical imperfection territory and h_ε is the transformed probability density function. Calculating $P(\lambda)$ with several load parameters we get a probability function. If we know the probability density function $h(\lambda)$ of the maximum load acting during a given period, then we can compute the probability of the instability of the structure in this period:

$$P_{\text{instab}} = \int_{-\infty}^{\infty} P(\lambda)h(\lambda)d\lambda.$$

We call attention to the limits of these calculations. Catastrophe theory describes functions only locally, so large imperfections may give wrong results. GÁSPÁR [1983a] gives a numerical example for calculation of the probability of the instability.

5

Buckling of frames

Josef APPELTAUER, Lajos KOLLÁR

5.1 GENERAL THEORY OF FRAME BUCKLING

Josef APPELTAUER

5.1.1 Description of the phenomena

The buckling of frame structures will be elucidated on bar structures with rigid nodes, which may buckle spatially. Since the skeleton of a building mostly consists of plane frames arranged parallel to each other, the direction parallel to the frames will be called *transverse* direction, while the *longitudinal* direction is perpendicular to this. We assume that the cross section of the column of the frame has two axes of symmetry, and the greater moment of inertia stiffens the frame in the transverse direction.

During fabrication and erection *accidental imperfections* arise in both principal directions. These can be *geometric imperfections*, characterized by some geometric quantity, as crookedness or skewness of the bar axis, the introducing of loads deviating from the planned position (e.g. eccentricity), furthermore *mechanical imperfections* as residual (initial) stresses, or unequal distribution of the physical characteristics (e.g. yield limit). The columns of the frame thus undergo a spatial deformation, and they are subjected not only to normal forces, but also to bending moments and shearing forces in two directions, and furthermore to torsional moments.

As the loads increase, the distribution of internal forces in the frame varies continuously, because the bending stiffnesses of the individual bars decrease in different degrees. This decrease is caused by the 'softening' (secondary) effect of the normal forces and by the spreading of the zone of plastic deformations, which can be followed reliably above all in steel frames. In reinforced concrete frames, add to all these the appearing of cracks in the tension zone. Thus the state of equilibrium of the comparatively more loaded (so-called *active*) bars of the frame becomes unstable, and the equilibrium of the whole structure is only assured by the supporting effect of the less loaded (*passive*) bars.

The whole frame buckles when the increasing deformation tendency of the active bars consumes the decreasing supporting capacity of the passive bars. Hence the buckling of frames is, as a rule, a process spreading to the whole bar structure, but in the case of local loading it may be practically confined to a part of the structure only, since the supporting effect ot the farther passive bars 'dies out' with the distance.

Engineers have tried more and more to approach this rather complicated phenomenon with computation models. First, mechanical models have been used which could take only bifurcation into account. Later, models describing loss of stability with limit point (divergence or snapping through, see Chapter 1) were developed, which were closer to reality. Snapping of frames occurs, however, very seldom (e.g. with very slender three-hinged portal frames), so that we treat this phenomenon only very briefly in Section 5.1.5.

To all these mechanical models, exact or approximate mathematical models may belong.

5.1.2 Mechanical models describing the various kinds of loss of stability

Bifurcation model

Apart from special cases (e.g. when the effect of structural and loading imperfections is contrary to each other [ROORDA, 1965]), the bifurcation model greatly simplifies reality, since its validity requires the introduction of several idealizations. Thus the bar axes are perfectly straight and the columns are exactly vertical, so that no geometric imperfections are present. The material undergoes elastic deformations only, and the corresponding moduli are everywhere identical, so that initial stresses and thus mechanical imperfections are absent. Though CHWALLA [1959] introduces also variable moduli into his bifurcation model, nevertheless this generalization has to be interpreted differently, according to the structural materials. In the case of steel, the straight line of the stress-strain diagram becomes curved before reaching the yield stress because of the initial (residual) stresses [BEEDLE and TALL, 1960], since in some fibres the stresses reach the yield limit sooner than in the case of a uniform stress distribution. Hence the generalization of *Chwalla* 'smuggles' a group of imperfections into the bifurcation model. This would be not bad in itself, but causes difficulties in practical applications, see Section 5.1.6. On the other hand, with aluminium (and also with reinforced concrete) plasticity does not occur suddenly, but gradually, and the stress-strain diagram is a curved line from the beginning. In this case the generalization of *Chwalla* means the extension of the bifurcation model to plastic deformations. Theoretical and experimental investigations at Lehigh University [GALAMBOS, LU and DRISCOLL, 1968] have applied the bifurcation model also in the domain of elastic-plastic deformations, when symmetrically loaded symmetric frames lose their equilibrium after developing several plastic hinges.

The most basic simplifications have to be introduced, however, with respect to the loading and internal force system of frames. These belong to the criterion of *Klöppel* and *Lie* [KOLLBRUNNER and MEISTER, 1955], which mathematically formulates the possibility of bifurcation. The primary deformation (caused by the loading and still stable) can be developed into the series of the eigenfunctions of the frame. The criterion of *Klöppel* and *Lie*, ensuring the possibility of bifurcation, requires that in the series of the primary deformation the functions of the secondary (buckling) deformation do not appear. This means that the primary and secondary deformations are orthogonal to each other. We can also say that, with respect to the primary deformation, qualitatively new (bending or torsional) deformation components have to come into the buckling deformation [MATEESCU, APPELTAUER and CUTEANU, 1980]. Hence, bifurcation really can occur only on ideal, geometrically

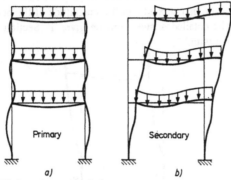

Fig. 5-1   Bifurcation of equilibrium of a symmetric frame

symmetric and symmetrically loaded frames (*Fig. 5-1*). Here the primary deformation is symmetric, and the buckling shape is antisymmetric, i.e. orthogonal to the primary one. In the general case (*Fig. 5-2*) we have to concentrate the loads acting on the beams to the nodes, and on sway frames we have to neglect the horizontal forces (*Fig. 5-2a*). Furthermore we have to neglect the primary deformation too, which seemingly consists of the compression of the columns only, but since the shortening of the individual columns is not necessarily equal, this may result in bending and transverse shear. The buckling shape similarly may contain compression, shear and bending of general distribution, so that the primary and secondary deformations cease to be orthogonal to each other [APPELTAUER, 1970a]. If, however, we introduce the simplifications mentioned above, the stable state of equilibrium of the frames shown in *Fig. 5-2* does not contain deformations, and the buckling shape bifurcates from this (theoretical) state.

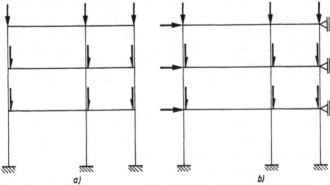

Fig. 5-2   Bifurcation models of general frames. *a*) Unbraced (sway) frame, *b*) braced (no-sway) frame

In the case of sway frames the normal forces change anyway due to the horizontal displacements of the constant directional (conservative) loads, but if we neglect the horizontal forces, then the normal force distribution may essentially deviate from that of the real frame. We thus may rightly ask: is the critical load parameter still relevant to the buckling of real frames? The answer also depends on the magnitude of the horizontal forces related to the vertical ones. The detailed answer will be given in Section 5.1.6 where practical applications will be dealt with.

Fig. 5-3 Additional forces and moments due to sway

Returning to the classical bifurcation case of *Fig. 5-1* we have to remark that when calculating the critical load parameter we always concentrate the distributed loads into the nodes. By so doing we neglect the primary (symmetric) deformation, so that the (antisymmetric) buckling deformation starts from the undeformed state. The consequences of this simplification have been dealt with by many authors, mostly on two- or three-hinged portal frames with beams of straight or broken line, assuming elastic behaviour. (At Lehigh University this investigation has been extended to the domain of elastic-plastic deformations of multi-storey frames.) The concentration of the loads into the nodes results in most cases in a somewhat greater critical load parameter than with distributed loads, because – as mentioned before – the normal forces in the columns change due to sway, and in the two columns additional forces ΔN of opposite signs arise (*Fig. 5-3*). These influence the primary deformation in the same sense, by increasing the sway to some extent. If we concentrate the loads into the nodes, the primary deformation ceases to develop, and so does the increasing influence of the additional forces ΔN. This concentration means also, as a rule, that the influence of the forces ΔN on the bending stiffnesses of the columns is neglected, so that from this point of view the normal forces in the columns can be considered as unchanged.

The simplifications of the bifurcation model are also valid for the secondary (buckling) deformation. Here mostly the bending deformations are considered, i.e. (with plane frames) the deformations due to the normal and shearing forces are neglected. The influence of shear has to be taken into account with built-up columns with bracing or batten plates, and the influence of compression is relevant only with frames having a great storey height [CHWALLA and JOKISCH, 1941], i.e. when the length of the columns is much greater than that of the beams, or more generally: when the bending stiffness of the beams is much greater than that of the columns. Due to this effect the buckling shape of single-bay multi-storey frames deviates from that of the 'normal' frames and approaches the deformation of the column with solid cross section (*Fig. 5-4*).

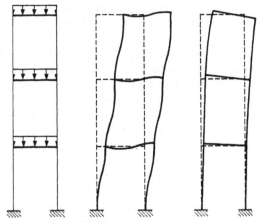

Fig. 5-4 Buckling shapes of a slender frame

As can be seen, the disadvantage of the bifurcation model is that in some cases it poorly approaches reality. Its advantage is, however, that it can give a clear picture of even the most complicated buckling phenomena, for which we also have to take into account the various kinds of post-buckling behaviour treated by KOITER (see Chapter 1).

The bifurcation model also simplifies the mathematical model when determining the critical load parameter, since a buckling shape infinitely close to the equilibrium position has to be described by a linearized differential equation, which leads to the so-called linear stability theory. (This circumstance made it possible for CHWALLA to use the tangent modulus in the plastic zone.) However, this same linearization prevents us from predicting the post-critical behaviour of frames. Thus the horizontal line obtained by the bifurcation theory must not be considered as the description of the post-critical behaviour of the structure, see Section 1.1.5.

The asymptotic application of the linear elastic theory (when we determine the critical load as the asymptote of the load-deflection curve) has also an uncertain validity, since at larger deformations it is not possible to use the linearized differential equation, and we have to resort to the higher order (nonlinear) theory.

The bifurcation model has a further advantage, although this also contradicts reality: it makes it possible to treat the in-plane buckling of the transverse frames and the longitudinal instability of the hall separately. The first kind of instability entails, as a rule, only bending deformation, while the second involves also torsional deformation. We shall deal with the coupling of both buckling modes in Section 5.1.6.

Divergence model

The divergence model eliminates all shortcomings of the bifurcation model, since it idealizes neither the frame, nor the structural material or the loading. The divergence model starts, in fact, from the real frame, is able to take into account all the imperfections, and can follow the plastic deformations. The divergence model does not know two separate buckling modes in the two principal directions, since it describes the complete, spatial buckling from the beginning on. Only in special cases, when the buckling mode gets very close to the transverse or longitudinal direction, does the buckling decompose into two plane deformations.

Whether the frame loses its stability by divergence depends on several factors; of these the stiffness ratios of the bars are of great importance. If the beams are much stiffer than the columns, then plastic deformation develops first of all in the columns, reducing their stiffnesses. In no-sway frames, individual columns may become unstable, while in sway frames whole storeys can lose their stability at once. In both cases, the increasing internal moments are carried mainly by the beams, and their failure entails the collapse of the whole frame. On the other hand, if the beams are less stiff, then plastic hinges develop in them, so that the columns above each other behave 'independently' from each other: they become continuous beams in no-sway frames, cantilevers clamped at the bottom in sway frames, and in both cases these columns buckle [WOOD, 1958].

As is well known, the notion of divergence has been introduced by DUTHEIL [1966]. In single steel bars divergence appears when, due to plastic deformation, the internal moment increases more slowly than the external one, and cannot maintain equilibrium with the latter. In reinforced concrete bars, the same phenomenon occurs due to plastification of concrete and steel bars and due to cracking which reduces bending stiffness as well. With thin-walled tubes, flattening of the cross section may also reduce the bending stiffness and thus causing divergence.

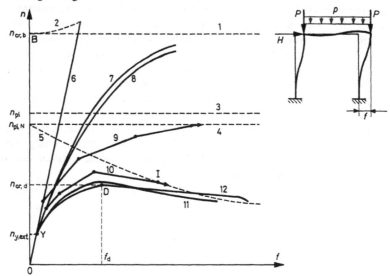

Fig. 5-5 Limit curves plotted on the basis of various assumptions

In the case of frames, divergence occurs when the decreasing stiffness of the passive (supporting) bars diverges from the increasing 'buckling tendency' of the active (unstable)

bars. As a two-hinged bar collapses before the most critical cross section becomes entirely plastic, frames also collapse before all plastic hinges develop that are necessary to produce an unstable mechanism of hinged bars. All these are shown in *Fig. 5-5.*

First we treat the *limit curves.* Curve 1 is a horizontal line at the height of the critical bifurcation parameter ($n_{cr,b}$), and, as has been said already, can be interpreted only mathematically, but not mechanically. Line 2 is the 'exact' curve of the post-critical load-bearing capacity of bifurcation (with elastic bars), showing symmetric-stable behaviour. (It should be mentioned that in these two cases no horizontal load is acting on the frame, i.e. $H = 0$. If either the frame or the loading cease to be symmetric, then the post-critical curve becomes 'asymmetric', see Section 1.1.3.) Line 3 shows the limit parameter of the rigid-plastic computation. In case 4 we considered the influence of the normal force on the development of the plastic hinge. Curve 5 has been constructed also on the basis of rigid-plastic behaviour, but taking into account sway deformation, hence this is the result of the second-order theory.

The *equilibrium paths* showing the behaviour of the frame according to various assumptions are those marked 6 to 12. Line 6 describes the linearly elastic behaviour, i.e. the first-order theory. Line 7 is constructed by the second-order elastic theory. As can be seen, the curve 8 of the eigenfunction of the buckling deformation is very close to 7. Curve 9 shows the rigid-plastic behaviour, according to the first order theory. Line 10 depicts the same, but using the second-order theory. In this case the places and sequence of the development of plastic hinges may change as compared with the previous case [HALÁSZ, 1969]. Line 11 is obtained by the elasto-plastic second-order theory, and line 12 shows experimental results. It is remarkable that this experimental curve intersects the limit curve 5 of plastic mechanism. HORNE [1963a] explained this by pointing out that, due to the large deformations, in the plastic hinges strain-hardening of steel becomes effective. HALÁSZ [1967] draws the attention to the fact that in some cases this strain-hardening may compensate the whole second-order effect, the limit curve 5 may lose its descending character and become ascending, representing stable equilibrium states. The point Y in *Fig 5-5* marks the appearance of yield stress in the extreme fibre.

The second-order theory suffices to determine the critical load parameter of divergence ($n_{cr,d}$), but large post-critical deformations can be described by the third-order theory only.

5.1.3 Bifurcation

Classification of the computation methods

Engineers trust exact methods above all, so that the second-order generalizations of the three main computation procedures (force, deformation, and mixed methods) have all been worked out. In analytical form all three result in a system of homogeneous linear equations with deformation or force quantities of the buckling shape as unknowns. Buckling can occur only if the determinant of this equation system becomes equal to zero. Developing the determinant we obtain a 'characteristic' equation containing trigonometric functions which represent the second-order influence of the normal force on the deformations and the bending moments. The characteristic equation has several roots, yielding the critical load parameters, of which we need the smallest one. The deformation method has the advantage over the force method that its determinant varies much more smoothly, so that it is easier to find the smallest root.

Before the computer age several procedures have been found to reduce the order of the stability determinant or to eliminate it completely. Also an immense number of approximate methods have been developed, and one is inclined to wonder how such a complicated problem can be solved in some cases so simply, even to reduce it to closed formulas.

For a better overview we classify the approximate methods into the following groups:
– methods obtained from the exact theory by simplifications,
– equilibrium methods,
– energy methods,
– methods based on cutting out subsystems,
– reducing the problem to known simple cases,
– replacing the discrete frame by an equivalent continuum.

In the computer age the determination of the critical load by exact methods no longer presents difficulty, so that most of these approximate methods, above all the iterative procedures, have lost their importance, and only those remain useful which enable the design engineer to find the critical load parameter very quickly. In the following we shall present some of these approximate methods which, to our judgement, may be useful.

The *exact methods*, furthermore the *approximate methods obtained from the exact theory by simplifications and the equilibrium method* have been extensively treated in the classical literature and can be found easily, hence we will not deal with them in the following. We also omit the equivalent continuum method (which 'smears out' the beams and coalesces the columns into a single one) because it can be found in [ZALKA and ARMER, 1992]. We only mention that, considering the whole frame as one girder, as the continuum method does, the deformation of the frame can be characterized by four 'basic' types of deformation: local bending, global bending, local shear and global shear deformations [ZALKA, 1992]. Characteristic stiffnesses can be attached to these deformations, and the stiffnesses can then be associated with the corresponding partial critical loads. By the application of the continuum method and using the summation theorems treated in Chapter 2, simple design formulae can be produced which make possible to carry out stability analysis in minutes [ZALKA, 1998].

A more general treatment of the continuum method is made possible by the theory of sandwich bars, presented in Chapter 6.

In the following we shall treat the other approximate methods which we deem practically very useful, such as the energy method, or which are difficult to find in the literature, as those using subsystems or based on simple cases.

Approximate methods
Energy methods
All energy methods require the knowledge of the buckling shape. According to the way this deformation is determined, they can be classified into two groups. Those belonging to the

first group find the deformation by more or less complicated static calculations, while those of the second group determine this shape very simply (e.g. by computing the deformation of the sway frame due to a single horizontal concentrated force acting on the top beam) without sensibly impairing the accuracy of the result. The reason for this phenomenon most probably is that the strain energy stored in the frame is not very sensitive to the exactness of the assumed buckling shape. We will call the horizontal loads causing bending deformation which approximates the buckling shape the 'equivalent loading'.

The most characteristic procedure of the first group is the so-called bending method of SATTLER [1953], which has been generalized by PENELIS [1968] to space frames whose columns are connected by rigid floors. We still have to mention the methods of JOHNSON [1960] and of BERNARDINIS [1968], who started from the energy expressions of TIMO-SHENKO [TIMOSHENKO and GERE, 1961]; the second author arrived at closed formulas.

The most remarkable method of the second group is that of PUWEIN [1936-38]. He approximates the buckling shape by the deformation of the sway frame due to a horizontal force acting on the uppermost beam, calculates the strain energy stored, and in determining the work done by the vertical forces assumes that the deformations of the columns are affine to those connected to infinitely rigid beams. The second-order bowing out of the columns is considered by the usual correction factor $\pi^2/12$, which compensates the difference between the vertical displacements of the upper end of a column bent by a horizontal force and of a column buckling under a compressive force. SIEVERS slightly modified the method of PUWEIN and obtained an explicit (but rather complicated) formula for multi-storey frames [BÜRGERMEISTER, STEUP and KRETSCHMAR, 1963]. SAHMEL [1955] essentially simplifies the method of PUWEIN in the case of multi-storey frames by determining the horizontal displacements of the beams without any previous static analysis, assuming the points of inflexion at mid-height of the storeys.

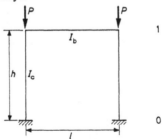

Fig. 5-6 Notations for one bay

APPELTAUER [1961] developed, by generalizing the ideas of PUWEIN and SAHMEL, a weighted energy method which yields rather simple explicit formulas for multi-storey and/or multi-bay frames. He approximates the buckling shape by the deformation caused by unit horizontal forces, acting at the levels of the beams, except for multi-storey frames rigidly clamped at the bottom, whose rigidity at the ground floor, defined by the relation (E denoting the modulus of elasticity):

$$\kappa = \frac{E_c I_c l}{E_b I_b h}, \tag{5-1}$$

is greater than $k = 0.5$ (*Fig. 5-6a,*). On these – in accordance with the original idea of PUWEIN – it is sufficient to let the horizontal force act at the level of the uppermost beam. If this rigidity is less than 0.5, and also for frames elastically clamped or hinged at the bottom, it is necessary to apply the horizontal load at every level, mainly because of the greater sway of the ground floor. It is permissible to draw the bending moment diagram using the simplification of SAHMEL, mentioned above, except for frames which are elastically clamped or hinged at the bottom, or which have a very weak beam between two very rigid beams.

On the basis of [APPELTAUER and BARTA, 1964], STEVENS [1967] investigated the appropriateness of the simplification of SAHMEL. The weighted method starts from the second-order moment

$$M_i = N_i \frac{\Delta f_i}{2} = N_i \frac{f_i - f_{i-1}}{2}, \tag{5-2a}$$

see *Fig. 5-7*, and after summing these up at the nodes, weighting them by the curvatures of the columns at the nodes, and introducing the correction factor $12/\pi^2$ mentioned above, leads to the general formula

$$\frac{12}{\pi^2} \sum_n N_i \frac{f_i - f_{i-1}}{2} \frac{M_i}{E_i I_i} = \sum_n \frac{M_i^2}{E_i I_i}, \tag{5-2b}$$

which equalizes, in fact, the external work and internal energy referred to unit length close to the node. This notion has been later introduced also by ORAN [1967] who called it the 'density of the strain energy'

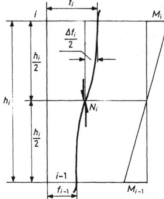

Fig. 5-7 The second-order bending moment at the node of a multi-storey and/or multi-bay frame

For multi-storey frames (*Fig. 5-8a*) we show the bending moments caused by a unit force applied at the uppermost level in *Fig. 5-8b*, and those caused by forces applied at all levels in *Fig. 5-8c*. If.e.g. the stiffness, defined by Eq. (5-1), of the ground floor of the n storey symmetric frame, clamped at the bottom, is greater than 0.5, then the explicit form of Eq. (5-2b) becomes:

$$P_{n,\mathrm{cr}} = \frac{\pi^2 E_n I_n}{h_n^2} \frac{\sum_n \overline{h}_i \overline{h}_i'}{\sum_n \overline{f}_i \left(\overline{N}_i \overline{h}_i' - \overline{N}_{i+1} \overline{h}_{i+1}' \right)}, \tag{5-3}$$

where the following notations have been used:

$$\overline{h}_i = \frac{h_i}{h_n}, \tag{5-4}$$

$$\overline{h}'_i = \overline{h}_i \frac{E_n I_n}{E_i I_i}, \tag{5-5}$$

$$\overline{N}_i = \frac{N_i}{N_n}, \tag{5-6}$$

$$\overline{f}_i = \frac{24 E_n I_n}{h_n^3} f_i. \tag{5-7}$$

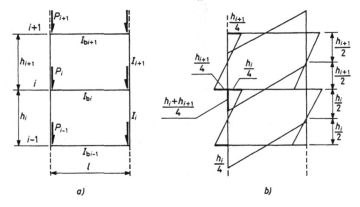

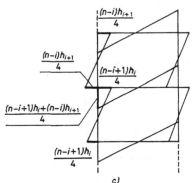

Fig. 5-8 Bending moments caused by equivalent horizontal loading on multi-storey frames

The recursion formula of the reduced displacements of the beams [Eq. (5-7)] is:

$$\overline{f}_i = \overline{f}_{i-1} + \overline{h}_i^2 \overline{h}'_i + 0.411 \overline{l}'_{i-1} \overline{h}_i \left(\overline{h}_{i-1} + \overline{h}_i \right) + 0.411 \overline{l}'_i \overline{h}_i \left(\overline{h}_i + \overline{h}_{i+1} \right), \tag{5-8}$$

where also the notation

$$\overline{l}'_{i-1} = \frac{l_i}{h_n} \frac{E_n I_n}{E_{bi} I_{bi}} \tag{5-9}$$

has been used.

The frames hinged at the bottom are mostly equistable with their ground floor parts. For example, in the case of the asymmetric two-bay frame (*Fig. 5-9*) the explicit form of Eq. (5-2*b*) is the following:

$$P_{\mathrm{cr}} = \frac{\pi^2 EI}{h^2} \frac{1 + 4m + r}{\bar{f}\left(1 + 2\bar{P}_m m + \bar{P}_r r\right)}. \tag{5-10}$$

The notations are:

$$m = \frac{EI}{E_m I_m}, \tag{5-11}$$

$$r = \frac{EI}{E_r I_r}, \tag{5-12}$$

$$\bar{P}_m = \frac{P_m}{P}, \tag{5-13}$$

$$\bar{P}_r = \frac{P_r}{P}, \tag{5-14}$$

$$\bar{f} = \frac{48EI}{h^3} f. \tag{5-15}$$

The expression of the reduced beam displacement becomes:

$$\bar{f} = \frac{1}{4}\left[1 + 4m + r + 0.822\left(\bar{l}' + \bar{l}'_r\right)\right], \tag{5-16}$$

where the abbreviations

$$\bar{l}' = \frac{l}{h} \frac{EI}{E_b I_b} \tag{5-17}$$

$$\bar{l}'_r = \frac{l_r}{h} \frac{EI}{E_{br} I_{br}} \tag{5-18}$$

were introduced.

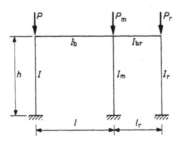

Fig. 5-9 Asymmetric two-bay frame

APPELTAUER and BARTA [1960 and 1961] generalized the approximate method of PUWEIN to frames of industrial halls loaded also by cranes. The cross section of the columns varies stepwise (*Fig. 5-10a*). They approximated the buckling shape by the deformation line caused by horizontal forces proportional to the vertical ones (*Fig. 5-10b*). In the case of the symmetric frame depicted in *Fig. 5-10a* the energy condition of *Puwein* runs, assuming a unit horizontal force:

$$\frac{1}{2}\left[(1 - k)f_2 + kf_1\right] = P_1 w_1 + P_2 w_2, \tag{5-19}$$

where

$$k = \frac{P_1}{P_1 + P_2} 1kN = \frac{P_1}{P} 1kN \qquad (5\text{-}20)$$

and

$$w_i = \frac{1}{2}\overline{\psi}_i \frac{f_i^2}{h_i}, \qquad (5\text{-}21)$$

as the vertical displacements of application of the forces.

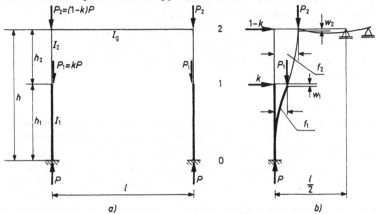

Fig. 5-10 Equivalent horizontal loading on the frame of an industrial hall

Since the multiplication factor ψ_i also depends on the elastic line of the column, they introduced, using the results of PUWEIN, the values

$$\psi_1 = \frac{\overline{\psi}_1}{1.2} = 1.0, \qquad (5\text{-}22)$$

$$\psi_2 = \psi = \frac{\overline{\psi}_2}{1.2}, \qquad (5\text{-}23)$$

and – after modifying the correction factor to $\pi^2/10$ – they arrived at the critical force

$$P_{cr} = \frac{\pi^2 E_1 I_1}{h^2} \frac{1 - k\left(1 - \frac{\overline{f}_1}{\overline{f}_2}\right)}{2\psi\overline{f}_2\left\{1 - k\left[1 - \frac{1}{\alpha_1\psi}\left(\frac{\overline{f}_1}{\overline{f}_2}\right)^2\right]\right\}}. \qquad (5\text{-}24)$$

The reduced displacements and α_1 can be computed by the formulas

$$\overline{f}_i = \frac{6E_1 I_1}{h^3} f_i \qquad (5\text{-}25)$$

and

$$\alpha_1 = \frac{h_1}{h}. \qquad (5\text{-}26a)$$

The computation formulas for the reduced displacements and the numerical values of the factors ψ as functions of α_1 and

$$\beta = \frac{E_2 I_2}{E_1 I_1} \qquad (5\text{-}26b)$$

in the cases of rigid and hinged beam-to-column connections are to be found in [MA-TEESCU, APPELTAUER and CUTEANU, 1980]. These data are valid also for columns elastically clamped or hinged at the bottom. For multi-bay frames we refer to [APPELTAUER and BARTA, 1961]. It should be mentioned that in [ROZENBLAT, 1962] an energetic solution for industrial frames with columns having a cross section varying in one step can be found.

The crane bridge loads locally, as a rule, a group of three frames. Due to the longitudinal stiffening of the hall, the less loaded frames stiffen this group, increasing the stability of the whole hall. This problem has been treated by LIPTÁK [1963], APPELTAUER [1970b] and AGENT [1970]. The second of these papers applied the generalization of the method of PUWEIN, described earlier, to industrial halls stiffened by the roof. [MATEESCU, APPELTAUER and CUTEANU, 1980] also treats the case of longitudinal stiffening bracings, giving programs for computers, too.

Methods using cut-out subsystems

The bifurcation problem can be simplified by considering only a part of the frame, taking the stiffening (or loading) effect of the other parts approximately into account. This method will be extensively treated in Section 5.2, so that we confine us only to some remarks.

SLAVIN [1950] considers five bars of the no-sway frame, taking into account the restraints of the other ends of the bars stiffening the column investigated. BOLTON [1955a, b] reduces the iteration on no-sway frames to one step by rotating the (mostly extreme) node of which the stiffness is the least but the loading is the highest. With sway frames BOLTON considers three storeys above each other, and includes, like MERCHANT, the influence of the shearing forces into the correction (or stability) functions [MERCHANT and SALEM, 1960], which represent the second-order effects on the general displacements of the bar end sections.

SCARLAT [1969] found that the critical load of a bar depends mainly on the sum of the restraints at both ends, and much less on their ratio at the two bar ends. He dissects the frame into parts which contain one overloaded column each, and conects these parts by hinges. It is easy to determine the critical loads of these parts, and also which ones are active and passive during buckling. The connecting hinges can then be replaced by elastic restraints in a recursive way. His method yields realistic results for multi-bay and single-storey frames.

KIRSTE [1956] and SNITKO [1963] distribute the stiffnesses of the beams between the storeys above each other, so that they dissect the frame into independent storeys.

HANGAN [1967] solves the stability problem of one column of the frame by integrating its deformation line, considering the rigidities of the joining bars in the boundary conditions. He also gives diagrams for practical use. WOOD [1958] also presents diagrams for solving the stability problem of an elastically restrained column.

APPELTAUER [1961] gives a very simple formula for determining the critical load of a column picked out from a no-sway frame

$$N_{cr} = \frac{\pi^2 EI}{(\beta l)^2}, \tag{5-27}$$

where

$$\beta = 1 - 0.61 (t_{ik} + t_{ki}) + 0.43 t_{ik} t_{ki}, \tag{5-28}$$

with t_{ik} and t_{ki} as the carry-over factors for bending moments between the nodes i and k.

Methods using the solutions of known simple cases

The obvious idea of reducing complicated problems to known simple cases originates most probably from KORNOUKHOV [1948] who reduced multi-bay and multi-storey frames to portal or symmetric multi-storey ones. It is more expedient to follow the train of thought in the reversed situation, see *Fig. 5-11*, where the three frames are equistable. In a similar way the frame shown in *Fig. 5-12a* can be replaced by that of *Fig. 5-12b*, where the sum of the moments of inertia of the columns and that of the loads acting at the nodes are equal in both frames at all levels. (LIGHTFOOT called this 'Principle of Multiples' [HORNE and MERCHANT, 1965].)

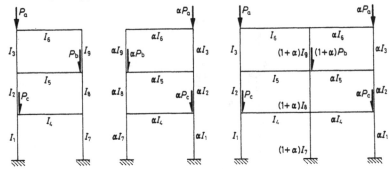

Fig. 5-11 Equistable frames

The frame of *Fig. 5-12b* is equistable with that of *Fig. 5-12c*. The moments of inertia of the columns and the loads acting at the nodes of the replacement multi-bay, multi-storey frame (*Fig. 5-12b*) can be determined by the formulas

$$I_{I,II} = \frac{\sum I}{2n}, \tag{5-29}$$

$$P_{I,II} = \frac{\sum P}{2n}, \tag{5-30}$$

where n is the number of the bays [BOWLES and MERCHANT, 1958]. These summations are justified by energy considerations [JOHNSON, 1960], saying that the sum of the work done by the vertical forces acting at the nodes does not depend on the distribution of these forces along one level, since the rotations of the nodes of one level are practically equal. This statement is similar to that made by DULÁCSKA, see Section 5.2.3. POPOV [1959] reduced multi-storey frames up to four bays to a built-up column with batten plates of finite stiffness. KERKHOFS [1965] condenses the frame into one cantilever with the aid of the horizontal displacements, which can be computed taking plastic deformations and the influence of normal forces into account. He also proposes that if $\kappa \leq 0.5$, we should consider the beams as infinitely rigid. Here κ is the frame stiffness according to Eq. (5-1).

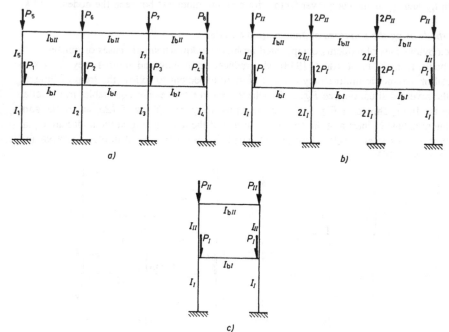

Fig. 5-12 A series of equistable frames

We have to mention here also the various formulas yielding the critical loads of simple frames, although these give mostly the buckling length. The relations of the German code DIN 4114 [1952] are based on the simplifications of PUWEIN [1936-38], according to which the buckling shape can be approximated by the deformation line caused by a unit horizontal force applied at the uppermost beam. If we denote the horizontal displacement of the uppermost beam by Δ, then the product $P_{cr}\Delta$ hardly varies with different frame types. Hence the equality

$$P_{cr}\Delta = \overline{P}_{cr}\overline{\Delta} \qquad (5\text{-}31a)$$

approximately holds, where the quantities with a bar refer to a known simple frame type. From this relation we obtain:

$$P_{cr} = \overline{P}_{cr}\frac{\overline{\Delta}}{\Delta}. \qquad (5\text{-}31b)$$

We also refer to the diagrams published in [PETERS, 1963]. The relations and nomograms contained in the French Règles CM [1966] and in the American AISC Commentary [1978], furthermore the relations of MURASEW [1958], all refer to the typical case when all columns are equally active with respect to buckling [BECK and ZAR, 1963]. The relations of FREI-HART and FREIHART [1968] valid for single- and multi-bay frames are based on the replacement cantilever. CHU and CHOW [1969] determine the buckling length of the most loaded column of an asymmetric frame with the aid of the nomograms of the American Building Code, and correct it by a factor corresponding to the absence of symmetry in geometry and loading. It should be remarked that the formulas and tables referring to columns with stepwise varying cross sections of industrial frames originated mostly from the Russian Building Code (SNIP). The formulas referring to industrial and other high-rise buildings, contained systematically in [APPELTAUER and BARTA, 1961], also belong to this group.

Special problems of elastically supported sway frames

The basic question is: what is the necessary stiffness of the bracing in order to allow the frame to be considered as no-sway? This problem will be treated in Chapter 7, so that we make here only some remarks.

GOLDBERG [1960] investigated the effectiveness of the bracing stiffness on portal frames with stiff and soft beams, and calculated the minimum necessary stiffness which prevents sway buckling. HOLMES [1961] determined, on the basis of experiments by himself and Wood, the critical loads of closed frames stiffened by concrete or brick filling walls, substituting the compressive stiffness of a diagonal strip for the shear rigidity of the wall. The filling wall may contain also openings. SMITH [1962] calculates the effective width of the filling wall and substitutes this for the compressed diagonal. GOLDBERG [1968] determined the minimum stiffness of the beams necessary for preventing sway buckling in the cases when the column and bracing stiffnesses are given and the characteristics of the regular frame and of the bracing vary along the vertical direction according to a geometric progression.

It should be remarked that, according to the American Building Code, the buckling lengths of the columns of multi-bay, multi-storey frames can be reduced to the geometric lengths if the sway of the frame is limited [BECK and ZAR, 1963].

5.1.4 Divergence

Buckling analysis taking plasticity into account

If plastic deformations occur, stability analysis has to be performed only if great compressive forces act in slender, sway columns, as e.g. in the case of industrial or multi-storey frames [GENT, 1962]. If the nonsway frame has short columns and is loaded symmetrically, that is if the columns undergo a deformation with an inflexion point, the second-order effect, characteristic for stability investigation, can be mostly neglected even if the normal force is high [HALÁSZ, 1967]. With decreasing column slenderness not only the critical bending moment approaches the plastic limit moment of the cross section, but also the rotation of the cross section increases without any decrease of the corresponding moment, so that the plastic rotation capacity of the structure increases [HALÁSZ and IVÁNYI, 1979].

The stability of frames is mostly depreciated by the additional moment of the vertical load due to sway (the so-called $P - \Delta$ effect). This is small in the domain of elastic deformations [KAZAZAEV, 1960], but gains importance with increasing plastic deformations, since up to plastic failure the horizontal displacement of multi-storey frames may reach even 2% of the height of the structure [WRIGHT and GAYLORD, 1968]. HEYMAN [LU, 1965a] determined analytically the number of storeys as a function of the number of bays, the ratio of vertical and horizontal loads, and the ratio of span and storey height, from which on stability analysis becomes necessary. According to the European Steel Commission the frame can be considered braced if its horizontal stiffness is at least five times that of the unbraced frame [MASSONNET, 1976], see also Chapter 7. Finally it should be mentioned that in the domain of plastic buckling, the role of imperfections also increases. FEY [1966] takes the imperfections into account by assuming a certain inclination α of the whole frame, and derives the value of α from the building codes.

Exact methods of analysis

As with first-order plastic analyses, there is a fundamental difference between checking and design methods. In design methods, optimization must also be included. The checking methods can be divided, according to OSTAPENKO [1965], into two groups: the compatibility and the second-order elasto-plastic methods. The compatibility methods take into account the elasto-plastic zones developing between the plastic hinges, while in the other methods plastic deformation is confined to the discreet plastic hinges. The design methods will be treated in Section 5.1.6.

Checking the stability by compatibility methods

These methods are rather complicated and mostly applied to simple frames. An elasto-plastic stress-strain diagram is taken as a basis, and initial stresses are considered; furthermore the decrease of the compressive stresses in the tension zone of the buckling deformation and the strain-hardening of the steel are neglected [OSTAPENKO, 1965]. Using these assumptions the moment-curvature diagram is constructed, with the normal force as a parameter. In the following we treat only the pioneer methods; the overview of the further developments would exceed the scope of this chapter.

OXFORT [1961, 1963] was one of the first researchers who applied the force method to the half frame shown in *Fig. 5-13*. When integrating the differential equations he divided the deformation line into several sections, according to the elastic or elasto-plastic stress distribution in the rectangular cross section. The two curves of the nodal rotation as functions of the column and beam moments intersect, as a rule, at two points, denoting the stable and unstable positions. Divergence occurs if these two points coincide and the two curves are tangent to each other.

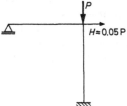

Fig. 5-13 A half frame computed exactly

CHU and PABARCIUS [1964] resort to the moment-curvature relation and solve the portal frames of *Fig. 5-14a* and *b* with a wide flange I section, considering either P or H as constant. MOSES [1964] solves similar frames, but he approximates the moment-curvature diagrams, depending also on the normal forces, by a polygon consisting of three straight lines. The computation of YURA and GALAMBOS, and furthermore that of ADAMS was the most exact: they determined the maximum value of H to a given P for the frame of *Fig. 5-14a*, and also the whole equilibrium path $H(\delta)$ [OSTAPENKO, 1965]. All these computations use the successive approximation: e.g. for certain values of P and horizontal displacement δ they assumed various nodal rotations, and satisfied both the compatibility of rotations at the nodes and the vertical equilibrium of the whole frame; finally they obtained H from the horizontal equilibrium of the frame. KLÖPPEL and UHLMANN [1968] applied the displacement method for frames similar to that in *Fig. 5-14* and considered the effect of plastic deformation by transfer matrices. They divided the bars into 40 sections each, and used the so-called *quantified plastification*: they increased the height of the plastic zone stepwise by $h_S/100$, with h_S as the height of the cross section. In this way they followed the variation of the critical force in dependence of the height of the wide-flange I shaped cross section, of the ratios h/l and h/i (h is the height of the column, l the span of the beam, and i the radius of inertia of the cross section), of the horizontal load and of the unequal vertical loading of the columns.

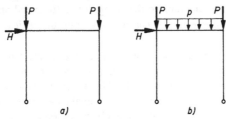

Fig. 5-14 The portal frame computed exactly

Check of stability by second-order elasto-plastic methods

MERCHANT and SALEM [1960] also computed the frame depicted in *Fig. 5-14a* (but assuming rigid clamping) by the second-order elasto-plastic method, for the ratio $H/P = 0.2$. In the case of $N_{\lim}/N_E = 0.232$ – where $N_{\lim}$ is the limit normal force of the column obtained by assuming a rigid-plastic mechanism and N_E is its *Euler* load – the critical force of divergence proved to be by 20% less than the simple plastic limit load. Similar calculations have been performed at Lehigh University and at the University of Nagoya [OSTAPENKO, 1965]. In their displacement method SAWKO and WILDE [1967] took also the effect of strain-hardening into account when writing the rigidity matrix.

The procedure of BEER [1966] is rather peculiar. It starts from a general stress-strain curve which may vary in the cross section and along the length of the bar as well, and thus it can follow the behaviour of structures made of various materials.

Approximate methods

The approximate methods try to avoid, first of all, the complicated computations of the exact methods, and – with remarkable intuition – they increased the simplifications up to developing explicit formulas. They can be divided into two groups. The methods belonging to the first group approach the critical load of the divergence. These are:

– methods using cut-out subsystems,

  – second-order rigid-plastic computations,
  – approximate explicit formulas.

Into the second group only the second-order elastic computation belongs which can be performed either by exact or approximate methods, but which approaches the critical load of divergence from underneath in every case, i.e. it always remains on the safe side.

Methods using cut-out subsystems

These procedures have been developed at Lehigh University for braced and unbraced frames [TALL, BEEDLE and GALAMBOS, 1964]. The subsystem contains the column investigated and all its neighbouring bars, but the practical computation confines itself to one column only whose ends are restrained against rotation by (sway or no-sway) nodes.

In the case of no-sway frames we know, on the one hand, the functions of clamping exerted by the bars joining the column and, on the other hand, the equilibrium shapes of the column considered as hinged-hinged and loaded by a compressive force and two different end moments. Using the equations of equilibrium and compatibility, the critical load of the column can be obtained by iteration: if the compressive force and one end moment are given, then we obtain the other end moment characterizing the critical state [GALAMBOS, 1965a].

In the case of sway frames we have, in addition, the moment-rotation curves, with the deviation of the chord connecting the bar end points from the pressure line of the compressive force as a parameter [GALAMBOS 1965b]. If the normal force, one end moment and the horizontal (sway) force are given, we can obtain the other critical end moment by double iteration; or if both end moments are given, we obtain the critical sway force [GALAMBOS, 1965c]. This computation is simplified by sway force-displacement diagrams containing the restraining moment exerted by the beam as a parameter [LU, 1965b].

Second-order rigid-plastic computation

If the computation proper of the previous case is exact, then the second-order rigid-plastic procedures do not lead to the critical state of divergence with partial plastification, but to a second-order rigid-plastic mechanism [VOGEL, 1963, 1965]. Thus the critical load determined in this way corresponds to the point of intersection I of *Fig. 5-5*. In the case of statically determinate structures, the approximation 'from above' is very good if the relative eccentricities are large ($e/i > 2$, with i as the radius of inertia) and the normal forces are small ($N_{cr}/N_y < 0.4$, with N_y as the central compressive force causing yielding of the whole cross section). On the other hand, with small relative eccentricities, the error in the approximation 'from underneath' is 5% [VOGEL, 1963]. In the case of statically indeterminate structures, the approximation may be from above or from underneath [VOGEL, 1965], but if the horizontal forces are medium or great ($H/P \geq 0.05$) the error is very small [KNOTHE, 1963].

Neglecting the plastification between the plastic hinges may lead in high portal frames with a medium ratio h/i and loaded by a horizontal force and very unequal vertical forces to an overestimation of the critical force of divergence by 20% [KLÖPPEL and UHLMANN, 1968]. The trigonometric equation system of the method results from the equilibrium of the whole structure written on the basis of virtual work, from the equilibrium equations of the elements of the statically determinate node configuration, and from the compatibility written for every rigid node and the last plastic hinge. This equation system is solved by iteration on a previously selected bar configuration.

KÄRCHER [1968] established more severe criteria of convergence, since the iteration of the above system of equations is equivalent to the approximation of the normal forces by a geometric series. The *method of the last plastic hinge* was applied also by HALÁSZ [1977], who obtained an expression similar to that of MERCHANT, given below, and into which the influence of imperfections can also be included. It should be mentioned that the possibility of bifurcation cannot be excluded.

Explicit expressions for the critical load

The most important formula is that of MERCHANT [HORNE and MERCHANT, 1965], which is a further development of the well-known *Rankine* formula, i.e. actually the *Föppl-Papkovich* theorem presented in Chapter 2:

$$\frac{1}{n_{cr,d}} = \frac{1}{n_{cr,b}} + \frac{1}{n_{pl}}, \tag{5-32a}$$

where $n_{cr,d}$ is the critical load parameter of divergence, $n_{cr,b}$ is the critical parameter of elastic bifurcation, and n_{pl} is the parameter of the first-order rigid-plastic limit state. HORNE [1963b], LIGTENBERG [1965] and MAJID [1967] tried to prove the *Merchant* formula theoretically. The basic assumptions of HORNE are interesting:

- the normal forces of the bars of the frame hardly vary during the elastic-plastic deformation, so that they remain approximately proportional to the load parameter, as in the first-order rigid-plastic calculation (this condition is fulfilled by multi-storey frames loaded mainly by vertical forces);
- in the vicinity of the elastic bifurcation the first eigenfunction of the frame closely approximates the curve of the second-order elastic deformation (see the curves 7 and 8 of *Fig. 5-5*);
- the curve of the rigid-plastic deformation is also close to the first eigenfunction of the frame (see the curves 8 and 9 of *Fig. 5-5*).

If the above deformation curves are not similar to each other, then also higher eigenfunctions play a role, and Eq. (5-32a) yields a critical load too high.

LIGTENBERG extrapolates the second-order effect from the elastic to the elasto-plastic range, and rescribes Eq. (5-32a) to the form

$$\frac{1}{n_{cr,d}} = \frac{1}{n_{n-1}} + \frac{1}{n_{pl}}, \tag{5-32b}$$

where n_{n-1} is the load parameter corresponding to the last but one plastic hinge. SALEM calculated more than 300 cases [HORNE, 1963b] and showed that in the case of rectangular frames the *Merchant* formula yields very good values, since its results remain by at the most 4.3% under the exact critical loads. However, WOOD [1958] points out that this is true only if buckling occurs perpendicularly to the axis of minimum moment of inertia. In other cases even deviations of 24% (to the safe side) have been found [HORNE, 1963a]. The exactness of the *Merchant* formula deteriorates, as a rule, if the critical parameter of bifurcation is small [WOOD, 1958], if the frame fails as a mechanism caused by the development of plastic hinges in the beams [HORNE, 1963b], and if the horizontal loads are either very small [Discussion (1967) of the paper of DAVIES, 1966], or very great [MAJID, 1967]. WOOD [1958] included also the influence of strain-hardening into the *Merchant* formula in the following 'average' way:

$$\frac{1}{n_{\mathrm{cr,d}}} = \frac{1}{n_{\mathrm{cr,b}}} + \frac{0.9}{n_{\mathrm{pl}}}. \tag{5-32c}$$

This formula is recommended by the European Steel Commission in the range of

$$4 \le \frac{n_{\mathrm{cr,b}}}{n_{\mathrm{pl}}} \le 10.$$

If

$$\frac{n_{\mathrm{cr,b}}}{n_{\mathrm{pl}}} > 10,$$

then, on the basis of Eq. (5-32c),

$$n_{\mathrm{cr,d}} > n_{\mathrm{pl}},$$

and there is no need of checking the stability. If, on the other hand,

$$\frac{n_{\mathrm{cr,b}}}{n_{\mathrm{pl}}} < 4,$$

then the European Steel Commission requires a second-order elastic-plastic calculation [MASSONNET, 1976].

RAEVSKI [1964] generalized the *Merchant* formula as follows:

$$\frac{1}{n_{\mathrm{cr,d}}} = \frac{\alpha}{n_{\mathrm{cr,b}}} + \frac{1}{n_{\mathrm{pl}}}, \tag{5-32d}$$

where the factor α depends on the critical deformation. Here we have to distinguish between frames with infinitely rigid and with finitely rigid beams. In the first case the critical deformation is characterized by the 'storey mechanism' (when the hinges develop at the column ends), and

$$\alpha = 1 - (1.0 + 0.178 k \bar{e}) \frac{n_{\mathrm{cr,d}}}{n_{\mathrm{y}}}, \tag{5-33a}$$

where k is the factor of partial plastification,

$$\bar{e} = \frac{e}{r_{\mathrm{c}}} \tag{5-34}$$

is the relation of the eccentricity to the radius of the core of the cross section, and n_{y} is the load parameter corresponding to the plastic load-bearing capacity of the section. In the second case the critical deformation is characterized, in dependence of the ratios l/h (*Fig. 5-10*), H/P and $W_{\mathrm{beam}}/W_{\mathrm{column}}$ (W being the section modulus), either by a 'storey mechanism' (mostly that of the ground floor), to which

$$\alpha = 1 - \frac{n_{\mathrm{cr,d}}}{n_{\mathrm{y}}}, \tag{5-33b}$$

belongs; or by a 'mixed mechanism' (mostly produced by two hinges of the lowest beam), when

$$\alpha = 1 - \frac{M_{\mathrm{max}}^{\mathrm{beam}}}{2 M_{\mathrm{pl}}^{\mathrm{beam}}}, \tag{5-33c}$$

where

$$M_{\mathrm{pl}}^{\mathrm{beam}} = W_{\mathrm{pl}}^{\mathrm{beam}} \sigma_{\mathrm{y}} \tag{5-35}$$

is the plastic load-bearing capacity of the beam. If we plot Eq. (5-32d) for various values of $\alpha < 1$ in the co-ordinate system $n_{cr,d}/n_{pl}$, $n_{cr,d}/n_{cr,b}$, we obtain curves which lie higher than the straight line of *Merchant* ($\alpha = 1$).

The second approximate formula of MERCHANT is [STEINHARDT and BEER, 1968]:

$$n_{cr,d} = \beta n_{el,pl}, \tag{5-36}$$

where $n_{el,pl}$ is the ordinate of the intersection point of the second-order elastic behaviour (curve 7) and the second-order plastic mechanism (curve 5) in *Fig. 5-5*. According to KNOTHE [1963] this formula overestimates the critical load, since the maximum error was 17%. In some cases, however, n_{ep} comes that close to the parameter $n_{y,extr}$ characterizing yielding of the extreme fibre, that the contradiction $n_{cr,d} < n_{y,extr}$ appears [STEINHARDT and BEER, 1968]. Thus OXFORT [1963] proposed the following modified formula:

$$n_{cr,d} = n_{y,extr} + \gamma \left(n_{el,pl} - n_{y,extr} \right),$$

and assumed for the factor γ the value 0.5.

Second-order elastic computation

The second-order elastic computation follows the equilibrium curve of *Fig. 5-5* only up to the point Y representing yielding of the extreme fibre, which lies certainly lower than the point D of the divergence. The two points may come close to each other if small bending moments arise in the columns, i.e. if the horizontal loads are low. On the other hand, if the horizontal forces (and also the bending moments) are high, point Y may lie by even 50% under point D [KNOTHE, 1963]. DUTHEIL [1957] came closer to the divergence proper by allowing a slight plastification of the most unfavourable cross section (the plastic deformation cannot exceed 7.5% of the elastic one), because up to this value plastification does not change appreciably the elastic behaviour.

As mentioned earlier, the second-order elastic computation also can be performed exactly, using the same methods as for the determination of the critical bifurcation force. The difference is that now we start from the actual (vertical and horizontal) loads and we can consider the geometric imperfections of the structure and of the bars, too. (This latter effect will be treated in Section 5.1.6.) Consequently, the equations of the various procedures cease to be homogeneous. It should be noted that SCHEER [1966] gave the exact equations of virtual work starting from the elastic deformation line, since – except for multi-bay and multi-storey rectangular frames – the equations of virtual work related to the original deformation do not constitute equilibrium conditions, but excellent schemes of iteration.

As far as computer calculations are concerned, the continuous decrease of the frame rigidity is mostly described in the form

$$\overline{K} = K - G, \tag{5-37}$$

where K is the elastic rigidity matrix, and G the 'geometric' rigidity matrix containing the second-order effects, which thus continuously changes together with the deformation of the structure. Since the deformation of the frame can be determined only by previous computation, the second-order calculation is of iterative nature. If we begin the calculation with the full load, we can determine, in knowledge of the first deformation $\Delta^{(1)}$, the first expression of the 'secant geometric matrix' $G_{sec}^{(1)}$, and with the total matrix $\overline{K}^{(1)}$ we obtain the second deformation $\Delta^{(2)}$. Since the iteration changes the geometric matrix only, the procedure is simple, but the convergence is poor. The iterative approximation of the equilibrium path proper is demonstrated by *Fig. 5-15a*. If, however, we perform the calculation by increasing the loading stepwise by $\Delta P(i)$ (*Fig. 5-15b*), then in every step the 'tangent geometric matrix' $\mathbf{G}_t^{(i)}$ appears, and the procedure takes the form of a successive approximation which is more complicated than the previous method, but yields a better convergence. There are convergence improving methods in both cases, but their presentation would exceed the scope of this chapter.

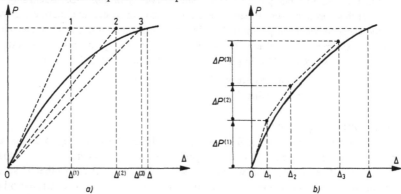

Fig. 5-15 The iterative variants of the second-order computation

Obviously, when using a computer, it is not necessary to determine the normal forces of the frame by a previous first-order computation. This was advantageous with hand calculations, since the second-order effect practically does not change the values of the normal forces and, furthermore, if the horizontal loads are small, the approximate values of the normal forces can be determined on statically determinate simply supported beams. However, an obvious difficulty of the second-order calculation has to be pointed out. The well-known principle of superposition can be applied only if the bar characteristics

$$\varepsilon_i = l_i \sqrt{\frac{N_i}{E_i I_i}} \tag{5-38}$$

remain unchanged. However, every load combination leads to different ε_i values, so that the second-order calculation has to be performed separately for every load combination.

According to KLÖPPEL and FRIEMANN [1964], the ratio of the second-order and first-order moments cannot be generally determined, but if the bar characteristics ε do not exceed the limit 1.5 to 2.0, then the second-order moments do not significantly deviate from the first-order ones [SCHABER, 1961]. The approximate methods of the second-order elastic computation are based, among others, also on this statement. In some cases the approximation hardly modifies the exact procedure. Thus SCHABER [1961] develops, in the frame of the *Kani* iteration, the correction factors of the second-order effect into series; BÜRGERMEISTER and STEUP [1957] determine the values appearing in the integral equation of the deformation by iteration; RÓZSA [1967] renders it unnecessary to perform a first-order calculation for every step of iteration; and STEVENS [1964a], using the well-known second-order formula of *Mohr-Maxwell*, eliminates the iteration (by the method of *Engesser-Vianello*) of the assumed elastic deformation line, and confines the computation to the determination of the critical (indifferent) point of the elastic deformation line only. The method of RESINGER and STEINER [1959] belongs here also: they neglected, in the case of $\varepsilon \leq 1.5$, the influence of the normal forces on the rotations of the nodes, which results in transforming the stability determinant into an algebraic equation. This method has been generalized for industrial hall frames by APPELTAUER, GIONCU and CUTEANU [1966].

Other approximate methods follow more or less independent trains of thought. KLÖPPEL and GODER [1957] repeat the first-order computation of the structure with the sway forces caused by the normal forces of the previous calculation. In calculating these forces, the crookedness of the columns can be neglected. One iteration step is in most cases sufficiently accurate for practical purposes; if not, then the fact can be taken into account that the increase of the first-order moments can be approximated by a geometric series. ROSMAN [1961, 1962] iterates also the *Klöppel-Goder* sway forces on a replacement cantilever. APPELTAUER and BARTA [1962] applied the method of KLÖPPEL and GODER to industrial halls and obtained closed formulas.

HERBER [1956] stated that the shearing forces due to the second-order effect of the normal forces are at every level proportional to the shearing forces of the wind load, so that the additional moments can be written with the aid of the moments of the wind load, without any second-order calculation.

DULÁCSKA and KOLLÁR [1960] presented closed formulas for the second-order bending moments of braced and unbraced frames using the well-known 'moment increase factor'

$$\Psi = \frac{1}{1 - \dfrac{N}{N_{cr}}}. \tag{5-39}$$

Most approximate methods determine the moments at the nodes. The maximum bending moment along the bar length can be computed either with the aid of diagrams [CASSENS, 1961; KETTER, 1961], graphically [BASLER, 1956] or analytically [APPELTAUER and BARTA, 1962].

Closed formulas have been derived using the method of KLÖPPEL and GODER for several frame types: for portal and two-bay frames [JUNG, 1961], for one- and two-bay industrial frames [APPELTAUER and BARTA, 1962]. Approximate formulas are to be found for portal frames in [OPLADEN, 1959], and for one and two storey, one- or two-bay frames in [HERBER, 1956].

Finally we mention the special method of SNITKO [1952] who approximates the divergence of equilibrium by a limit state at which in the beam or in the column the first plastic hinge appears, but the maximum load may also be limited by conditions concerning the stiffnesses of the frame. In this way we overestimate the critical load if the bending moments in the columns are small (KNOTHE [1963] found an excess of 11%), but with great column moments we understimate it (according to KNOTHE, the error may be even 20%).

5.1.5 Snapping through

The phenomenon of snapping is illustrated by HORNE [1961] on a polygonal portal frame (*Fig. 5-16*): the figures *a* to *d* correspond to the points B, C and D of the equilibrium path of *Fig. 5-16e.* Of these, the equilibrium states at B and D are stable, while that at C is unstable. HUDDLESTON [1967] shows, with the aid of the nonlinear theory of stability, the connection of snapping and bifurcation on the half of the two-hinged frame loaded by a concentrated force on the column. SCHINEIS [1960] clarified on the symmetric frame: when does symmetric snapping and when antisymmetric bifurcation occur, the condition depending on the geometric and mechanical characteristics of the frame. The two-hinged frame of CHWALLA [1938], loaded on the beam by two symmetrically arranged concentrated forces, snaps through if $\kappa \geq 3.5$ (Eq. (5-1)), but if we distribute the two concentrated forces uniformly on the beam, then snapping occurs only at $\kappa \geq 4.3$; and the three-hinged, uniformly loaded portal frame snaps through already at $\kappa \geq 1.5$. All this is true in the elastic range only, i.e. when the columns are very slender. Hence for steel frames, due to the development of plastic deformations divergence of the equilibrium will occur, instead of snapping.

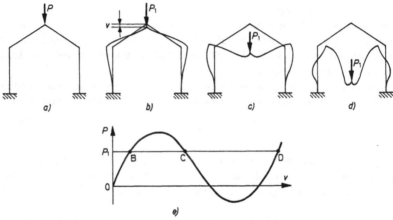

Fig. 5-16 Snapping through of the polygonal portal frame

In the case of two-hinged portal frames, the critical force of snapping decreases by 14% [JENNINGS, 1968], if the theory of large deformations is used. However, the influence of large deformations also depends on the shape of the frame: on polygonal frames it is somewhat less than on frames with straight beams [SAAFAN, 1963], because in polygonal frames the normal force of the beam decreases as related to the increasing peak force; this effect, however, cannot be percieved with small deformations. CHILVER [1956] states that the critical force of snapping may decrease even by 40% if the load is concentrated in the middle region of the beam. In the case of the half frame of HUDDLESTON [1967], the crookedness of the column, due to imperfections, significantly reduced the critical load of snapping.

The critical load causing snapping can be determined by the same methods as are used for bifurcation. HORNE [1962] arrived at an approximate closed formula for the frame of *Chwalla*, in which the parameters of the concentrated loads of the columns and of the loads acting on the beam appear in nonlinear and linear form, respectively.

5.1.6 Practical applications

Considerations on bifurcation loads: the general safety factor

When treating the bifurcation model we found that the simplifying assumptions used result in a sometimes rather distorted picture of reality. Nevertheless, HALLDORSSON and WANG [1968] emphasize that the safety factor related to the bifurcation load plays an as important role as the safety factor of a barrage against tilting. STEVENS [1967] states that properly constructed multi-storey frames hardly exhibited any elastic instability phenomena, but it is not advisable to make the bifurcation critical load too low, since the stiffness of the frame becomes low also, and thus the deformations increase and the eigenfrequency decreases: both effects can be disagreeable. Nevertheless, the direct determination of the safety factor against bifurcation proposed by CHWALLA [1959] did not gain general acceptance in engineering practice and, consequently, no systematic prescriptions on the minimum necessary values of the safety factor exist. These minimum values ought to compensate the deviations, varying from case to case, between bifurcation model and reality.

The bifurcation critical load can also be regarded, above all with larger horizontal loads, as a stiffness characteristic, see also Chapter 7. Thus the new German Building Code [DIN 18800, 1983] renounces the second-order calculation of the frame if the safety factor against bifurcation is greater than 10. Instead of this safety factor APPELTAUER and BARTA [1962] restrict the normal force of the most compressed column by the formula

$$N \leq \varphi m_b A \sigma_L, \qquad (5\text{-}40)$$

where N is the normal force increased by the overloading factor, σ_L is the limit stress, and m_b is a special safety factor which has to compensate the deficiencies of the bifurcation model. The buckling factor φ correlates the imperfections of the hinged-hinged bar to those of the most compressed column.

Apart from the fact that the values of m_b are rather uncertain, we have to make other remarks concerning Eq. (5-40). This formula replaces the imperfections of the frame by those of the most compressed column. It can be assumed, however, that the variation of the imperfections from column to column do not significantly influence the result of Eq. (5-40). This is true at least for the initial (residual) stresses. It is well known that the deformation modulus valid for the elastic-plastic section of the $\sigma(\varepsilon)$ curve represents the influence of the initial stresses, but HABEL [1958] and KERKHOFS [1965] showed that the variation of the deformation modulus over the frame does not significantly influence the bifurcation critical load. Including the deformation moduli into the bifurcation critical load causes trouble, since it unnecessarily separates mechanical and geometric imperfections (cf. Section 5.1.1). An even more important question is whether the frame can be correlated at all with the hinged-hinged bar. One difference is obvious: the hinged-hinged bar has no such structural imperfections as one bar of the frame has. Another question is whether the imperfection sensitivities of both kinds of bars are equal. As shown in Chapter 1, the imperfection sensitivity depends on the post-critical behaviour of the bar, characterized by the initial tangent of the bifurcated, secondary path. The hinged-hinged bar shows a symmetric-stable behaviour, so that the imperfection sensitivity of slender bars is comparatively low, and increases only if the bar becomes sturdy. Frames, however, may exhibit symmetric-unstable or even asymmetric bifurcation [BRITVEC and CHILVER, 1963; GODLEY and CHILVER, 1967; GIONCU, 1982; GIONCU and IVAN, 1983], resulting in an increased imperfection sensitivity.

As shown in Chapter 3, this sensitivity also increases due to the interaction of two or more buckling modes. This may be dangerous for the column of a frame constructed according to the principle of equal stiffnesses in both principal directions, since here the buckling in the plane of the bending moment may interact with the lateral buckling (combined bending and torsion). This is the reason why THOMPSON and HUNT [1973] call this method of design 'naive optimization'. In all cases when the imperfection sensitivity of the column of the frame is higher than that of the hinged-hinged bar, Eq. (5-40) results, due to the buckling factor φ, in an error of uncertain magnitude to the unsafe side.

LU [1965a,c] gives the diagram of *Fig. 5-17* for the elasto-plastic bifurcation load of symmetric frames. If

$$\frac{\left(n_{\mathrm{cr,b}}\right)_{\mathrm{el,pl}}}{\left(n_{\mathrm{cr,b}}\right)_{\mathrm{el}}} \leq 0.4, \tag{5-41a}$$

then the frame does not buckle, but after a symmetric deformation plastically collapses. If, on the other hand,

$$\frac{\left(n_{\mathrm{cr,b}}\right)_{\mathrm{el,pl}}}{\left(n_{\mathrm{cr,b}}\right)_{\mathrm{el}}} = 1.0, \tag{5-41b}$$

then elastic antisymmetric buckling occurs. Consequently, the buckling analysis proper has to be performed only between these two limits, using the formula

$$\frac{\left(n_{\mathrm{cr,b}}\right)_{\mathrm{el,pl}}}{\left(n_{\mathrm{cr,b}}\right)_{\mathrm{el}}} + 3\frac{\left(n_{\mathrm{cr,b}}\right)_{\mathrm{el,pl}}}{n_{\mathrm{pl}}} = 3.4 \tag{5-42}$$

which is a variant of the *Merchant* formula. We have to remark, however, that due to the bifurcation modelling of the imperfections only the initial stresses appear in Eq. (5-42), so that the problem of the safety factor remains unclear here also.

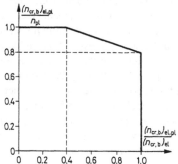

Fig. 5-17 The elasto-plastic bifurcation diagram of the symmetric frame (b: bifurcation, el: elastic, pl: plastic)

Dealing with optimization of plane frames, RAEVSKI [1968] started from the assumption of equal critical stresses and equal rotations of the end cross sections for the whole frame, and developed recursion relations to obtain the dimensions of the columns; furthermore, he established correlations between the dimensions of columns and bars. By so doing he achieves a saving of 16% in material. SWITZKY and WANG [1969] give, in the frame of the optimization proper, diagrams for the determination of the minimum own weight.

Introducing the buckling length
Using the buckling length

$$l_b = \beta l \tag{5-43}$$

we can extend the basic idea of Eq. (5-40) to all compressed bars of the frame, and thus we can consider every bar as hinged-hinged. The following relations hold between the factor β and the bar characteristic ε [Eq.(5-38)]:

$$\beta = \frac{\pi}{\varepsilon}, \tag{5-44}$$

and between the factors β of the individual bars (i,j):

$$\beta_j = \beta_i \frac{h_i}{h_j} \sqrt{\frac{N_i}{N_j} \frac{E_j I_j}{E_i I_i}}. \tag{5-45}$$

TALL, BEEDLE and GALAMBOS [1964] draw the attention to the fact that the buckling length determined in this way is not equal any more to the distance between the inflexion points, but the remark on the possibly different imperfection sensitivities of the frame bar and the hinged-hinged bar, made above, holds also here.

For that matter, the buckling length is, in the case of bar structures, not a consistent notion. According to Eq. (5-45), the shorter a column and the smaller its normal force, furthermore the greater its bending stiffness (i.e. the lower its propensity to buckle), the larger its buckling length. An infinitely small compressive force thus causes an infinitely large buckling length. This contradiction can be avoided if we resort to the notion of active and passive bars. If, e.g. the transverse rigidity of a sway frame is zero in the critical state, then the transverse rigidities of the active bars are negative, and only those of the passive ones remain positive. Logically we should determine buckling lengths for the active bars only, thus avoiding extremely large buckling lengths. The active and passive bars could be separated by comparing their normal forces resulting from the critical load to the critical force of the individual bar, which can be approximated, according to [DULÁCSKA and KOLLÁR, 1960], with no-sway frames by a hinged-hinged bar, and with sway frames by a clamped-clamped but sway bar.

The buckling length is determined, in knowledge of the critical force, i.e. of ε, by Eq. (5-44), but, as already mentioned before, the literature contains many data on the buckling length. It was perhaps DUTHEIL [1961] who first generalized the notion of buckling length by including the imperfections.

The buckling factor φ, determined by the buckling length or by the slenderness, is mostly introduced into the so-called *Iasinski* formula:

$$\frac{N}{\varphi A} + \psi \frac{M}{W} \le \sigma_L. \tag{5-46}$$

Here the influence of the horizontal loads, neglected so far, have also to be considered when computing the normal force and, above all, the bending moment. The factor Ψ (Eq. (5-39)) has been interpreted in various ways, depending on the fullness of the bending moment diagram as related to the 'full' rectangle, taking also partial plastification into account. However, Eq. (5-46) is not even theoretically correct, due to the simultaneous use of the factors φ and Ψ. In addition, this is not a bifurcation problem any more, but an investigation of divergence.

Considerations for solving divergence problems. Direct plastic design
STEVENS [1964b] concentrates the plastic deformations into the plastic hinges and replaces the stability analysis proper by the limitation of the deformations of the frame. If

$$\beta = \frac{\delta_a}{\delta_s} \tag{5-47}$$

(where δ_a denotes the displacement due to the additional loading and δ_s corresponds to the service state), then

$$\frac{\beta - 1}{n - 1} = 4. \tag{5-48}$$

If the whole frame buckles, WOOD [1958] proposes for the ratio n of the critical load to the service load the value $n = 3$, so that $\beta = 9$; but he also recommends reductions in the vertical loads of the lower storeys and possibly also in their horizontal loads. He confines the sway-angle of one level by 0.0025 rad (or at $\beta = 9$ by 0.0225 rad).

HOLMES and GANDHI [1965], further developing the method of HEYMAN, valid for plastic beams and elastic columns, condensed the buckling effects into a factor, depending on the bar characteristic ε and on the relative rigidites of the bars, expressing the increase of the deformation obtained by the simple plastic calculation. Thus multi-storey frames can be classified into three groups according to the magnitude of the horizontal forces (small, medium and large), characterized by beam, mixed, and column mechanisms, respectively, corresponding to whether the plastic hinges develop in the beams, in both the beams and columns, and in the columns only. MAJID and ANDERSON [1968] start from the dimensions obtained from the simple plastic calculation, correct them by iteration taking the elasto-plastic deformations into account, maintaining the prescribed load parameter, avoiding the development of plastic hinges in the beams under service load, and ensuring that the columns remain in the elastic state (to avoid torsional-flexural buckling). HALÁSZ [1972] published several theorems on plastic buckling, applicable to both 'rigid' (with no danger of buckling) and 'flexible' frames (which may buckle). He also considered the possibility of bifurcation occurring before divergence. In [HALÁSZ, 1969] he published an easy-to-use design procedure based on the notion of the *reduced critical load*, which appears at the nodes of frames whose rigidity is reduced by structural hinges replacing the plastic hinges of the real frame. In the case of single-bay, one- or two-storey sway frames, the second-order effect has been condensed into a reduction factor by HALÁSZ and IVÁNYI [1977].

LIGTENBERG [1965] developed a very simple formula for the buckling length, valid for the limit state characterized by the load parameter n_{n-1} appearing in Eq. (5-32b):

$$l_b = \frac{n - 5}{6} l + l_{b0} + \frac{100}{6} \frac{H}{V} l. \tag{5-49}$$

Here l_{b0} is the buckling length of the lowest column considered as rigidly clamped at top and bottom (in the case of a no-sway column $l_{b0} = 0.5l$), H and V are the total horizontal and vertical loads, respectively, acting on the frame. LIGTENBERG checked this formula using test results of LOW [1959], YEN, LU and DRISCOLL [1962], and his own results.

Reducing the buckling analysis to the computation of stresses
The divergence can be analysed by determining the stresses using the second-order elastic calculation, which always lies on the safe side. We can also include imperfections, but it is necessary to transform them into geometric ones, i.e. into crookednesses and inclinations. The amplitude of the crookedness can be most expediently calculated according to DUTHEIL [1957]:

$$f_i = c\frac{\sigma_{max}}{\sigma_E}r_c = c\frac{\sigma_{max}}{\pi^2 E}\lambda^2 r_c, \qquad (5\text{-}50a)$$

with r_c as the radius of the core, λ as the slenderness, and c as a constant which can be assumed in the case of normal structural steel approximately as 0.37. For sway frames this amplitude has to be assumed logically at the nodes. The German Building Code on stability [DIN 18800, 1983] prescribes numerical values for both the amplitude of the crookedness and the inclination of the bar axis. With small and medium bar characteristics ε, the crookedness can be disregarded in sway frames, and only the inclination has to be taken into account. The geometric imperfections have to be distributed over the frame in such a way that the initial shape be as close to the buckling shape as possible. This can be achieved with sway frames comparatively easily by visual means. The imperfections have a random character, so that their distribution is also random. A distribution could be also conceived in which the most fully used columns obtain the maximum imperfections, while the others obtain them according to random numbers. However, as has been said earlier, this distribution hardly affects the result of the analysis.

If M_x and M_y are the bending moments of the transverse and longitudinal frames respectively, then, according to the US Building Code, the calculation can be made by the following formula of the *Dunkerley* type:

$$\frac{N}{\varphi A} + \frac{c_x M_x}{\varphi_b\left(1 - \frac{N}{N_{Ex}}\right)W_x} + \frac{c_y M_y}{\left(1 - \frac{N}{N_{Ey}}\right)W_y} \leq \sigma_L, \qquad (5\text{-}51)$$

where φ_b is the factor of lateral buckling, N_{Ex} and N_{Ey} are the *Euler* critical forces in x and y directions, respectively, and c_x and c_y depend on the variation of the moments along the bar length. M_x and M_y are the respective first-order moments which are transformed to second-order moments by the terms in parentheses in the denominators, using the *Ayrton-Perry* approximation. The influence of imperfections appears in the buckling factors φ and φ_b only. If, however, we perform a second-order calculation, then $M_{x,max}^{II}$ and $M_{y,max}^{II}$ are known, and the terms in parentheses in the denominators have to be omitted. The factor φ_b represents the influence of the moment on the flexural-torsional buckling. For the factors c_x and c_y the US Building Code assumes the average value 0.85 for sway frames, but they can be expediently calculated by the formula of MASSONNET [CAMPUS and MASSONNET, 1955]:

$$c_{x,y} = \sqrt{0.3\left[1 + \left(\frac{M_2}{M_1}\right)^2\right] + 0.4\frac{M_2}{M_1}}, \qquad (5\text{-}52)$$

where M_1 and M_2 are the end moments, maximum and minimum in absolute value. This is more correct than the average value, because due to the condition min (φ_x, φ_y) prescribed for the buckling factor φ of central compression, this belongs to the phenomenon of lateral buckling, and although the transverse frame is very often unbraced, the longitudinal one is, as a rule, braced. In the denominator of the third term of Eq. (5-51), we omitted the factor φ_b, because the moments M_y, which are smaller anyway, do not cause lateral buckling in the direction where the maximum moment of inertia is effective. HALÁSZ and IVÁNYI [1979] report on special, equivalent factors $c_{x,y}$ valid for plastic zones. As far as the imperfections are considered, it is highly improbable that they appear in both directions with their maximum values. Consequently, we mostly assumed maximum imperfections in the transverse direction, and in the longitudinal direction we neglect their influence. Finally it should be remarked that the US Building Code extended Eq. (5-51) also to plastic deformations.

As can be seen from the short review of Eq. (5-51), the calculation taking interaction into account is rather complicated. Hence, further developing the relations valid for central compression of CSELLÁR and HALÁSZ [1961], APPELTAUER and BARTA [1962] set up the following formula for the stresses:

$$\sigma_{max} = \frac{N}{A} + \frac{M_x^{II}}{K_x W_x} + \frac{M_y^{II}}{K_y W_y} + \frac{B}{W_\omega} + \sigma_{N,M}^i \leq \sigma_L, \tag{5-53}$$

where K_x and K_y are the factors taking partial plastification into account, corresponding to an approximation of *Dutheil*, referred to at the beginning of the paragraph 'Second-order elastic computation' in Section 5.1.4, which can be computed by the approximate formula of the Soviet Building Code:

$$K = \frac{W_{el} + W_{pl}}{2 W_{el}} \tag{5-54}$$

and $\sigma_{N,M}^i$ contains the influence of the imperfections on the spatial deformation of the bar by substituting the initial curvature of the bar axis for all imperfections. Consequently:

$$\sigma_{N,M}^i = \frac{E c_{0,y}}{10^3 K_y} \frac{1}{\bar{n} - 1} f(z), \tag{5-55}$$

where c_{0y} comes from rewriting the expression of the imperfection amplitude (5-50a) into the form:

$$f_{0y} = c_{0y} \left(\frac{\lambda_y}{100}\right)^2 \frac{W_y}{A}, \tag{5-50b}$$

$\bar{n}$ is the parameter of flexural-torsional buckling (determined on the perfect bar), and $f(z)$ describes the variation of the stress $\sigma_{N,M}^i$ along the bar axis. Since (5-50b) also is valid for normal structural steel, c_{0y} is identical with the factor c of Eq. (5-50a). The function $f(z)$ describes a sine or cosine half wave in the cases of braced and unbraced frames respectively. The values of the imperfections have to be assumed according to what has been said so far, but when computing M_x^{II}, the structural imperfections have to be taken into account. Eq. (5-53), besides to be rather accurate, can be handled as easily as any other stress formulas. In the 1970s the European Steel Commission recommended a similar formula, but only for no-sway frames, without taking flexural-torsional buckling into account.

5.1.7 Conclusions

As can be seen, stability of frames is a rather complicated phenomenon which can be presented here only in outline. Our main aim has been to give a survey which was possible only by referring to the literature concerning questions which could not be treated in detail. We tried to fulfil different levels of exactness, so that some older methods have also been included. We also tried, on the one hand, to clarify the phenomena and problems in principle and, on the other hand, to consider the requirements of practice. We hope to have given a general orientation for engineers interested in frame buckling.

5.2 APPROXIMATE STABILITY ANALYSIS OF FRAMES BY THE BUCKLING ANALYSIS OF THE INDIVIDUAL COLUMNS

Lajos KOLLÁR

5.2.1 Basic principles of the method

On the basis of what has been said in Section 5.1, strictly speaking we cannot look for the critical loads of the individual columns, only for that of the whole structure, since – due to our assumption of perfectly elastic material – the buckling of one bar means the buckling of all other bars, so that the whole frame structure will buckle at once. It is true that some bars are loaded comparatively more than others, but the less loaded bars stiffen (support against displacement and restrain against rotation) the more loaded ones against buckling.

However, if we can determine exactly, to which extent the two ends of a bar are stiffened by the others, then the stability analysis of a frame performed by the buckling analysis of the individual bars becomes justified. This method can be expediently used in practical design, and is contained in several building codes. In the following we intend to investigate systematically the conditions of application of this method, the errors committed with it, and the principles which follow from it and can be used in practice.

The basic question to be clarified is: what are the conditions which allow us to dissect the frame into individual columns and to determine their critical loads separately? Let us investigate this problem on the simple two-legged frame of *Fig. 5-18a* by analysing the stiffening effect of the adjoining bars on the loaded one.

- The adjoining bars elastically restrain the end points of the loaded column against rotation. This effect is characterized by the rotational spring constant c_φ: in *Fig. 5-18b* we depicted only the restraining bar acting as a spring and the arising bending moment $M = -c_\varphi \varphi$.
- The adjoining bars elastically hinder the displacement of a bar end perpendicularly to the bar axis as a translational spring with the constant c_δ. *Fig. 5-18c* shows the bar

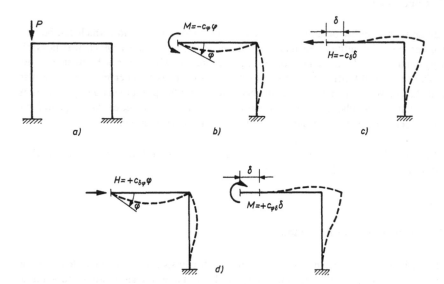

Fig. 5-18 The stiffening effects of the adjoining bars on the loaded column. *a*) The frame investigated, *b*) elastic restraint, *c*) elastic support, *d*) 'mixed' support

horizontally supporting the loaded column with the arising force $H = -c_\delta \delta$.

– The rotation of the end cross section of the loaded column causes also a supporting force in the adjoining bars ($H = c_{\delta\varphi}\varphi$), and the displacement of the same cross section causes also a restraining moment in the adjoining bars ($M = c_{\varphi\delta}\delta$). Due to *Betti*'s theorem, the two spring constants are equal: $c_{\delta\varphi} = c_{\varphi\delta}$. We indicated this double effect of the third spring constant in *Fig. 5-18d*.

It should be remarked that to exactly describe the stiffening effect of the soil on the column we need also this three-spring model [LIPTÁK, 1973].

The determination of the critical load of a column stiffened by such three springs at both ends leads to a rather complicated transcendental equation [LIPTÁK, 1960]. Hence we want to investigate whether we could use a simpler model.

If the nodes of a frame are rigidly supported against displacement (*braced* or *no-sway frames*), then of the three springs we have to take only that restraining against rotation, c_φ, into consideration. That is, the displacement δ is now zero, and the force H arising from the rotation φ is taken by the supporting structure, so that it does not influence the buckling of the column investigated.

On the other hand, if we have to deal with *unbraced* or *sway frames* (whose nodes are free to displace horizontally), then the condition of developing a simpler model is that no force H arises. This is fulfilled if the ratios of the loads of neighbouring columns are such that the columns 'use' only the restraining effect of the beams, and by so doing they want to buckle simultaneously, so that neither is able to support the other with a force H. Of course, the restraining effects of the two ends of each beam have to be determined in both cases according to the corresponding deformations, see Sections 5.2.2 and 5.2.3.

The three-spring model describing the stiffening effect of the soil simplifies, in fact, only in the cases of rigid clamping or hinged support. However, even in the case of an elastic restraint we will not take the exact, three-spring model into account, but we content ourselves with the rotational stiffness c_φ. This approximation is justified by the fact that the horizontal displacements of the foundations are, as a rule, negligibly small, and if only a rotation can come about, the spring $c_{\varphi\delta}$ has no sense either.

On the basis of what has been said so far, it is sufficient to determine the critical load of the column restrained at the lower end by a spring c_l against rotation, and at the upper end restrained by a spring c_u against rotation and by a spring c_d against displacement (*Fig. 5-19*), because with $c_d = \infty$ we obtain the case of the braced frame, and with $c_d = 0$ the case of the unbraced frame.

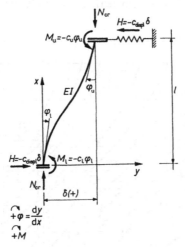

Fig. 5-19 The simplified column model

Starting from the differential equation of the elastic bent bar and taking the boundary conditions described above into account, we can derive the transcendental equation determining the critical load N_{cr} of the column. Introducing the notations:

$$k = \sqrt{\frac{N_{cr}}{EI}} \tag{5-56}$$

$$\frac{c_l}{EI/l} = \rho_l, \tag{5-57}$$

$$\frac{c_u}{EI/l} = \rho_u, \tag{5-58}$$

$$\frac{c_d}{EI/l} = \rho_d,$$ (5-59)

the transcendental equation for the general case runs:

$$\left[\cos kl - 1 - \frac{(kl)^2}{\rho_l} + \frac{(kl)^4}{\rho_l \rho_d}\right]\left[\rho_u kl \cos kl - \rho_u kl - (kl)^2 \sin kl\right] +$$

$$+ \left[\rho_u kl \sin kl + \frac{\rho_u}{\rho_l}(kl)^2 + (kl)^2 \cos kl\right]\left[\sin kl - kl + \frac{(kl)^3}{\rho_d}\right] = 0.$$ (5-60)

The roots kl of this equation yield, according to (5-56), the critical load N_{cr}. We obtain the case of a braced frame with $\rho_d = \infty$, and that of an unbraced frame with $\rho_d = 0$.

It is more practical to use the buckling length l_0 related to N_{cr} by the expression

$$N_{cr} = \frac{\pi^2 EI}{l_0^2},$$ (5-61)

or the ratio of the buckling length to the geometric (network) length of the column, which is, making also use of (5-56):

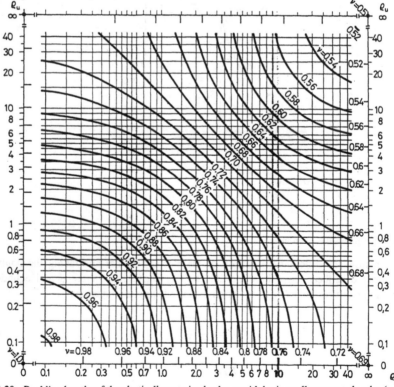

Fig. 5-20 Buckling lengths of the elastically restrained column with horizontally supported nodes (no-sway column)

$$v = \frac{l_0}{l} = \frac{\pi}{l}\sqrt{\frac{EI}{N_{cr}}} = \frac{\pi}{kl}. \qquad (5\text{-}62)$$

In *Figs 5-20* and *5-21* we plotted, on the basis of [NEMESTÓTHY, 1972], the values $v = l_0/l$ for columns with various end restraints in the cases of braced and unbraced frames respectively.

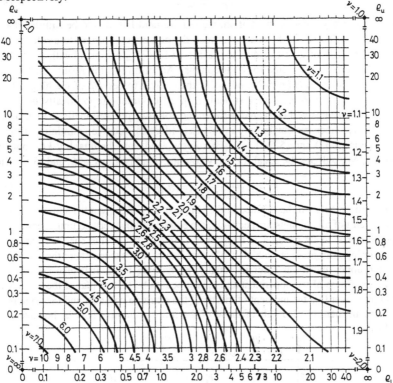

Fig. 5-21 Buckling lengths of the elastically restrained column with horizontally unsupported nodes (sway column)

For the stability investigation of frames by buckling analysis of the individual columns, we still need to know how to 'share' the restraining and supporting stiffnesses of the beams between the columns in such a way as to obtain the same ratio of the actual compressive force of every column to its critical load computed with these 'shared' stiffening effects of the beams. The requirement mentioned earlier, namely that all columns buckle at the same time, can be fulfilled only in this case.

This problem could be solved in the general case by a large system of linear equations. (This stiffness sharing may yield negative stiffnesses for some columns in cases when the column is comparatively less loaded, so that it also has to stiffen the more loaded ones.)

We want to avoid this rather complicated calculation. Let us start from the simplest case when we know in advance how to share the stiffnesses of the beams between the columns.

This simplest case comes about when the ratio of the actual load N to the *Euler* critical load N_E of the hinged-hinged no-sway column,

$$N_E = \frac{\pi^2 EI}{l^2},\tag{5-63}$$

is the same for every column. In addition, the requirement has also to be fulfilled that, after sharing the stiffnesses of the beams between the columns according to a rule known in advance, the ratio of the actual load N to the critical load N_{cr} computed with the stiffening effect of the beams be the same for each column. According to the results for the elastically restrained columns (*Figs 5-20* and *5-21*), the critical forces N_E of the hinged-hinged columns increase, due to the elastic stiffening, in all columns proportionally only, if the stiffnesses of the beams will be shared between the columns proportionally to their bending stiffnesses EI/l, since we obtain only in this case identical relative restraining stiffnesses ρ_l and ρ_u [Eqs (5-57) and (5-58)] for each column, and thus the quantity kl, containing the ratio N_{cr}/N_E, will also be equal for every column. That is, Eqs (5-56) and (5-63) yield:

$$kl = \pi\sqrt{\frac{N_{cr}}{\pi^2 EI/l^2}} = \pi\sqrt{\frac{N_{cr}}{N_E}}.\tag{5-64}$$

All these requirements will be fulfilled in the simplest way if the lengths and cross sections of all beams and of all columns, respectively, are equal, and in every column a compressive force proportional to its bending stiffness acts (*Fig. 5-22*). In this case we can assume equal restraining stiffnesses at both ends of every beam, and we can share these stiffnesses between the upper and lower columns equally, by halving them. It is also necessary that the bending stiffness of the uppermost and lowest beams be half of that of the intermediate ones, and moreover that the moment of inertia of the extreme columns and the compressive forces acting in them be also half of those of the inner columns. This is, however, not a practical case.

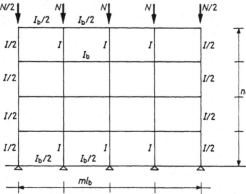

Fig. 5-22 A frame loaded on top, whose columns can be separated simply and exactly

We obtain an arrangement closer to the real frames if the stiffnesses of the columns and of the beams, respectively, and also the compressive forces, linearly increase downwards (*Fig. 5-23*). The stiffnesses of the uppermost and lowest beams, those of the extreme columns, and the compressive forces acting in the latter be also half those of the intermediate ones. In this case we can also assume equal stiffnesses at both ends of the beams, and share them between the upper and lower columns in the proportion of their stiffnesses EI/l.

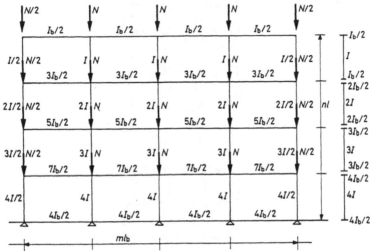

Fig. 5-23 A frame loaded at every storey, whose columns can be separated simply and exactly

In the following we will treat the stability of frames starting from these basic cases: first that of the braced, second that of the unbraced frames. We make the following assumptions.

- The material of the frame is perfectly elastic.
- If not said the contrary, we shall deal only with frames with a rectangular network.
- The loading be 'one-parametric', i.e. the ratio between the individual loads be constant during increase of the load intensity.
- We shall confine our investigation to loads which do not cause bending in the bars (central load). In the cases of eccentrically loaded frames we will determine the critical value of a central loading case which causes compressive forces equal to those caused by the eccentric loading.
- We will disregard the change in length of the columns during buckling. This causes a negligible error in the case of multi-legged frames; with two-legged frames, however, we may commit an error of 20–30%, in extreme cases even more, to the unsafe side, i.e. we obtain higher critical loads than with the exact calculation (see more detailed in [ZALKA and ARMER, 1992]). In Section 6.6.4 we find a method to take this effect into account.

5.2.2 Stability investigation of braced frames

The columns of braced frames buckle as shown in *Fig. 5-24*, because in the geometrically possible other way (*Fig. 5-25*) the beams would exert a much greater bending resistance

(Section 5.2.3). We will thus determine the bending resistance of the beams to the deformation shown in *Fig. 5-26*.

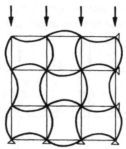

Fig. 5-24 The buckling shape of the braced frame

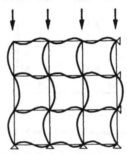

Fig. 5-25 A less dangerous buckling shape

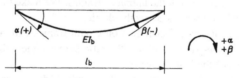

Fig. 5-26 Deformation of the beam of the braced frame

Denoting the bending rigidity of a beam end, with the other end rigidly clamped, by s_0, its moment carry-over factor by a, the bending rigidity of the left end of the beam, s_α, can be written as the function of the rotations of the two beam ends, α and β, as follows:

$$s_\alpha = \frac{M_\alpha}{\alpha} = \frac{\alpha s_{0\alpha} + a_{\beta\alpha} s_{0\beta}}{\alpha} = s_{0\alpha} + \frac{\beta}{\alpha} a_{\beta\alpha} s_{0\beta}. \tag{5-65}$$

For a beam with constant cross section:

$$s_{0\alpha} = s_{0\beta} = s_0 = \frac{4EI_b}{l_b} \tag{5-66}$$

and

$$a_{\alpha\beta} = a_{\beta\alpha} = a = \frac{1}{2}, \tag{5-67}$$

so that

$$s_\alpha = \frac{4EI_b}{l_b}\left(1 + \frac{\beta}{2\alpha}\right), \tag{5-68}$$

and, similarly:

$$s_\beta = \frac{4EI_b}{l_b}\left(1 + \frac{\alpha}{2\beta}\right).$$ (5-69)

On the beams of braced frames α and β have opposite signs (*Fig. 5-26*), so that the rigidities of the beam ends will always be smaller than $4EI_b/l_b$.

If we know the rotation of one end of the beam (or the moment arising here) in advance, then the rigidity of the other end assumes the values indicated in *Table 5-1*.

Table 5-1 Rigidities of the beams of braced frames

One end of the beam is	Rigidity of the other end
hinged	$3EI_b/l_b$
rigidly clamped	$4EI_b/l_b$
elastically clamped (e.g. continuous beam)	$\sim 3.5EI_b/l_b$
connected to a compressed column	$\sim 2EI_b/l_b$
Rigidity of a foundation resting on the soil	$cI_{\text{foundation}}$

In the table we also gave the rigidity of a beam which is connected at the other end to a compressed column (deforming as shown in *Fig. 5-26*); furthermore the restraining rigidity of the soil (considered as consisting of elastic springs) exerted onto a foundation (considered as rigid as compared with the soil), where c is the modulus of elastic bedding, which gives the magnitude of the reaction force between foundation and soil if the deflection is equal to unity, and $I_{\text{foundation}}$ is the moment of inertia of the (horizontal) surface of contact between foundation and soil. Of these, the case when the beam is connected at both ends to compressed columns needs a more detailed investigation.

If both ends of the beam join compressed columns, then the rotations α and β of the two bar ends are indeterminate, since, in the indifferent equilibrium state, the buckling amplitudes of the columns are also arbitrary. Hence the ratio of the rotations of the two end points is determined by the aforementioned static requirement that the two columns stiffened by the beam ends start to buckle simultaneously. This requirement determines the necessary ratio of the exerted rigidities of the two beam ends, which are, according to Eqs (5-68) and (5-69), uniquely related to the rotations α and β. In order to have a better overview we plotted the end stiffnesses s_α and s_β, made dimensionless, against the ratio of the two end rotations, α/β, in *Fig. 5-27a*. In the case of braced frames, α and β have opposite signs, so that the left-hand side of the diagram refers to our case. The diagram shows that one of the stiffnesses s_α and s_β can be zero, and even negative (if the rotation of one end is half that of the other, or less). This comes about if a less loaded column stiffens the beam, and through it the other, more loaded column.

The sum of the two end stiffnesses is:

$$s_\alpha + s_\beta = \frac{4EI_b}{l_b}\left[2 + \frac{1}{2}\left(\frac{\alpha}{\beta} + \frac{\beta}{\alpha}\right)\right].$$ (5-70)

This relation has been plotted in *Fig. 5-27b*. As can be seen, the curve has – in the domain of α and β of opposite signs – a local maximum at $\alpha = -\beta$. This means that the beam can, when undergoing the deformation according to *Fig. 5-26*, exert the maximum restraining effect at both ends together when its two ends rotate by the same amount. In this case:

$$s_\alpha = s_\beta = \frac{2EI_b}{l_b}. \tag{5-71}$$

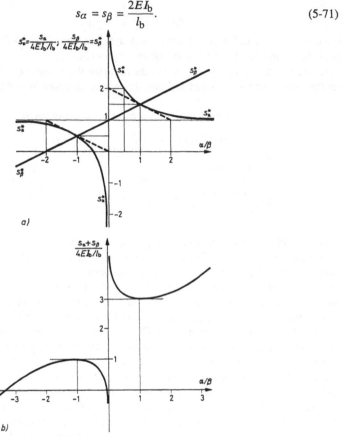

a)

Fig. 5-27   *a*) The stiffnesses of the two beam ends as functions of the ratio of the end rotations, *b*) the sum of the stiffnesses of the two beam ends as a function of the ratio of the end rotations.

Now, if the frame corresponds to *Figs 5-22 or 5-23*, and thus the rigidities EI/l of the columns standing beside each other and also their compressive forces are equal to each other, then the two ends of the beams has to exert identical restraints at both ends in order to cause the *Euler* forces N_E of the columns to increase proportionally. The fact that the extreme columns have half the moment of inertia and half the compressive forces of the inner ones ensures that the *Euler* loads N_E of these columns, being stiffened by one beam end only, increase at the same ratio as those of the inner columns.

We still have to share the beam end rigidities between upper and lower columns. Using the same consideration as before, we arrive at the conclusion that the beam end rigidities have to be shared proportionally to the rigidities EI/l of the upper and lower columns.

As can be seen, for the frames shown in *Figs 5-22* and *5-23*, we obtain their critical load exactly when we 'dissect' them into individual columns, provided we shared the beam end rigidities as discussed before.

If the stiffnesses EI/l of the columns being beside each other are not equal, but the compressive forces acting in them are proportional to their *Euler* forces, then the rigidity ratio s_α/s_β of the two ends of each beam should be determined in such a way that the sum of the restraining rigidities of all beams stiffening each column be proportional to the rigidities EI/l of the columns. This could be achieved either by a large set of linear equations or by a lengthy trial-and-error method. Thus let us investigate what error we commit if we choose a simpler way: if we take the two end rigidities of the beams equal to each other also in this case.

The error we commit is twofold: first, we consider some columns as being more stiffened and the others less stiffened than in reality they are; and, second, since the sum of the two end stiffnesses is maximum when they are equal, in the whole we assume stronger restraints than the real ones.

If, in addition, the compressive forces acting in the columns are not proportional to the *Euler* forces, we commbit a further error by assuming equal beam end rigidities.

Consequently, if we want to assume equal beam end rigidities in every case, it is necessary to investigate the magnitude of the error committed in the general case.

For columns being beside each other a detailed investigation has been carried out by KORONDI [1974]. In *Table 5-2* we present some of his results. The data refer to the two columns of the frame shown in *Fig. 5-28*. The table gives, besides the exact and the approximate values of the critical loads (the latter being computed by taking the two beam end rigidities equal to each other), the error committed in percentage. A positive percentage means an error to the unsafe side, a negative one to the safe side.

Table 5-2 Exact and approximate critical loads of the frame shown in *Fig. 5-28 a*)$I_2/I_1 = 1$

$\dfrac{I_b/l_b}{I_1/l}$	$\dfrac{N_2}{N_1}$	$\dfrac{N_{1,cr}^{exact}}{\pi^2 EI_1/l^2}$	$\dfrac{N_{1,cr}^{approx}}{\pi^2 EI_1/l^2}$	$\Delta\%$	$\dfrac{N_{2,cr}^{exact}}{\pi^2 EI_1/l^2}$	$\dfrac{N_{2,cr}^{approx}}{\pi^2 EI_1/l^2}$	$\Delta\%$
	0.2	1.276	1.14	−11	0.255	1.14	+347
0.2	1	1.156	1.14	−1	1.156	1.14	−1
	5	0.255	1.14	+347	1.276	1.14	−11
	0.2	1.971	1.62	−18	0.394	1.62	+311
1.0	1	1.668	1.62	−3	1.668	1.62	−3
	5	0.394	1.62	+311	1.971	1.62	−18
	0.2	3.154	2.72	−14	0.631	2.72	+331
5.0	1	2.854	2.72	−5	2.854	2.72	−5
	5	0.631	2.72	+331	3.154	2.72	−14

Buckling of frames

Table 5-2 (continued) $b) I_2/I_1 = 5$

$\dfrac{I_b/I_b}{I_1/l}$	$\dfrac{N_2}{N_1}$	$\dfrac{N_{1,cr}^{exact}}{\pi^2 E I_1/l^2}$	$\dfrac{N_{1,cr}^{approx}}{\pi^2 E I_1/l^2}$	$\Delta\%$	$\dfrac{N_{2,cr}^{exact}}{\pi^2 E I_1/l^2}$	$\dfrac{N_{2,cr}^{approx}}{\pi^2 E I_1/l^2}$	$\Delta\%$
	0.2	1.294	1.16	-10	0.259	5.15	$+1888$
0.2	1	1.293	1.16	-10	1.293	5.15	$+298$
	5	1.045	1.16	$+11$	5.225	5.15	-14
	0.2	2.065	1.64	-21	0.413	5.78	$+1299$
1	1	2.048	1.64	-20	2.048	5.78	$+182$
	5	1.217	1.64	$+35$	6.085	5.78	-5
	0.2	3.209	2.73	-15	0.642	8.22	$+1180$
5	1	3.182	2.73	-14	3.182	8.22	$+158$
	5	1.879	2.73	$+45$	9.395	8.22	-13
	0.2	$-$			$-$		
25	1	3.827	3.56	-7	3.827	13.65	$+257$
	5	2.780	3.56	$+28$	13.890	13.65	-2

The results of *Table 5-2* show that:
- if the moments of inertia of the two columns are equal to each other, then in the case of equal compressive forces the approximate method yields almost exact results; and if the compressive forces are different, then the approximate method yields for the stronger loaded column lower critical loads, and for the less loaded one higher critical loads than the exact ones;
- if the moments of inertia of the two columns differ from each other then, in the case of compressive forces proportional to the moments of inertia, the approximate method yields for the column loaded by the greater force lower critical loads, and for the column loaded by the smaller force higher critical loads than the exact ones; and if the ratio N_2/N_1 differs from the ratio I_2/I_1, then the approximate method yields for the comparatively stronger loaded column lower critical loads, and for the comparatively less loaded ones higher critical loads than the exact ones. In the case of load ratios equal to the ratios of the moments of inertia $(N_2/N_1 = I_2/I_1)$, the approximate method does not yield exact results (as contrasted to the case of columns being above each other), since we did not adjust the rigidities of the two beam ends to the ratio I_2/I_1, but considered them as equal.

We can summarize these results by stating that *in the case of the comparatively stronger loaded columns we are always on the safe side if we use the approximate method*. This can be easily explained by considering that, bby taking the rigidities of the two bar ends equal to each other, we restrain the comparatively stronger loaded columns less than it would be necessary, resulting, in fact, in lower critical loads than the exact ones.

This statement is also true for columns being above each other. That is, in this case the approximate method is closer to the exact one, because the beam end rigidities are shared between the columns proportional to their stiffnesses EI/l, but if the compressive forces are not proportional to these stiffnesses, then the same statement holds as has been said on the columns being beside each other.

The (sometimes rather large) error to the unsafe side, committed with the approximate method used in computing the critical loads of the comparatively less loaded columns, is reduced by the following circumstances.

If, after determining the critical loads, we *dimension* the columns, then we assign greater moments of inertia to the more loaded ones, and smaller moments of inertia to the less loaded ones. By so doing, we approach the actual compressive forces by the moments of inertia (i.e. by the *Euler* loads) more and more, so that the approximate method becomes more and more exact. In fact, the results of *Table 5-2* show that, in the cases $N_2/N_1 \neq 1$, the error in the critical load of the less loaded column becomes a minimum when $I_2/I_1 = N_2/N_1$.

For properly evaluating the data of *Table 5-2* we also have to consider that the two-legged frame of *Fig. 5-28* only imperfectly reflects the behaviour of the multi-legged frames (*Fig. 5-24*), that we, in fact, want to investigate. The difference between them is, first, that the ratio I_b/I_{col} of the two-legged frame is not equivalent to that of the multi-legged frame. That is, for proper equivalence, the ratio of the restraining effect of the beam(s) to the stiffness of the column should be equal in both structures. In the two-legged frame, however, one column is restrained by one beam only, whereas in the multi-legged ones by two beams (except for the extreme columns), thus to the ratio I_b/I_{col} of the two-legged frame a multi-legged frame with half this value would correspond. The second difference is that, in the case of very different column stiffnesses, the deformation of the beam in the two-legged frame very strongly deviates from the assumed symmetric shape, while in the multi-legged one this deviation is less, due to the influence of the neighbouring columns. Since the sum of the rigidities of the beam ends is maximum if the beam deforms symmetrically (cf. *Fig. 5-27a*), we can state that the less the deviation from the symmetric shape, the smaller the error committed to the unsafe side.

Most building codes take the limit eccentricity as a basis for the dimensioning of compressed columns. This means an additional safety for the comparatively less loaded ones, since they bow out, actually, to a smaller extent, and thus only a fraction of the limit eccentricity develops in them.

The error committed in the elastic critical load manifests itself in the critical load only in the cases of very slender columns. The smaller the slenderness, the more accentuated the reducing effect of plasticity, which depends only on the plastic load-bearing capacity of the cross section and is independent of the slenderness in which we committed the error. Thus the error will be less with stubbier columns.

Finally, we have also to consider that the error committed in the critical loads of the comparatively less loaded columns means, in fact, an error to the unsafe side only if we want to vary or choose the loads acting on the columns independently of each other ('multi-parametric loading'), and we want to fully use the computed load-bearing capacity of the individual columns. If, however, we deal with 'one-parametric loading', i.e. if the ratio of the loads acting on the individual columns is set, then we cannot fully use the computed critical loads of the less loaded columns, since the intensity of the load system acting on the frame is determined by the load-bearing capacities of the more loaded columns, which are, according to the approximate calculation, less than the exact values. Consequently, we cannot use the less loaded columns even up to their exact critical loads.

In summary we propose to consider the aforementioned points of view, and in the case of unequal loads and column stiffnesses we may use a higher factor of safety for the comparatively less loaded columns.

Let us briefly deal with *frames with triangular network* which generally appear as plane trusses (*Fig. 5-29a*) or chord planes of double-layer space frames (*Fig. 5-29b*). As contrasted to frames with rectangular networks where all bars can bow out in a single half-wave (*Fig. 5-24*), here, due to the odd number of sides, either one of the bars must bow out in two half waves (*Fig. 5-29c*), or two bars have to deform in a shape as if they were rigidly clamped at one end (*Fig. 5-29d*). (Buckling modes between these two shapes may also occur.) The bars bowing out in this way have a much higher critical load, and also a much greater rigidity by which they stiffen the other, adjoining bars. We do not want to analyse the possible cases in detail; we are content to simply state that the buckling lengths of the bars of frames with triangular network will be smaller than those of rectangular frames determined on the basis of what has been said in the foregoing paragraphs.

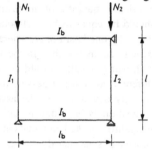

Fig. 5-28 The nonsway frame investigated

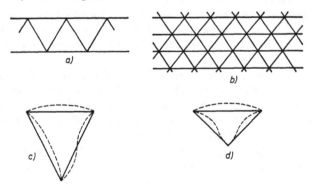

Fig. 5-29 Frames with triangular network. *a*) Plane truss, *b*) chord plane of a space frame, *c*) and *d*) buckling modes

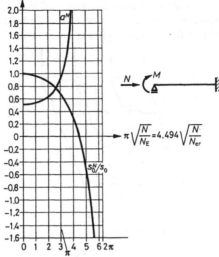

Fig. 5-30 The influence of the compressive force on the stiffness and carry-over factor of the beam

Finally we would like to mention that if the *compressive forces acting in the beams* are not negligibly small, then we have to take into account that these forces reduce the bending stiffnesses of the beams. BLEICH [1952] has given the stiffness factor $s_0^{(N)}$ and the carry-over factor $a^{(N)}$ of a beam rigidly clamped at the other end and loaded by a compressive force N, as functions of the ratio N/N_{cr}, in *Fig 5-30*, where N_{cr} is the critical force of the beam hinged at one end and rigidly clamped at the other end. The diagram gives the value of $a^{(N)}$ and the ratio of $s_0^{(N)}$ to the stiffness s_0 of the beam without a compressive force. The notation N_E appearing also in *Fig. 5-30* denotes the critical load of the hinged-hinged bar, as before, see Eq. (5-63).

DULÁCSKA [1987] found that the stiffness $s_0^{(N)}$ and carry-over factor $a^{(N)}$ of a beam, loaded by a compressive force N and arbitrary supported at the other end, can be computed with sufficient accuracy by the relations

$$s_0^{(N)} = s_0 \left(1 - \frac{N}{N_{cr}}\right) \tag{5-72}$$

$$a^{(N)} = a \frac{1}{1 - \dfrac{N}{N_{cr}}} \tag{5-73}$$

where N_{cr} denotes the critical force of the beam taking the actual support of the other end into account.

5.2.3 Stability investigation of unbraced frames

Unbraced frames buckle according to *Fig. 5-31* with laterally displacing nodes. In the meantime the beams deform (approximately) antisymmetrically (*Fig. 5-32*).

Accordingly, their two ends exert the restraining stiffnesses given by Eqs (5-68) and (5-69), but since the signs of α and β are now equal, the beam end stiffnesses will be greater than $4EI_b/l_b$ [Eq. (5-66)], see the right-hand side of *Fig. 5-27a*.

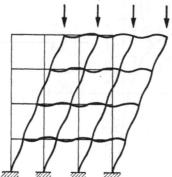

Fig. 5-31 The buckling shape of the unbraced frame

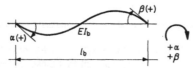

Fig. 5-32 The deformation of the beam of the unbraced frame

In the case of antisymmetric deformations, i.e. in the domain of α and β of equal signs, the sum of the end stiffnesses of the beam becomes a minimum if $\alpha = \beta$ (*Fig. 5-27b*). In this case the two end stiffnesses are equal, and their value is

$$s_\alpha = s_{0\beta} = s_0 = \frac{6EI_b}{l_b}. \tag{5-74}$$

If we use these end stiffnesses, then in the cases of $\alpha \neq \beta$ we commit an error to the safe side, since – as shown also in *Fig. 5-27b* – the sum of the two end stiffnesses is, actually, greater than the double of that given by Eq. (5-74).

According to what has been said in Section 5.2.1, we have to investigate: what conditions have to be fulfilled in order to obtain no horizontal forces H between the neighbouring columns during buckling.

It is obvious that if all columns beside each other have the same stiffness EI/l and the same compressive forces (and the extreme columns have half these values), then every column wants to use the same beam end stiffnesses and buckle simultaneously, without supporting each other. This happens in the cases of *Figs 5-22* or *5-23*. Hence in these cases we obtain the critical load of the frame with the approximate method exactly, assuming the beam end stiffnesses according to Eq. (5-74), and sharing them between the upper and lower columns in the proportion of their stiffnesses EI/l.

If the rigidities of the *columns above each other* do not follow the patterns of *Figs 5-22* or *5-23*, but the compressive forces are proportional to their *Euler* loads, we still have to share the beam end stiffnesses between the columns proportional to their stiffnesses EI/l. We can maintain this sharing even if the compressive forces are not proportional to the *Euler* loads. About the error committed and the circumstances reducing this error, we can use the same reasoning as with the braced frames.

If, however, the rigidities of the *columns beside each other* differ and, moreover, the compressive forces acting in them are also not proportional to their *Euler* loads, then the comparatively less loaded columns support horizontally the more loaded ones, and we should not only take these forces H into account, but we also have to return to the three-spring model, which would rather complicate our task. It would thus be expedient to reduce the problem to that without H forces. This is made possible by the following load rearrangement theorem of DULÁCSKA [DULÁCSKA and KOLLÁR, 1960]:

A thorough investigation of the phenomenon shows *that the critical total load of one storey of unbraced frames hardly depends on the distribution of the loads between the individual columns.* The reason for this is the following.

Let us consider the frames shown in *Fig. 5-33*. These differ from each other in that the unloaded columns of *Fig. 5-33a* elastically support the loaded ones, while the three columns on the right-hand side of the frame of *Fig. 5-33b* do not support the left one but they also take part in the load-bearing.

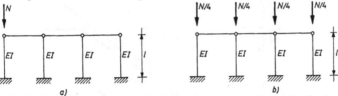

Fig. 5-33 The four-legged frame investigated. *a*) Only one column is loaded, *b*) all columns are equally loaded

The investigation of the column rigidly clamped at the bottom and elastically supported at the top (see in detail in Section 7.5.1) shows that its critical load increases – at least in the range of small spring constants – proportionally with the spring constant, see *Fig. 5-34*. In the figure we plotted to the exact curve also its initial tangent which represents this linear relation. N_E is defined by Eq. (5-63), and c is the spring constant of the support.

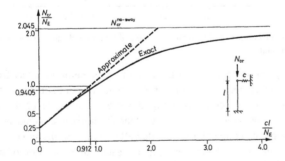

Fig. 5-34 The critical load of the column rigidly clamped at the bottom and elastically supported at the top, as a function of the spring constant

It can be also shown that the spring constant exerted by an unloaded column, effective for supporting the loaded one, is proportional to EI/l^2 as well as its critical load (cf. with *Fig. 5-34* where on the horizontal axis the product of the spring constant c, proportional to EI/l^3, with the length l is plotted). Thus it makes almost no difference whether some columns support the more loaded other ones, or they also take part in the load-bearing.

We obtain similar results for the frame with columns rigidly clamped at both ends (*Fig. 5-35*). The diagram of the critical load, together with its initial tangent, is plotted in *Fig. 5-36.*

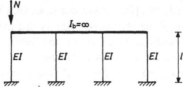

Fig. 5-35 A four-legged frame with an infinitely stiff beam

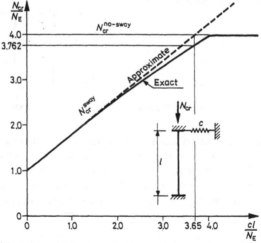

Fig. 5-36 The critical load of the column rigidly clamped at top and bottom and supported elastically, as a function of the spring constant

It may seem surprising that the columns of unbraced frames, being far away from each other, can stiffen each other to such an extent, above all if we compare their stiffening effect to that, much weaker, of the columns being above each other, or of the columns of unbraced frames. The explanation lies in the fact that the horizontal beam provides a *rigid connection* between the buckling columns, so that all columns on the same level must bow out to the same measure. The columns being above each other (or those of unbraced frames) are compelled only by the bending rigidities of the adjoining beams (or columns) to buckle simultaneously with the comparatively most loaded column, and this connection is by far not rigid. The less loaded columns thus bow out less than the most loaded one, their buckling amplitude being only proportional to that of the most loaded one.

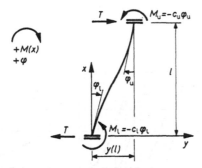

Fig. 5-37 The column elastically restrained at both ends

The distribution of the loads between the columns hardly affects the magnitude of the total critical load of one storey in cases of other clamping rigidities affecting the column ends either. That is, the change in the clamping rigidity modifies the supporting stiffness (against horizontal displacement) of a column and its 'own' critical load almost to the same measure. As a proof we develop the rigidity against displacement, c, of a column restrained against rotation at its two ends by the spring stiffnesses c_u and c_l, respectively, see *Fig. 5-37*, and compare it with the critical load of the column.

The solution of the differential equation of the bent bar yields, taking into account the boundary conditions expressing the elastic restraints and using the notations of Eqs (5-57) and (5-58), the ratio $T/y(l)$ which is equal to the spring constant c effective against horizontal displacement:

$$c = \frac{T}{y(l)} = \frac{EI}{l^3} \frac{12\,(\rho_l + \rho_u + \rho_l\rho_u)}{12 + 4\rho_l + 4\rho_u + \rho_l\rho_u}. \tag{5-75}$$

We compared this spring constant with the critical load taken from the diagram of *Fig. 5-21*, in *Fig. 5-38*. To be more visual, we plotted the quantity

$$\kappa = \frac{2\sqrt{3}}{\sqrt{\dfrac{12\,(\rho_l + \rho_u + \rho_l\rho_u)}{12 + 4\rho_l + 4\rho_u + \rho_l\rho_u}}}, \tag{5-76}$$

proportional to the square root of the inverse of c given by Eq. (5-75), and the multiplicator of the buckling length, ν (5-62), proportional to the square root of the inverse of the critical force, in the figure; thus we have to deal with nondimensional quantities. We have obtained the factor $2\sqrt{3}$ in κ by setting the values of κ and ν equal to each other in the cases $\rho_l = \rho_u = \infty$ (clamped-clamped bar) and $\rho_l = \infty, \rho_u = 0$ (clamped-hinged bar). *Fig. 5-38a* shows the curves plotted against the restraining stiffnesses, ρ_l and ρ_u, and *Fig. 5-38b* the sections of the two surfaces ν and κ at $\rho_l = \rho_u = \infty$ and at $\rho_l = \rho_u = 0$. As can be seen, the spring constant and the critical fobrce vary, as functions of the restraining stiffnesses at the bar ends, almost proportional to each other.

Hence what we said on the frame of *Fig. 5-33*, namely that it makes practically no difference whether some columns support the more loaded ones or carry a part of the load, remains valid also when the columns are elastically restrained at both ends.

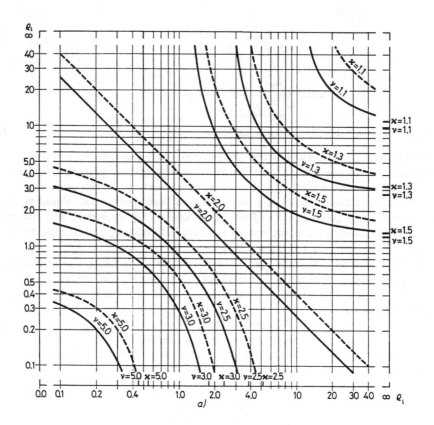

Fig. 5-38a Comparison of the stiffness against displacement and the critical load of the column elastically restrained at both ends. *a*) The square root of the inverse value of the spring constant and the buckling length.

It follows from the foregoing that the statement of DULÁCSKA on the rearrangement of the loads, referring to the columns of one storey, remains valid also if the individual columns are restrained to different degrees, even if they have different lengths. That is, the change in the column length changes the *Euler* load [Eq. (5-63)] proportionally to its second power, and the stiffness against displacement proportionally to its third power [Eq. (5-75)]; consequently, the the quotient cl/N_E, characterizing the supporting stiffness (*Figs 5-34* and *5-36*), is independent of the column length.

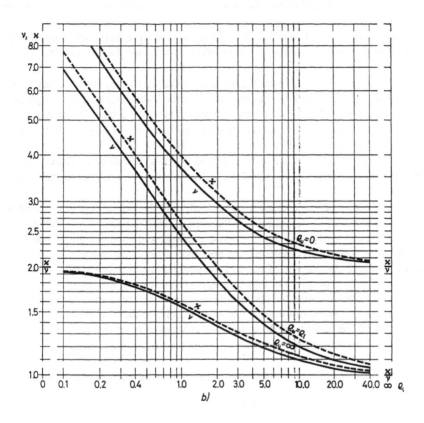

Fig. 5-38b Comparison of the stiffness against displacement and the critical load of the column elastically restrained at both ends. *b*) The sections of the surfaces κ and ν at the extreme values of the restraining stiffnesses

Since we can rearrange the load acting on the columns of one storey without significantly changing the critical total load, we can find a very easy way to determine this latter. We choose a load arrangement under which the individual columns buckle without exerting horizontal forces H on each other; that is, every column carries just a load which it can carry with the aid of the restraining effect of the beams only, without needing a horizontal supporting force. The critical loads computed in this way will be called the 'own critical loads' of the columns ($N_{cr}^{(c\,=\,0)}$). These can be simply obtained from the chart of *Fig. 5-21* in knowledge of the restraining effect of the beam ends. We could determine these by adjusting their values to the actual stiffnesses of the columns, but it is much simpler to assume equal stiffnesses at the two ends of the beams, which is anyway an approximation lying on the safe side.

Thus we still have to find out what error we commit by determining the total critical load of the storey with the aid of the 'own critical loads' of the columns, and how can we limit this error.

The error stems from the fact that although the supporting effect of the columns and their critical loads vary very similarly to each other, nevertheless they do vary not identically.

This error is obviously the greater, the more the actual load arrangement deviates from that corresponding to the distribution of the 'own' critical loads. This deviation is greatest if only one column is loaded and all others are supporting it. We can put it also in another way: the deviation is greatest if the summarized spring constants of the less loaded columns supporting the more loaded one(s) is highest (see *Figs 5-34* and *5-36*). In this case the loaded column must be overloaded with respect to its 'own' critical load. Hence we can confine the error by limiting the overloading of the columns with respect to their 'own' critical loads.

On the basis of numerical computations we shall limit the overloading of one column in the *fourfold* of its 'own' critical load. If the column were restrained by infinitely rigid clampings at top and bottom, this value would correspond to the ratio of the critical loads in the states with horizontally supported and unsupported nodes, so that it would become superfluous to check the stability of the column with horizontally supported nodes. The restraints are, however, actually not infinitely rigid. In these cases the ratio of the critical loads in the two states is higher, but the restraining effect of the beam ends is considerably lower in the case of fixed nodes than in that of free sway, so that it is practically always necessary to check the *stability of the overloaded columns as non-sway columns*.

Let us check the error committed by rearranging the loads if we overload one column fourfold.

In the case of *Fig. 5-33a* the summarized spring constant of the three unloaded columns is

$$c = 3\frac{3EI}{l^3} = \frac{0.912\,\pi^2 EI}{l}\frac{1}{l^2} = \frac{0.912}{l}N_E,$$

and the corresponding (exact) critical load of the loaded column is obtained from *Fig. 5-34* as

$$N_{cr}^{sway} = 0.9405 N_E = 0.9405 \times 4N_{cr}^{(c\,=\,0)}.$$

According to the theorem of load rearranging we can take the fourfold of the 'own' critical load of the loaded column as the critical total load of the column row $(N_{cr}/N_E = 1)$. This is by 6.3% higher than the exact value, i.e. we are on the unsafe side.

In the case of *Fig. 5-35* the summarized spring constant is

$$c = 3\frac{12EI}{l^3} = \frac{3.65\,\pi^2 EI}{l}\frac{1}{l^2} = \frac{3.65}{l}N_E,$$

with which we obtain the exact critical load of the loaded column from *Fig. 5-36* as

$$N_{cr}^{sway} = 3.762 N_E = 0.9405 \times 4N_{cr}^{(c\,=\,0)}.$$

Thus the error committed by the load rearranging is again 6.3%, on the unsafe side.

We also have investigated the four-legged frames shown in *Figs 5-39* and *5-40* under the indicated two load arrangements with both the exact and the approximate method. (The latter, using the load rearrangement, yields the same result in both cases.) We set the ratio of the rigidities of beam and column, $(EI_b/l_b)/(EI_c/l_c)$, to 1/3, 1 and 3, respectively. The results are given in *Tables 5-3* and *5-4*. In order to have a better overview, we related the critical loads obtained by the exact and the approximate methods to the *Euler* load of the hinged-hinged column [Eq. (5-63)]. The data of the two tables show that the critical loads obtained by the load rearrangement method are lower than the exact ones, and when they are higher, the error is less than 3%.

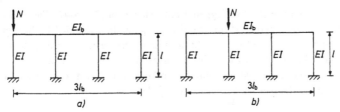

Fig. 5-39 The four-legged frame, rigidly clamped at the bottom, investigated in detail

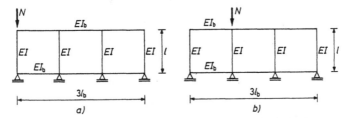

Fig. 5-40 The four-legged frame, stiffened at the bottom by a beam, investigated in detail

Table 5-3 Critical loads of the frame of *Fig. 5-39*

$\frac{I_b/l_b}{I/l}$	Exact method			Approximate method		
	a)	b)	Δ %		Δ %	
	$\frac{N_{cr}}{\pi^2EI/l^2}$	$\frac{N_{cr}}{\pi^2EI/l^2}$	related to a)	$\frac{N_{cr}}{\pi^2EI/l^2}$	related to a)	to b)
1/3	2.37	2.42	+2.3	2.40	+1.4	−0.9
1	3.24	3.14	−3.0	3.16	−2.4	+0.5
3	3.72	3.62	−2.6	3.53	−5.2	−2.7

Table 5-4 Critical loads of the frame of *Fig. 5-40*

$\frac{l_b/l_b}{I/l}$	Exact method			Approximate method		
	a) $\frac{N_{cr}}{\pi^2 EI/l^2}$	b) $\frac{N_{cr}}{\pi^2 EI/l^2}$	Δ % related to a)	$\frac{N_{cr}}{\pi^2 EI/l^2}$	Δ % related to a)	to b)
1/3	1.55	1.51	−2.6	1.53	−1.5	+1.2
1	2.65	2.51	−5.3	2.58	−2.5	+2.9
3	3.43	3.28	−4.4	3.37	−1.7	+2.8

Let us also investigate the two-legged frame shown in *Fig. 5-41*. The exact and approximate critical loads for some stiffness ratios are given in *Table 5-5*. In the approximate method we assumed equal stiffnesses at both ends of the beam, according to Eq. (5-74). The ratios of the column stiffnesses are, in the cases investigated, such as to ensure that the loaded column is overloaded just four times as related to its 'own' critical load computed with the beam end stiffnesses according to Eq. (5-74).

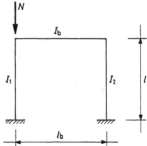

Fig. 5-41 The two-legged frame investigated

Table 5-5 Critical loads of the frame of *Fig. 5-41*

$\frac{l_b/l_b}{I_1/l}$	I_2/I_1	$\frac{N_{cr}^{exact}}{\pi^2 EI_1/l^2}$	$\frac{N_{cr}^{approx}}{\pi^2 EI_1/l^2}$	Δ%
0	3.0	0.94	1.00	+6.3
1	5.22	2.97	3.02	+1.7
2	4.31	3.13	3.39	+8.3
5	3.4	3.35	3.62	+8.1
7	3.3	3.49x	3.68	+5.5
∞	3.0	3.76	4.00	+6.3

The results of *Table 5-5* show that the error to the unsafe side (with a + sign) is somewhat higher than in the previous case, but less than 10%. For the evaluation we have, however, to take into account the two points of view what we have said on the difference between two- and multi-legged frames in connection with *Table 5-2*. (In the present case the deformation of the beam deviates not from the symmetric, but from the antisymmetric shape, and since the sum of the two end stiffnesses is a minimum in the case of antisymmetric deformation, this deviation has an opposite effect as with the braced frames.)

Since in the approximate method for unbraced frames we sum up the load-bearing capacity of columns standing beside each other, the circumstances, which reduce the error in the case of unbraced frames under one-parametric loading, do not become effective; they are effective only for the total loads of storeys being above each other.

In summary, we can state that if we actually overload some columns to the fourfold in relation to their 'own' critical load, then we may commit at most an error of 10% to the unsafe side when determining the critical total load of the storey with the load rearrangement method based on the theorem of DULÁCSKA. In such cases it is thus expedient to reduce the computed critical total load by 10%

If the column stiffnesses of storeys being above each other greatly differ, then, according to what has been said in Section 5.2.2, we may consider a somewhat higher safety factor to the columns of the less loaded storey.

Summing up the approximate procedure, we can determine the critical load of an unbraced frame by assuming the restraining stiffnesses of the bar ends according to Eq. (5-74) and sharing them between the columns above each other proportionally to their stiffnesses EI/l. We determine the 'own' critical loads $N_{cr}^{(c\,=\,0)}$ of the columns with the aid of the chart of *Fig. 5-21*, and check whether no column is more than four times overloaded in relation to its 'own' critical load. We then compute the total critical load of each storey as the sum of the 'own' critical loads, and reduce this by 10% or less, according to whether the most loaded columns are four times or less overloaded. Finally we must check whether no overloaded column buckles as a non sway column, according to Section 5.2.2.

If the compressive forces acting in the beams are not negligibly small, then we again have to use the diagrams of *Fig. 5-30* or Eqs (5-72) and (5-73) to determine the reduced values of s_0 and a appearing in Eq. (5-65). Since the beams deform in two half waves instead of one, as contrasted to braced frames, by using these equations we are on the safe side, because for N_{cr} we should use, in fact, a higher value than that of the hinged-clamped bar.

In order to dimension the individual columns we have to know their 'computational buckling lengths', with the aid of which we can determine their actual critical loads (taking into account that the less loaded ones support the more loaded ones), and compare them with their actual loads. This can be done in the following way:

We know the ratio α_i of the load Q_i of the ith column to the total load of the storey, Q^{storey}:

$$Q_i = \alpha_i Q^{\text{storey}}. \tag{5-77}$$

The share of the column of the total critical load of the storey is the same:

$$Q_{\text{cr},i} = \alpha_i Q_{\text{cr}}^{\text{storey}}. \tag{5-78}$$

Since the computational buckling length l_0 is defined by the expression

$$Q_{\text{cr},i} = \frac{\pi^2 EI_i}{l_{0i}^2}, \tag{5-79}$$

the computational buckling length is:

$$l_{0i} = \pi \sqrt{\frac{E I_i}{Q_{cr,i}}}.$$

(5-80)

This length will be, of course, greater for more loaded, and smaller for less loaded columns than the buckling lengths of their 'own' critical loads, but it is just this deviation by which the fact that they can buckle only simultaneously is taken into account.

6

Application of the sandwich theory in the stability analysis of structures

István HEGEDŰS and László P. KOLLÁR

Sandwiches consist of layers which have different stiffness and strength characteristics. The most common sandwiches have three layers: two faces (or face sheets) and the core between them. The face sheets are usually thinner, but significantly stiffer (and stronger) than the core.

Structural sandwiches are usually sandwich panels with a total thickness much smaller than the other dimensions (*Fig. 6-1*). The face sheets can be flat or corrugated plates, the core can be low-density solid material or some kind of structure, such as honey-comb (*Fig. 6-1c*) or corrugated sheets (*Fig. 6-1d*). The material can be manifold, the face sheets are commonly made of metal, or fibre reinforced plastics, the solid core is often made of foam or balsa wood, while the honey-comb core can be made of aluminium, resin-impregnated paper or plastic.

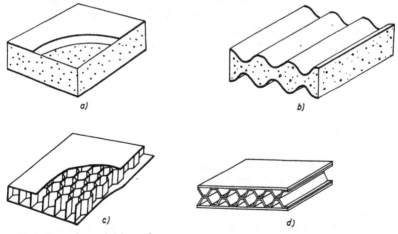

Fig. 6-1   The built-up of a sandwich panel

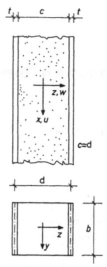

Fig. 6-2 Dimensions of a sandwich beam

In most of this chapter we will deal with *sandwich beams*, in which both the thickness $(c + 2t)$ and the width (b) are much smaller than the third dimension (*Fig. 6-2*). The most important difference (but not the only one) between the analysis of a sandwich beam and a classical beam is that the shear deformation of the latter is neglected.

An important application of the sandwich beam theory is the analysis and design of regular structures assembled of identical elements in a regular pattern, such as frames or trusses [KOLLÁR and HEGEDŰS, 1985]. The analysis of these structures can be replaced by that of continua, identified as '*replacement continua*'. On the basis of the continuum analysis we can derive approximate results 'on the back of the envelope'. This method was applied e.g. to the stability analysis of trusses [TIMOSHENKO and GERE, 1961], frames [CSONKA, 1961a; ZALKA, 1976, 1977, 1979; ZALKA and ARMER, 1992], space frames and trusses [KOLLÁR and HEGEDŰS, 1985].

6.1 ASSUMPTIONS, DEFINITIONS

We assume that the materials behave in a linearly elastic manner, strains and deflections are small. The axes of the beams throughout this chapter are vertical, and the load which causes buckling is also vertical. For simplicity, we will refer to these structures as beams.

We mainly deal with sandwiches that consist of three layers in a symmetric built-up, namely two faces of thickness t and a core of thickness c, which are continuously connected to each other (*Fig. 6-2*).

The faces behave according to the classical (*Bernoulli-Euler*) beam theory (for sandwich plates the classical plate theory), i.e. their transverse shear deformations are neglected.

The material of the core may be isotropic or orthotropic. The elastic moduli of the core are much smaller than those of the faces. Consequently, in most cases, the core will be treated as an 'antiplane' material, which means that it is considered to be infinitely soft in the direction of the axis of the beam (or in the plane of the plate), however its shear stiffness is finite. In most cases (except those in Section 6.3.2), the core will be treated as incompressible in the thickness direction. This means that we model the core as a strongly anisotropic material: it has a very low stiffness in the direction of the axis of the beam (or in the plane of the plate), very high stiffness through the thickness, and it has a finite shear stiffness. This is a very good model for a honeycomb core (*Fig. 6-1c*). It is not so obvious that this is also a reasonable approximation for isotropic cores, as it will be illustrated in Section 6.3.2.

For the layers of the sandwich we will assume that the cross sections remain plane after deformation, however the rotations of the cross sections of the layers could be different from each other (*Fig. 6-3a*).

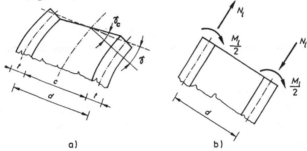

a)   b)

Fig. 6-3 Deformation of the cross section of a sandwich beam *a*) and the forces in the faces *b*)

If the sandwich is subjected to *pure bending*, equal and opposite forces (N_l) and equal bending moments ($M_l/2$) arise in the faces (*Fig. 6-3b*). The resultant of these is:

$$M = M_0 + M_l, \qquad (6\text{-}1)$$

where

$$M_0 = dN_l. \qquad (6\text{-}2)$$

We define three stiffnesses for a sandwich beam. The bending stiffness consists of two parts: the 'local bending stiffness', D_l, which is due to the bending of the face sheets:

$$D_l = 2\frac{E_f t^3 b}{12}, \qquad (6\text{-}3)$$

the 'global bending stiffness', D_0, which is due to the compressibility of the face sheets (from the parallel axes theorem):

$$D_0 = \frac{E_f b t d^2}{2}, \qquad (6\text{-}4)$$

and the shear stiffness

$$S = G_c b \frac{d^2}{c}, \qquad (6\text{-}5)$$

which will be discussed later. In these equations E_f is *Young*'s modulus of the faces, G_c is the shear modulus of the core, b is the width of the beam, t is the thickness of the faces, d is the distance between the middle planes of the faces and c is the width of the core.

In the case of pure bending (if the shear force is zero), M_0 and M_l are related to the bending stiffnesses D_0 and D_l by $M_0 = -D_0 w''$ and $M_l = -D_l w''$, where w'' is the second derivative of the horizontal displacement of the beam. (Prime denotes the derivative with respect to x.)

Sandwich beam with thin faces

We refer to a sandwich as a *sandwich with thin faces* if the following relationship holds:

$$t \ll c. \tag{6-6}$$

In this case c and d are approximately equal and, consequently, the shear strain in the core (γ_c) and γ defined in *Fig. 6-3a* can be considered identical:

$$c \approx d, \qquad \gamma_c \approx \gamma. \tag{6-7}$$

For a sandwich beam with thin faces both the local bending stiffness (D_l) and the bending moment (M_l) are zero:

$$M_l = 0, \qquad D_l = 0. \tag{6-8}$$

This model is often referred to in the literature as a 'beam with shear deformation', or as a '*Timoshenko-beam*'.

Sandwich beam with thick faces

We refer to a sandwich as a *sandwich with thick faces* if Eq. (6-6) does not hold. For this case we have to consider the difference between the shear strain in the core (γ_c) and γ defined in *Fig. 6-3a*:

$$\gamma_c = \gamma \frac{d}{c} \tag{6-9}$$

and neither the local bending stiffness (D_l) nor the bending moment (M_l) are neglected:

$$M_l \neq 0, \qquad D_l \neq 0. \tag{6-10}$$

6.2 SANDWICH BEAM WITH THIN FACES (TIMOSHENKO-BEAM)

In this section we consider sandwiches with thin face sheets, where Eq. (6-6) holds. This assumption will significantly simplify the analysis. Sandwiches with thick faces will be treated in Section 6.3.

Since the core has a finite shear stiffness, the shear deformation of the sandwich beam must be taken into account.

Governing equations

A deformed element of the sandwich beam is shown in *Fig. 6-4*. It can be observed that the slope of the beam $\left(w' \right)$ is not equal to the rotation of the cross section (χ). In fact, the slope consists of two terms:

$$w' = \chi + \gamma, \tag{6-11}$$

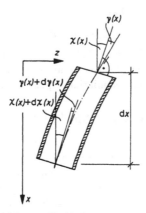

Fig. 6-4 Deformation of a sandwich beam with thin faces

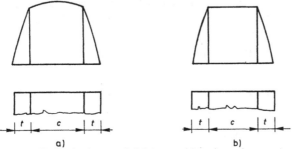

Fig. 6-5 Shear stress distribution in a sandwich beam with an isotropic core a), and with an antiplane core b)

where γ, the shear strain, is defined in *Figs 6-4* and *Fig. 6-3a*. (In the classical beam theory, where the shear deformations are neglected, there is only bending deformation and $w' = \chi$.) Similarly to the deformations, the displacements of the beam could be defined as the sum of the 'bending displacement' and the 'shear displacement':

$$w = w_D + w_S, \tag{6-12}$$

where

$$w'_D = \chi, \qquad w'_S = \gamma. \tag{6-13}$$

The shear force in the beam is the integral of the shear stresses. The distribution of the shear stresses is illustrated in *Fig. 6-5a* for an isotropic core, and in *Fig. 6-5b* for an antiplane core [ALLEN, 1969]. Hence the shear force is (assuming constant stress distribution in the core) as follows:

$$V \approx G_c b d\gamma_c = S\gamma, \tag{6-14}$$

where S is defined by Eq. (6-5). Using Eq. (6-7) we may have $S = G_c bc$. The bending moment is the moment resultant of the forces arising in the faces. (The local bending moments in the faces are negligible, due to Eq. (6-8)). The axial strain in the faces is equal to $\varepsilon = -\chi' d/2$, and, consequently, we have

$$M = M_0 = -\left(E_f \chi' \frac{d}{2} bt\right) d = -D_0 \chi'. \tag{6-15}$$

The stiffnesses D_0 was defined by Eq. (6-4). The equilibrium equations for a beam is independent of the deformations. For a beam subjected to loads perpendicular to the axis and to axial loads, the horizontal and moment equilibrium equations are as follows [TIMOSHENKO and GERE, 1961]:

$$-V' + q - \left(N\beta w'\right)' = 0, \tag{6-16}$$

$$M' - V = 0, \tag{6-17}$$

where q is the horizontal load and $N\beta$ is the normal force in the beam. Here N is independent of x, and can be considered as the load parameter, while β depends on x. The vertical load, p, is the first derivative of the normal force: $p = N\beta'$. Eq. (6-12) through (6-17) form the differential equation system of a sandwich beam with thin faces with the unknown functions $w, w_D, w_S, \chi, \gamma, M, V$. These equations can be simplified by algebraic manipulations. Two forms of these simplified equations have practical interest.

If all the unknown functions are expressed in terms of displacements w_D and w_S, Eqs (6-12) through (6-17) result in:

$$D_0 w_D'''' = q - \left(N\beta \left(w_D' + w_S'\right)\right)', \tag{6-18}$$

$$S w_S = -D_0 w_D''. \tag{6-19}$$

If all the unknown functions are expressed in terms of w and χ, we obtain:

$$D_0 \chi''' = q - \left(N\beta w'\right)', \tag{6-20}$$

$$S \left(w' - \chi\right) = -D_0 \chi''. \tag{6-21}$$

Boundary conditions

Both groups of the above equations form fourth order differential equation systems. Consequently, two boundary conditions are required at each end of the beam. The boundary conditions are

for a built-in boundary:

$$w_D + w_S = 0, \qquad w_D' = 0; \quad \text{or} \quad w = 0, \qquad \chi = 0, \tag{6-22}$$

for a hinged support:

$$w_D + w_S = 0, \qquad w_D'' = 0; \quad \text{or} \quad w = 0, \qquad \chi' = 0, \tag{6-23}$$

while for an (unloaded) free end:

$$w_S' = 0, \qquad w_D'' = 0; \quad \text{or} \quad w' - \chi = 0, \qquad \chi' = 0. \tag{6-24}$$

Note that for a built-in boundary the rotation of the cross section (χ) is zero, however the slope of the beam (w') is not.

Buckling of simply supported beams

We consider a simply supported beam (*Fig. 6-6a*) subjected to an axial load N. In the differential equations (Eqs (6-18) and (6-19)) we introduce $q = 0$ and $\beta \equiv 1$. These equations, and the pertaining boundary conditions:

$$w_D(0) + w_S(0) = 0, \qquad w_D''(0) = 0, \tag{6-25}$$

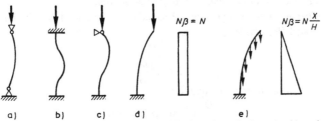

a)   b)   c)   d)   e)

Fig. 6-6 Sandwich beam subjected to a concentrated load *a*) through *d*), and subjected to a uniformly distributed load *e*).

$$w_D(H) + w_S(H) = 0, \qquad w_D''(H) = 0 \qquad (6\text{-}26)$$

can be satisfied by the trigonometrical functions:

$$w_{Dk} = A_k \sin\left(\frac{k\pi}{H}x\right) \quad \text{and} \quad w_{Sk} = -B_k \sin\left(\frac{k\pi}{H}x\right). \qquad (6\text{-}27)$$

By introducing these functions into Eqs (6-18) and (6-19), after algebraic manipulations, we obtain

$$N_{cr,k} = \left(N_{0,k}^{-1} + S^{-1}\right)^{-1} \quad \text{and} \quad B_k = \frac{D_0}{S}\frac{k^2\pi^2}{H^2}A_k, \qquad (6\text{-}28)$$

where $k = 1, 2, \ldots$, A_k is arbitrary, and

$$N_{0,k} = k^2\frac{\pi^2 D_0}{H^2}. \qquad (6\text{-}29)$$

The smallest critical load is obtained by setting $k = 1$. This gives:

$$N_{cr} = \left(N_0^{-1} + S^{-1}\right)^{-1}, \qquad (6\text{-}30)$$

where

$$N_0 = \frac{\pi^2 D_0}{H^2} \qquad (6\text{-}31)$$

is the critical load of a simply supported column with bending stiffness D_0 undergoing bending deformations only. Note that S is the critical load of a simply supported column undergoing shear deformations only as it will be shown in Section 6.4.2.

Buckling of beams built-in at both ends

We consider a beam (*Fig. 6-6b*) built-in at both ends subjected to an axial load N. The smallest buckling load (omitting the derivation) is as follows:

$$N_{cr} = \left(N_0^{-1} + S^{-1}\right)^{-1} \qquad (6\text{-}32)$$

where

$$N_0 = \frac{\pi^2 D_0}{\left(\dfrac{H}{2}\right)^2}. \qquad (6\text{-}33)$$

Buckling of beams built-in at one end and hinged at the other
We consider the beam shown in *Fig. 6-6c* subjected to an axial load N. The buckling load N_{cr} can be calculated [PLANTEMA, 1961] as

$$N_{cr} = \frac{\alpha^2 D_0}{1 + \alpha^2 \frac{D_0}{S}}, \tag{6-34}$$

where α is the root of the following equation:

$$\left(1 + \frac{D_0}{S}\alpha^2\right) \tan \alpha H = \alpha H. \tag{6-35}$$

Buckling of a cantilever beam
We consider a beam built-in at one end and free at the other.

The beam is subjected to an axial load N which results in a *constant normal force* (*Fig. 6-6d*). The smallest buckling load is:

$$N_{cr} = \left(N_0^{-1} + S^{-1}\right)^{-1}, \tag{6-36}$$

where

$$N_0 = \frac{\pi^2 D_0}{(2H)^2}. \tag{6-37}$$

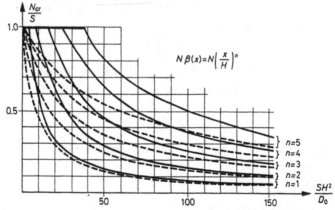

Fig. 6-7 Buckling of a sandwich beam subjected to a distributed load resulting in a normal force $N\left(\frac{x}{H}\right)^n$. The accurate solutions are presented by solid lines. The approximate solutions calculated by the *Föppl-Papkovich*'s theorem are shown by dashed lines

Critical force of a cantilever beam subjected to a *distributed load* which results in a normal force

$$N\beta = N\left(\frac{x}{H}\right)^n. \tag{6-38}$$

was numerically determined by [HEGEDŰS and KOLLÁR, 1984a], and for $n = 1$ by [ZALKA, 1979]. Some results are presented in *Fig. 6-7* by solid lines. For $n = 1$ the distribution of the normal force is linear, hence the load is uniformly distributed along the cantilever (*Fig. 6-6e*). For uniformly distributed load the first column of *Table 6-1* can also be used.

6.3 SANDWICH BEAM WITH THICK FACES

In many practical cases the local bending stiffness, D_l plays an important role in the analysis of sandwiches and it is not negligible. These structures are called 'sandwiches with thick faces'. For these cases the equations derived in Section 6.2 for sandwiches with thin faces must be modified.

6.3.1 Incompressible core

In this section the core is treated as incompressible perpendicular to the axis of the beam, consequently the distance between the face sheets is constant. The effect of compressibility will be treated in Section 6.3.2.

The bending moment due to the load is equilibrated partly by the compressive forces of the faces (M_0), which is given by Eq. (6-15), and partly by the bending moments of the faces (M_l), which, according to the classical beam theory, gives $M_l = -D_l w''$, where w is the horizontal displacement of the beam (identical to the horizontal displacement of the face sheets). By taking both bending moments into account we obtain:

$$M = M_0 + M_l = -D_0 \chi' - D_l w''. \tag{6-39}$$

The shear force is given by Eq. (6-14), but we have to take into account that $\gamma_c = \gamma \frac{d}{c}$ (*Fig. 6-3 a*). Hence the shear force can be calculated as $V = G_c b d \gamma_c = G_c b d \gamma \frac{d}{c}$, and we have:

$$V = G_c b d \gamma_c = G_c b d \gamma \frac{d}{c} = S\gamma. \tag{6-40}$$

where S is identical to Eq. (6-5).

Equations (6-12), (6-13), (6-40), (6-39), (6-16), (6-17) form the differential equation system of a sandwich beam with thick faces with the unknown functions w, w_D, w_S, χ, γ, M, V. These equations can be simplified by algebraic manipulations. Three forms of these equations have important practical interest.

If all the unknown functions are expressed in terms of displacements w_D and w_S, the above six equations result in

$$-\frac{D_l D_0}{S} w_D'''''' + (D_0 + D_l) w_D'''' = q - \left(N\beta \left(w_D' + w_S' \right) \right)', \tag{6-41}$$

$$S w_S = -D_0 w_D''. \tag{6-42}$$

If the unknown functions are expressed in terms of w and χ, we obtain

$$D_l w'''' + D_0 \chi''' = q - \left(N\beta w' \right)', \tag{6-43}$$

$$S \left(w' - \chi \right) = - \left(D_0 \chi' \right)'. \tag{6-44}$$

Both groups of the above equations form sixth-order differential equation systems.

If the sandwich is statically determinate, we can determine the shear force directly by integrating the load. In this case the differential equation (not considering vertical loads) can be written as:

$$-\frac{D_l D_0}{S} w_D''''' + (D_0 + D_l) w_D''' = -V,$$

(6-45)

$$S w_S = -D_0 w_D''.$$

(6-46)

By introducing V_D, defined as:

$$V_D = -(D_0 + D_l) w_D''',$$

(6-47)

into Eq. (6-45), we obtain the following second-order differential equation:

$$V_D'' - \frac{(D_0 + D_l) S}{D_0 D_l} V_D = -\frac{(D_0 + D_l) S}{D_0 D_l} V.$$

(6-48)

This differential equation is well-known in the (continuum) theory of coupled shear walls and rectangular frames, and was derived independently by many authors [CSONKA, 1962; ROSMAN, 1965; SZERÉMI, 1984, STAFFORD SMITH and CROWE, 1986].

Boundary conditions

Six boundary conditions are required for the above sixth-order differential equation systems. Consequently three boundary conditions must be specified at each end of the beam. We can prescribe:

– the horizontal displacement (w) or the shear force (which is the first derivative of the bending moment, $-D_0\chi'' - D_l w'''$),
– the global bending moment ($-D_0\chi'$) or the rotation of the cross section (χ),
– the local bending moments in the faces ($-D_l w''$) or the rotation of the cross section of the faces (w').

The boundary conditions are:

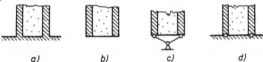

a)   b)   c)   d)

Fig. 6-8 Boundary conditions of sandwich beams with thick faces

for a built-in boundary (*Fig. 6-8a*):

$$w = 0, \qquad w' = 0, \qquad \chi = 0 \quad \text{or}$$

(6-49)

$$w_D + w_S = 0, \qquad w_S' = 0, \qquad w_D' = 0,$$

(6-50)

while for a free end (*Fig. 6-8b*):

$$w'' = 0, \qquad \chi' = 0, \qquad -D_0\chi'' - D_l w''' = 0 \quad \text{or}$$

(6-51)

$$w_S'' = 0, \qquad w_D'' = 0, \qquad -D_0 w_D''' - D_l \left(w_D''' + w_S''' \right) = 0,$$

(6-52)

for a both locally and globally hinged support (*Fig. 6-8c*):

$$w = 0, \qquad \chi' = 0, \qquad w'' = 0 \quad \text{or}$$

(6-53)

$$w_D + w_S = 0, \qquad w_D'' = 0, \qquad w_S'' = 0,$$

(6-54)

for a locally hinged and globally built-in support (*Fig. 6-8d*)

$$w = 0, \qquad \chi = 0, \qquad w'' = 0 \quad \text{or}$$

(6-55)

$$w_D + w_S = 0, \qquad w_D' = 0, \qquad w_S'' + w_D'' = 0.$$

(6-56)

For a built-in boundary both the rotation of the cross section (χ) and the slope of the beam (w') are zero. Note that at a free end only the resultant of the shear forces is zero, but equal and opposite shear forces may arise in the core and in the faces.

Buckling of simply supported beams

We consider a simply supported beam (*Fig. 6-6a*) subjected to an axial load N which results in a constant normal load. The critical loads, omitting the derivation, are as follows:

$$N_{cr,k} = \left(N_{0,k}^{-1} + S^{-1}\right)^{-1} + N_{l,k}, \qquad \text{where} \qquad k = 1, 2, \ldots \tag{6-57}$$

and

$$N_{0,k} = \frac{k^2 \pi^2 D_0}{H^2}, \qquad N_{l,k} = \frac{k^2 \pi^2 D_l}{H^2}. \tag{6-58}$$

The smallest critical load is obtained by setting $k = 1$. This yields:

$$N_{cr} = \left(N_0^{-1} + S^{-1}\right)^{-1} + N_l, \tag{6-59}$$

where

$$N_0 = \frac{\pi^2 D_0}{H^2}, \qquad N_l = \frac{\pi^2 D_l}{H^2}, \tag{6-60}$$

are the critical loads of a simply supported columns undergoing bending deformations only with bending stiffnesses D_0 and D_l, respectively. Note that S is the critical load of a simply supported column undergoing shear deformations only, as it will be shown in Section 6.4.2.

Buckling of beams built-in at both ends

We consider a beam (*Fig. 6-6b*) built-in at both ends subjected to an axial load N. The smallest buckling load is as follows:

$$N_{cr} = \left(N_0^{-1} + S^{-1}\right)^{-1} + N_l, \tag{6-61}$$

where

$$N_0 = \frac{\pi^2 D_0}{\left(\dfrac{H}{2}\right)^2}, \qquad N_l = \frac{\pi^2 D_l}{\left(\dfrac{H}{2}\right)^2}. \tag{6-62}$$

Buckling of a cantilever beam

We consider a beam built-in at one end and free at the other. The beam is subjected to an axial load N which results in a *constant normal force* (*Fig. 6-6d*). The smallest buckling load is as follows:

$$N_{cr} = \left(N_0^{-1} + S^{-1}\right)^{-1} + N_l, \tag{6-63}$$

where

$$N_0 = \frac{\pi^2 D_0}{(2H)^2}, \qquad N_l = \frac{\pi^2 D_l}{(2H)^2}. \tag{6-64}$$

We consider now the cantilever beam subjected to a *uniformly distributed load*, p (*Fig. 6-6e*) which results in a normal force $N\beta$, where

$$N = pH, \qquad \beta = xH. \tag{6-65}$$

The solution was determined numerically by [HEGEDŰS and KOLLÁR, 1984b]. The buckling load is given by the following expression:

$$N_{cr} = p_{cr}H = c_1 \frac{D_0 + D_l}{H^2}, \tag{6-66}$$

where c_1 is presented in *Table 6-1*.

Table 6-1 Parameter c_1 in Eq. (6-66) for the calculation of buckling load of sandwich beams with thick faces subjected to uniformly distributed load

$\dfrac{SH^2}{D_l+D_0}$	$\dfrac{D_l}{D_l + D_0}$														
	0	0.001	0.005	0.01	0.05	0.1	0.2	0.3	0.4	0.5	0.6	0.7	0.8	0.9	1.0
0	0	0.008	0.039	0.078	0.392	0.784	1.567	2.351	3.135	3.918	4.702	5.486	6.269	7.053	7.837
0.05	0.050	0.099	0.161	0.211	0.535	0.928	1.712	2.496	3.279	4.062	4.844	5.626	6.405	7.178	7.837
0.1	0.100	0.171	0.255	0.320	0.668	1.064	1.850	2.632	3.414	4.195	4.974	5.750	6.519	7.267	7.837
0.2	0.200	0.304	0.412	0.500	0.904	1.314	2.102	2.882	3.658	4.432	5.021	5.957	6.698	7.385	7.837
0.5	0.500	0.665	0.815	0.933	1.465	1.917	2.717	3.486	4.238	5.025	5.691	6.378	7.015	7.552	7.837
1.0	1.000	1.260	1.403	1.536	2.142	2.642	3.449	4.185	4.887	5.551	6.179	6.757	7.265	7.658	7.837
2.0	2.000	2.289	2.396	2.519	3.094	3.589	4.366	5.026	5.618	6.178	6.679	7.111	7.473	7.729	7.837
π	3.141	3.201	3.278	3.373	3.869	4.313	5.000	5.583	6.098	6.556	6.960	7.303	7.575	7.762	7.837
5.0	4.211	4.230	4.279	4.339	4.709	5.057	5.637	6.117	6.532	6.892	7.202	7.458	7.655	7.787	7.837
10.0	5.597	5.600	5.626	5.655	5.861	6.080	6.457	6.773	7.052	7.279	7.466	7.620	7.736	7.810	7.837
20.0	6.570	6.572	6.584	6.599	6.706	6.828	7.045	7.230	7.388	7.522	7.632	7.719	7.783	7.823	7.837
50.0	7.287	7.288	7.292	7.298	7.344	7.395	7.571	7.574	7.641	7.700	7.749	7.787	7.815	7.832	7.837
100	7.554	7.555	7.557	7.560	7.583	7.609	7.657	7.700	7.736	7.767	7.792	7.812	7.826	7.834	7.837
∞	7.837	7.837	7.838	7.837	7.837	7.837	7.837	7.837	7.837	7.837	7.837	7.837	7.837	7.837	7.837

Beam-column

A simply supported sandwich beam is considered subjected to both an axial force, N, and a horizontal load with arbitrary distribution, $q(x)$ (*Fig. 6-9*). Beams subjected to both horizontal and vertical loads are referred to in the literature as beam-columns.

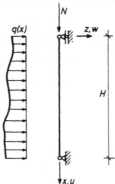

Fig. 6-9 Sandwich beam-column

The displacements of the beam, for $N = 0$, is as follows [ALLEN, 1969]:

$$w = \sum_{k=1}^{\infty} a_{0,k} \sin \frac{k\pi x}{H}, \qquad a_{0,k} = q_k \frac{H^2}{k^2 \pi^2} \frac{1}{N_{cr,k}}, \qquad (6\text{-}67)$$

where $N_{cr,k}$ is given by Eq. (6-57), and the q_k are the coefficients of the Fourier series expansion of the load:

$$q_k = \frac{2}{H} \int_0^H q(x) \sin \frac{k\pi x}{H} dx. \qquad (6\text{-}68)$$

(For a uniformly distributed load $q_k = 4q/(k\pi)$ if k is odd, and $q_k = 0$ if k is even.)

If the beam is subjected to both vertical and horizontal load, the following formula can be derived [ALLEN, 1969] for the displacement:

$$w = \sum_{k=1}^{\infty} \Psi_k a_{0,k} \sin \frac{k\pi x}{H}, \qquad (6\text{-}69)$$

where

$$\Psi_k = \frac{1}{1 - \dfrac{N}{N_{cr,k}}} \qquad (6\text{-}70)$$

is the 'amplification factor', taking the secondary displacements into account due to the normal force [TIMOSHENKO and GERE, 1961].

We may use simple formulas based on Eq. (6-69) to estimate the secondary displacements of sandwich beams due to normal loads, as follows. A sandwich beam is subjected to a normal load (the load parameter of which is N), and a horizontal load, perpendicular to the axis of the beam. The critical load parameter of the beam, under the action of the normal load only, is denoted by N_{cr}. The maximum horizontal displacement of the beam subjected to horizontal load only is denoted by w_0. The maximum displacement (w) of the beam under the combined action of the vertical and horizontal loads can be estimated as

$$w = \Psi w_0, \qquad \text{where} \qquad \Psi = \frac{1}{1 - \dfrac{N}{N_{cr}}}. \qquad (6\text{-}71)$$

The above formula is the more accurate the closer the buckling shape is to the deformed shape of the beam under horizontal loads only. (Eq. (6-71) is accurate, if the buckling shape and the deformed shape under horizontal loads only are identical.)

6.3.2 Compressible core

In the previous section we assumed that the core is incompressible perpendicular to the beam axis. This meant that the local buckling of the faces was not possible. In this section we investigate the local buckling of the faces which is referred to in the literature as *wrinkling*.

First we will treat the wrinkling of the faces independently of the column buckling (presented in the previous section) then we will show the interaction between local and global buckling, i.e. we will show that the wrinkling reduces the global buckling load (Eq. (6-57)).

Beam on elastic foundation

First we consider a sandwich beam with an isotropic core subjected at both faces to equal normal forces (*Fig. 6-10*). *Young*'s modulus and *Poisson*'s ratio of the core are denoted by E_c and v_c, respectively. We assume that the thickness of the core (c) is considerably bigger than the (half) buckling wave length (l). This means that the wrinkling of the two face sheets are independent of each other and we may consider, instead of the total beam, only one of the face sheets resting on an infinite medium (*Fig. 6-11*). This approximation is acceptable if $l < c/3$.

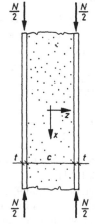

Fig. 6-10 Sandwich beam with thick faces with a compressible core

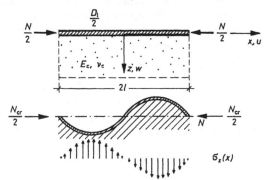

Fig. 6-11 Beam on an elastic foundation

The differential equation of a beam with bending stiffness $D_l/2$ subjected to a normal force $N/2$ is [TIMOSHENKO and GERE, 1961]:

$$\frac{D_l}{2}w'''' + \frac{N}{2}w'' = b\sigma_z, \tag{6-72}$$

where $b\sigma_z$ is the reaction force exerted by the elastic foundation. We assume that the displacement of the beam is given by $w = w_0 \sin(\pi x/l)$. If the displacement function of the upper boundary of the medium is identical to w, the σ_z stresses on the boundary are as follows [ALLEN, 1969]:

$$\sigma_z = -\frac{a}{l} w_0 \sin\frac{\pi x}{l}, \qquad \text{where} \qquad a = \frac{2\pi E_c}{(3 - v_c)(1 + v_c)}. \qquad (6\text{-}73)$$

These stresses act on the beam as 'loads' as shown on the right-hand side of Eq. (6-72). Note that the intensity of the stress depends both on the amplitude of the buckled shape, w_0, and on the buckling length, l. Introducing Eq. (6-73) into Eq. (6-72), we obtain the following expression for the buckling load

$$\frac{N_{cr}}{2} = \frac{D_l \pi^2}{2 \ l^2} + \frac{a}{l}\frac{l^2}{\pi^2}b. \qquad (6\text{-}74)$$

The smallest buckling load and the corresponding (half) buckling wave length can be obtained from the condition $dN_{cr}/dl = 0$. This equation results in

$$\frac{N_{cr}}{2} = 1.5\pi^{-\frac{2}{3}} D_l^{\frac{1}{3}} (ab)^{\frac{2}{3}}, \qquad \text{and} \qquad l_{cr} = \left(\frac{\pi^4 D_l}{ab}\right)^{\frac{1}{3}}. \qquad (6\text{-}75)$$

Interaction of wrinkling and global buckling

Several authors investigated the buckling of the sandwich beam taking the compressibility of the core into account [ALLEN, 1969; BOLOTIN, 1965; POMÁZI, 1980; BENSON and MAYERS, 1967; HEGEDŰS and KOLLÁR, 1989; CHONG and HARTSOCK, 1969].

We present the results of BENSON and MAYERS (published also in [ALLEN, 1969]). They investigated a sandwich beam with an isotropic core. They found that the compressibility of

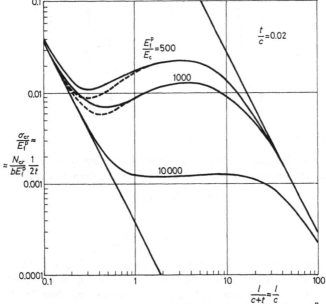

Fig. 6-12 Buckling of a sandwich plate subjected to unidirectional compressive load. E_f^p, $v_f = 1/3$, and E_c, $v_c = 1/3$ are the Young moduli and the Poisson ratios of the faces and the core, respectively. The solid lines were calculated assuming isotropic core [ALLEN, 1969], while the dashed lines by using Eq. (6-79).

the core reduces the buckling load. The values of the buckling load parameter are presented by solid lines in *Fig. 6-12* for a given geometry ($t/c = 0.02$) and for three different ratios of the elastic moduli of the faces and the core. These results are presented as functions of the buckling length (l). Note that if the compressibility of the core is negligible, the bucking load should be a monotonic function of the buckling length (Eq. (6-59)): the longer the buckling length, the lower the critical load. Interestingly, the curves in *Fig. 6-12* have local minima which means that buckling may occur with shorter buckling waves than the overall length of the beam. (Also note that the curves approach asymptotically two straight lines in the log-log scale. The left line, at short wave length, corresponds to the buckling load of the faces, i.e. $N_{cr} = \pi^2 D_l / l^2$, while the line on the right-hand side, at long wave length, corresponds to the buckling load of a beam without shear deformations, $N_{cr} = \pi^2 (D_0 + D_l)/l^2$). The calculation, leading to the above results, is straightforward but quite cumbersome, and not presented here.

Wrinkling of faces of sandwiches subjected to normal load

HEGEDŰS and KOLLÁR [1989] derived an expression for the buckling load of the sandwich beam subjected to equal normal loads on the faces (*Fig. 6-10*). They assumed that the core material of the sandwich behaves in an antiplane manner (i.e. the stiffness of the core is zero in the axial direction). Their result is accurate for honeycomb core, but also applicable, as a reasonable approximation, for isotropic core.

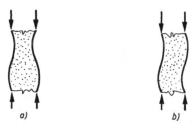

a)   *b)*

Fig. 6-13 Symmetrical and antisymmetrical buckling

They obtained that the buckling shape can be symmetrical or antisymmetrical (*Fig. 6-13*). The buckling load corresponding to the symmetrical buckling shape is as follows:

$$\frac{1}{2}N_{cr}^s = \frac{1}{2}N_l + 2R, \tag{6-76}$$

where

$$N_l = \frac{\pi^2 D_l}{l^2}, \qquad R = \frac{E_c b}{c}\frac{l^2}{\pi^2}. \tag{6-77}$$

The above buckling load is identical to that of a beam with bending stiffness $D_l/2$, resting on an elastic foundation with spring constant $E_c b /(c/2)$. Eq. (6-76) depends on the buckling length, l. The smallest buckling load and the corresponding buckling length are:

$$\left(\frac{1}{2}N_{cr}^s\right)_{min} = 2\sqrt{D_l\frac{E_c b}{c}}, \qquad l = \pi \left(\frac{D_l c}{4 E_c b}\right)^{\frac{1}{4}}. \tag{6-78}$$

The buckling load, corresponding to the antisymmetrical buckling shape, is:

$$N_{cr}^a = \left(N_0^{-1} + S^{-1} + (12R)^{-1}\right)^{-1} + N_l,$$ (6-79)

where

$$N_0 = \frac{\pi^2 D_0}{l^2}.$$ (6-80)

By setting E_c (and, consequently, R) equal to infinity in Eq. (6-79), the buckling load becomes identical to that of a sandwich beam with incompressible core and buckling length l (Eq. 6-59).

The smallest buckling load can be determined in dependence of the buckling length. This length is either identical to the length of the (simply supported) beam, or shorter if the critical load has a local minimum at a shorter buckling length.

This result, which was derived for an antiplane core, is applicable, as an approximation, for an isotropic core. This is demonstrated by a numerical example, the results of which are presented in *Fig. 6-12*. The solid lines were calculated assuming isotropic core, while the dashed lines show the results assuming an antiplane core. The properties of the antiplane core and the faces were calculated from those of the isotropic plate by the following formulas: $G_c = E_c/2$; $E_f = E_f^p / \left(1 - v_f^2\right)$. For $E_f^p / E_c = 10,000$, the results are practically identical, for the other two cases the results are reasonably close. It is important to mention that assuming an antiplane core instead of an isotropic one is always a *conservative* estimate for the critical load.

Wrinkling of faces of sandwiches subjected to normal force and bending moment
Let us consider a sandwich beam subjected to different normal forces on the faces (*Fig. 6-14*), which results in a normal force and a bending moment. The sandwich beam

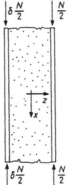

Fig. 6-14 Sandwich beam subjected to different normal forces on the faces

with a compressible antiplane core was analysed by HEGEDŰS and KOLLÁR [1989] assuming that the axis of the beam is straight prior to buckling, i.e. the pre-buckling deformations are neglected. It was obtained that the critical load, N_{cr}, can be calculated from the following second-order equation:

$$\left(\frac{N_{cr}}{2}\right)^2 \delta - \frac{N_{cr}}{2}\frac{1+\delta}{2}\left(\frac{N_{cr}^s}{2} + \frac{N_{cr}^a}{2}\right) + \frac{N_{cr}^s}{2}\frac{N_{cr}^a}{2} = 0,$$ (6-81)

where N_{cr}^s and N_{cr}^a were defined by Eq. (6-76) and Eq. (6-79), respectively, and δ is explained in *Fig. 6-14*.

Normal load only

If the two faces are subjected to identical compressive loads, i.e. $\delta = 1$, Eq. (6-81) results in:

$$\frac{N_{cr}}{2} = \begin{cases} \dfrac{N_{cr}^s}{2}, \\ \dfrac{N_{cr}^a}{2}, \end{cases}$$

(6-82)

which is identical to the expressions Eqs (6-76) and (6-79).

One face is loaded only

If only one of the faces is subjected to a compressive load, i.e. $\delta = 0$, Eq. (6-81) results in:

$$\frac{N_{cr}}{2} = 2\left[\left(\frac{N_{cr}^s}{2}\right)^{-1} + \left(\frac{N_{cr}^a}{2}\right)^{-1}\right]^{-1}.$$

(6-83)

Bending moment

If the faces are subjected to equal loads of opposite direction, i.e. $\delta = -1$, which is identical to a bending moment, Eq. (6-81) results in:

$$\frac{N_{cr}}{2} = \pm\sqrt{\frac{N_{cr}^s}{2}\frac{N_{cr}^a}{2}}.$$

(6-84)

One face is rigid

If one face is subjected to an infinite tensile force, while the other to a finite compressive load, i.e. $\delta = -\infty$, the face in tension will not deform during buckling. Equation (6-81) results in:

$$\frac{N_{cr}}{2} = \frac{1}{2}\left(\frac{N_{cr}^s}{2} + \frac{N_{cr}^a}{2}\right).$$

(6-85)

Expressions (6-83) through (6-85) are the harmonic, geometrical and arithmetic means of $N_{cr}^s/2$ and $N_{cr}^a/2$. This is a surprising general feature of nonsymmetrically loaded symmetrical structures [HEGEDŰS and KOLLÁR, 1990].

6.4 MODELS BASED ON THE SANDWICH BEAM WITH THICK FACES

The sandwich beam model was obtained by generalizing the classical *Euler-Bernoulli* beam. In the following we will discuss a few models which can be obtained by the simplification of the sandwich beam, and some others which are more complex than the sandwich beam. All these models have practical interest as it will be discussed in Section 6-6.

6.4.1 Beams with flexural deformations only ($S = \infty$ or $S = 0$)

We set the shear stiffness of the sandwich beam with thick faces equal to infinity, $S = \infty$. The differential equation system (Eqs (6-41) and (6-42)) becomes:

$$(D_0 + D_l)\, w'''' = q - \left(N\beta\left(w'\right)\right)', \tag{6-86}$$

$$w_S = 0. \tag{6-87}$$

These are identical to the (*Euler-Bernoulli*) beam equations, where the shear deformation is not considered [TIMOSHENKO and GERE, 1961]:

$$Dw'''' = q - \left(N\beta\left(w'\right)\right)'. \tag{6-88}$$

where $D = EI$ is the bending stiffness of the beam. The buckling load of a cantilever subjected to a concentrated force on the top (*Fig. 6-6d*, $\beta = 1$) is equal to

$$N_{cr} = \frac{\pi^2 D}{4H^2}, \tag{6-89}$$

while for uniformly distributed load (*Fig. 6-6e*, $\beta = x/H$) is equal to [TIMOSHENKO and GERE, 1961]:

$$N_{cr} = 7.837 \frac{D}{H^2}. \tag{6-90}$$

We obtained that a sandwich with infinite shear stiffness, $S = \infty$, is identical to an *Euler-Bernoulli* beam with bending stiffness $D = D_0 + D_l$. It can be shown that for zero shear stiffness, $S = 0$, the sandwich is also identical to an *Euler-Bernoulli* beam, but with a bending stiffness $D = D_l$.

6.4.2 Beams with shear deformations only ($D_0 = \infty$ and $D_l = 0$)

We set the global bending stiffness of the sandwich beam equal to infinity, $D_0 = \infty$, and its local bending stiffness equal to zero, $D_l = 0$. The differential equation system of the sandwich with thick faces (Eqs (6-41), (6-42)) becomes:

$$-Sw_S'' = q - \left(N\beta\left(w_D' + w_S'\right)\right)', \tag{6-91}$$

$$0 = w_D''. \tag{6-92}$$

From the second equation we obtain that w_D is a linear function: $w_D = w_{D0} + \varphi_0 x$, where w_{D0} and φ_0 are yet unknown constants. Note that w_S also contains a constant function ($w_S = w_{S0} + \dots$). If we add a constant to w_D and subtract the same constant from w_S, the results will not change. Hence we may choose w_{D0} or w_{S0} arbitrarily, so let w_{D0} be zero:

$$w_D = \varphi_0 x. \tag{6-93}$$

Three boundary conditions are needed to determine the unknowns in the solution of the differential equation system Eqs (6-91), (6-93).

For a cantilever, the equations simplify, because $\varphi_0 = 0$, and we have:

$$w = w_S, \tag{6-94}$$

$$-Sw_S'' = q - \left(N\beta\left(w_S'\right)\right)'. \tag{6-95}$$

If there is only vertical load on the beam, Eq. (6-95) becomes:

$$(S - N\beta) w_S' = 0. \tag{6-96}$$

This equation could have been also derived from the condition that the shear force resultant of the load, $N\beta w'$, must be equilibrated by the internal shear force, $S\gamma$.

Equation (6-96) has two solutions. Either w_S' or $(S - N\beta)$ is equal to zero at any point of the beam. If $w_S \equiv 0$, we obtain the trivial solution. We are looking for the nontrivial solutions of the beam. The normal load belonging to this solution is called the critical load, N_{cr}.

First, a *cantilever beam* subjected to a concentrated force on the top ($\beta = 1$) is considered. In this case we obtain the nontrivial solution if

$$N = N_{cr} = S. \tag{6-97}$$

In this case the multiplier of w_S' in Eq. (6-96) is identical to zero, consequently w_S' can be arbitrary along the entire length of the beam. This means that the beam does not have a definite buckling shape, as illustrated in *Fig. 6-15a*.

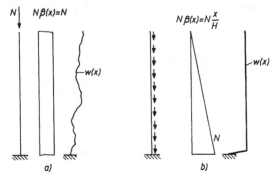

Fig. 6-15 Buckling of a cantilever beam with shear deformation only

If a distributed vertical load acts on the beam, (hence β is a function of x), w_S' must be equal to zero everywhere except for those points where $S - N\beta = 0$. If there is no point along the beam where the equation $S - N\beta = 0$ holds, only the trivial solution, $w_S \equiv 0$, exists. On the other hand, if there is at least one single point along the beam, where $N\beta$ is equal to S, then also a nontrivial solution exists, i.e. the beam buckles. The smallest critical load is:

$$N_{cr} = \min\left(\frac{S}{\beta}\right). \tag{6-98}$$

This expression is applicable even if S is a function of x. It is very important to notice that the critical load does not depend on the length of the beam. *The beam buckles if at any point the normal force reaches the value of the shear stiffness.*

A cantilever subjected to a *uniformly distributed load* has a linear normal force distribution, $\beta = x/H$. The maximum of the normal force is at the bottom, $x = H$. The critical load parameter is $N_{cr} = S$, and the buckling shape is a vertical line parallel to the axis of the beam, which has, at $x = H$, an infinitely short inclined part (*Fig. 6-15b*).

6.4.3 Sandwich beam with thin faces ($D_l = 0$)

If we set D_l equal to zero in Eqs (6-41), (6-42), we obtain the equations presented in Section 6.2 (Eqs (6-18), (6-19)) for a sandwich beam with thin faces, i.e. for a *Timoshenko-beam*. The buckling loads for the most important cases were presented in that section.

6.4.4 Beam on an elastic foundation which restrains the rotations – *Csonka-beam* ($D_0 = \infty$)

If the global bending stiffness of the sandwich beam is set equal to zero, $D_0 = \infty$, we obtain from Eqs (6-43), (6-44):

$$D_l w'''' - \left[S \left(w' - \chi \right) \right]' = q - \left(N\beta w' \right)', \tag{6-99}$$

$$0 = w_D''. \tag{6-100}$$

If at one cross section of the beam the rotation is zero, we obtain $\chi \equiv 0$, and Eq. (6-99) simplifies to

$$D_l w'''' - \left(S w' \right)' = q - \left(N\beta w' \right)'. \tag{6-101}$$

This is the differential equation of an (*Euler-Bernoulli*) beam, with bending stiffness D_l, the displacements of which are restrained by an attached beam which has shear deformation only and has shear stiffness S (*Fig. 6-16a*).

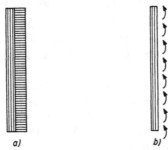

Fig. 6-16 Displacements of a beam restrained by a supporting beam (having shear deformations only) *a*), and the beam is supported by a foundation which restrains the rotations *b*)

The behavior of beams on classical elastic foundation is well known: the foundation provides supporting forces on the beam which are proportional to the displacements of the beam. The effect of the attached beam (with shear deformation only) is somewhat different, as it is discussed below. The shear force in the supporting beam (undergoing shear deformation only) is proportional to the first derivative of the displacement function:

$$V = S w'. \tag{6-102}$$

The load which results in the above shear force is as follows:

$$t = V' = \left(S w' \right)'. \tag{6-103}$$

These are the forces which will restrain the deformation of the *Euler-Bernoulli* beam, Eq. (6-101).

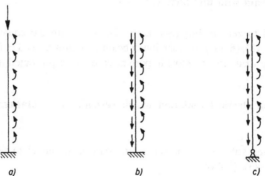

a)   b)   c)

Fig. 6-17 Buckling of cantilever beams on elastic foundation which restrains the rotation

The effect of the supporting beam with shear deformation only (on the buckling load, and on the displacements) is equivalent to the effect of an elastic foundation which restrains the rotations (*Fig. 6-16b*) [HEGEDŰS and KOLLÁR, 1988*b*]. This kind of foundation restrains the displacements of the beam by distributed moments which are proportional to the first derivative of the displacements:

$$m = kw',$$ (6-104)

where k is the spring constant of the foundation. Including this supporting moment in the moment equilibrium (Eq. 6-12), we obtain the following differential equation:

$$D_l w'''' - \left[kw' \right]' = q - \left(N\beta w' \right)'.$$ (6-105)

By comparing Eqs (6-101) and (6-105) we observe that they are identical if $k = S$. (Note that the shear forces calculated by the two models are different, however the shear forces do not cause deformation.)

This model is well known in the approximate analysis of frames and coupled shear walls, where the effect of the beams are considered as rotational springs which restrain the deformation of the columns. The model treated in this section will be referred to as *Csonka*-beam, because CSONKA introduced it in the approximate analysis of frames [CSONKA, 1961*a,b*, 1962, 1965*a,b*].

The critical load of the *Csonka*-beam can be calculated from the following theorem [HEGEDŰS and KOLLÁR, 1988*b*]: *The buckling load of the beam subjected to a normal force $N\beta$ and supported by a beam with shear stiffness S, is equal to the buckling load of an unsupported beam subjected to a normal force $N\beta - S$.*

We consider a (supported) cantilever subjected to a concentrated force on the top, N (*Fig. 6-17*). The critical load is equal to the sum of the critical load of the (unsupported) beam and the shear stiffness of the supporting beam:

$$N_{cr} = N_l + S = \frac{\pi^2 D_l}{4H^2} + S.$$ (6-106)

For a uniformly distributed load (*Fig. 6-17b*), $\beta = x/H$, ZALKA [1980, 1987] determined the solution numerically:

$$N_{cr} = \alpha \frac{7.84 D_l}{H^2},$$ (6-107)

where α is given for a cantilever in *Table 7-1*, and for a hinged cantilever – *Fig. 6-17c* – by ZALKA [in: KOLLÁR, 1991]. The α parameter can be determined in the function of the parameter $\beta = SH^2 / (7.84D_l)$.

6.4.5 Sandwich beam with thick faces on an elastic foundation which restrains the rotations

In the previous sections we investigated models which were obtained by simplifying the model of a sandwich beam with thick faces. In the next sections we consider a few cases which are obtained by generalizing the sandwich model: sandwiches on elastic foundation, multi-layered sandwiches and sandwich plates.

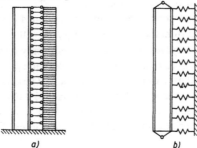

a)   *b)*

Fig. 6-18   Sandwich beam on an elastic foundation which restrains the rotation *a*), and the displacements *b*)

First, we consider a sandwich beam which is supported by a beam undergoing shear deformation only, with shear stiffness S_a (*Fig. 6-18a*). (The effect of this support is equivalent to the effect of a foundation which restrains the rotations [HEGEDŰS and KOLLÁR, 1988b].)

The theorem of the previous section on the calculation of the critical load of beams supported by beams (with shear deformation only) is applicable for sandwich beams with thick faces as well. Consequently, the critical load of a cantilever beam subjected to a concentrated load on the top is as follows:

$$N_{\text{cr}} = \left(N_0^{-1} + S^{-1}\right)^{-1} + N_l + S_a, \tag{6-108}$$

where N_0 and N_l are given by Eq. (6-64).

For hinged supports Eq. (6-108) is also applicable, but N_0 and N_l must be calculated from Eq. (6-60).

6.4.6 Sandwich beam with thick faces on an elastic foundation which restrains the displacements

A sandwich beam with thick faces is considered supported by an elastic foundation which restrains the horizontal displacements of the beam (*Fig. 6-18b*). The spring constant of the foundation is denoted by ρ, hence the supporting force is equal to ρw, where w is the horizontal displacement of the beam. The differential equations of the sandwich beam

(6-43, 6-44) must be modified according to these forces:

$$D_l w'''' + D_0 \chi''' + \rho w = q - \left(N \beta w' \right)', \qquad (6\text{-}109)$$

$$S \left(w' - \chi \right) = -D_0 \chi''. \qquad (6\text{-}110)$$

For a simply supported sandwich beam subjected to a concentrated force at the top (*Fig. 6-6a*) the buckling shape is sinusoidal, $w = w_0 \sin \pi x / l_k$, where $l_k = H/k$ ($k = 1, 2, \ldots$). The pertaining buckling load is:

$$N_{cr} = \left(N_0^{-1} + S^{-1} \right)^{-1} + N_l + N_\rho, \qquad (6\text{-}111)$$

where

$$N_0 = \frac{\pi^2 D_0}{l_k^2}, \qquad N_l = \frac{\pi^2 D_l}{l_k^2}, \qquad N_\rho = \rho \frac{l_k^2}{\pi^2}. \qquad (6\text{-}112)$$

It must be determined, which buckling shape ($k = 1, 2, \ldots$) gives the smallest critical load. In the case of long sandwich beams, l_k may be assumed continuously varying, in this way N_{cr} can be determined from the condition $dN_{cr}/dl_k = 0$.

Thin *isotropic sandwich cylinders* subjected to uniformly distributed axial load buckle in an axi-symmetrical pattern. Hence, the partial differential equations of their linear critical load can be reduced to ordinary differential equations. These equations consist of terms which are perfect analogues of those in Eqs (6-109) and (6-110). Also the formulas for the linear critical load are the same as Eq. (6-111) and (6-112), provided the spring constant is the ratio of the tensile stiffness to the square of the radius of the cylinder [HEGEDŰS, 1979].

6.4.7 Isotropic sandwich plate

A sandwich plate (*Fig. 6-19*) is considered where the face sheets are isotropic and the core is isotropic in the plane of the plate. This means that the faces can be characterized by Young's modulus, E_f, and *Poisson*'s ratio, ν_f, while the core by the shear modulus G_c. (For a core, compressible in the transverse direction, E_c must be specified as well.)

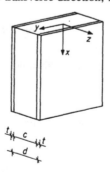

Fig. 6-19 Sandwich plate

First, we consider a *long plate* subjected to unidirectional compression in the shorter direction. The buckling shape of the plate will be *cylindrical*. We consider a strip of unit width, $b = 1$, of the plate. The buckling load of this strip is equal to that of a beam with unit width, and Eqs (6-59) or (6-63) or (6-66) can be used if we replace D_0, D_l, S by $\tilde{D}_0$, $\tilde{D}_l$, $\tilde{S}$ defined as follows:

$$\tilde{D}_0 = \frac{E_f t d^2}{2\left(1 - v_f^2\right)}, \qquad \tilde{D}_l = \frac{2 E_f t^3}{12\left(1 - v_f^2\right)}, \qquad \tilde{S} = G_c \frac{d^2}{c}. \qquad (6\text{-}113)$$

Note that the only difference is the multiplier $\left(1 - v_f^2\right)$ in the denominator.

Second, we consider a *rectangular plate simply supported along all four edges* and

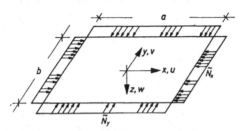

Fig. 6-20 Buckling of a simply supported sandwich plate

subjected to uniformly distributed compressive loads, $\widetilde{N}_x$ and $\widetilde{N}_y$ in the x and y directions, respectively (*Fig. 6-20*). We assume that the core is incompressible in the transverse direction. The buckling shape of the plate is [ALLEN, 1969]:

$$w(x,y) = w_0 \sin\left(\frac{\pi}{l_x}\left(x + \frac{a}{2}\right)\right) \sin\left(\frac{\pi}{l_y}\left(y + \frac{b}{2}\right)\right), \qquad (6\text{-}114)$$

where $l_x = a/k_a$ and $l_y = b/k_b$ are the buckling lengths in the x and y directions, respectively, and k_a and k_b are the numbers of half waves in the x and y directions, respectively ($k_a = 1, 2, 3, \ldots, k_b = 1, 2, 3, \ldots$).

The buckling load can be calculated from the following equation:

$$\lambda_{\text{cr}} = \left(\frac{1}{l_x^2} + \frac{1}{l_y^2}\right)\left[\left(\widetilde{N}_0^{-1} + \tilde{S}^{-1}\right)^{-1} + \widetilde{N}_l\right], \qquad (6\text{-}115)$$

where

$$\lambda_{\text{cr}} = \frac{\widetilde{N}_x}{l_x^2} + \frac{\widetilde{N}_y}{l_y^2}, \qquad (6\text{-}116)$$

$$\widetilde{N}_0 = \pi^2 \tilde{D}_0 \left(\frac{1}{l_x^2} + \frac{1}{l_y^2}\right), \qquad \widetilde{N}_l = \pi^2 \tilde{D}_l \left(\frac{1}{l_x^2} + \frac{1}{l_y^2}\right), \qquad (6\text{-}117)$$

and $\tilde{D}_0$, $\tilde{D}_l$, and $\tilde{S}$ are given by Eq. (6-113). The buckling load depends on the buckling lengths, l_x, l_y. We have to substitute various values of the buckling lengths (or the number of the half buckling waves) to find the smallest buckling load.

Eq. (6-115) results in a combination of various values of the normal forces, $\widetilde{N}_x$, $\widetilde{N}_y$. In practical problems usually the ratio of the normal forces is given, i.e.:

$$\widetilde{N}_x = \mu \widetilde{N}_{x0}, \qquad \widetilde{N}_y = \mu \widetilde{N}_{y0}, \qquad (6\text{-}118)$$

where $\widetilde{N}_{x0}, \widetilde{N}_{y0}$ are prescribed values, and μ is the load parameter the critical value of which can be calculated from Eq. (6-116) as follows:

$$\mu_{cr} = \lambda_{cr} \left(\frac{\widetilde{N}_{x0}}{l_x^2} + \frac{\widetilde{N}_{y0}}{l_y^2} \right)^{-1}. \tag{6-119}$$

As an example we present the results for a rectangular simply supported sandwich plate (*Fig. 6-20*) with thin faces ($\widetilde{D}_l = 0$) subjected to a uniformly distributed compressive load in the x direction ($\widetilde{N}_{y0} = 0$). The buckling occurs with a single half wave in the y direction ($k_b = 1$, $l_y = b$) and the buckling load from Eqs (6-115) through (6-119) is as follows:

$$\widetilde{N}_{x\,cr} = \mu_{cr}\widetilde{N}_{x0} = \left(\widetilde{N}_0^{-1} + \widetilde{S}^{-1} \right)^{-1} \left[1 + \left(\frac{l_x}{l_y} \right)^2 \right] = \frac{\pi^2 \widetilde{D}_0}{b^2} K, \tag{6-120}$$

where the parameter K is given in *Fig. 6-21* [ALLEN, 1969].

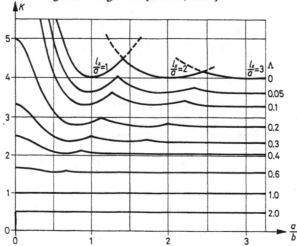

Fig. 6-21 Parameter K of Eq. (6-120) to calculate the buckling load of a sandwich plate;
$\Lambda = \frac{\pi^2}{2(1-\nu_f^2)} \frac{E_f}{G_c} \frac{td}{b^2}$

HEGEDŰS and KOLLÁR [1988a] generalized these results for *sandwich plates with compressible core*. Similarly to the buckling of sandwich beams we obtain symmetrical and antisymmetrical buckling shapes. The corresponding buckling loads are

$$\frac{1}{2}\lambda_{cr}^S = \left(\frac{1}{l_x^2} + \frac{1}{l_y^2} \right) \left[\frac{1}{2}\widetilde{N}_l + 2\widetilde{R} \right], \tag{6-121}$$

$$\lambda_{cr}^a = \left(\frac{1}{l_x^2} + \frac{1}{l_y^2} \right) \left[\left(\widetilde{N}_0^{-1} + \widetilde{S}^{-1} + \left(12\widetilde{R} \right)^{-1} \right)^{-1} + \widetilde{N}_l \right], \tag{6-122}$$

where λ_{cr}, $\tilde{N}_0$, and $\tilde{N}_l$ are given by Eqs (6-116), (6-117), and

$$\tilde{R} = \frac{E_c}{c\pi^2 \left(\dfrac{1}{l_x^2} + \dfrac{1}{l_y^2}\right)}. \tag{6-123}$$

6.4.8 Orthotropic sandwich plate

The face sheets of sandwich plates are often made of fibre reinforced plastics or corrugated plates, consequently these plates do not show isotropic behavior. Results are presented here only for orthotropic sandwich plates. The axes of orthotropy coincide with the co-ordinate axes.

Orthotropic sandwich plates show different stiffness characteristics in the x and y directions. The material law for the global bending is as follows:

$$\begin{Bmatrix} \widetilde{M}_{0,x} \\ \widetilde{M}_{0,y} \\ \widetilde{M}_{0,xy} \end{Bmatrix} = \begin{bmatrix} \tilde{D}_{0,11} & \tilde{D}_{0,12} & \\ \tilde{D}_{0,12} & \tilde{D}_{0,22} & \\ & & \tilde{D}_{0,66} \end{bmatrix} \begin{Bmatrix} -\dfrac{\partial \chi_x}{\partial x} \\ -\dfrac{\partial \chi_y}{\partial y} \\ -\left(\dfrac{\partial \chi_x}{\partial y} + \dfrac{\partial \chi_y}{\partial x}\right) \end{Bmatrix}, \tag{6-124}$$

where the bending moments per unit length, (similarly to M_0) comes from the normal

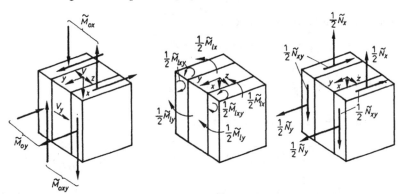

Fig. 6-22 Forces acting on an element of sandwich plates or shells

forces arising in the faces (*Fig. 6-22*), and χ_x, χ_y are the rotations of the cross section perpendicular to the x and y axis, respectively. The material law for the local bending is:

$$\begin{Bmatrix} \widetilde{M}_{l,x} \\ \widetilde{M}_{l,y} \\ \widetilde{M}_{l,xy} \end{Bmatrix} = \begin{bmatrix} \tilde{D}_{l,11} & \tilde{D}_{l,12} & \\ \tilde{D}_{l,12} & \tilde{D}_{l,22} & \\ & & \tilde{D}_{l,66} \end{bmatrix} \begin{Bmatrix} -\dfrac{\partial^2 w}{\partial x^2} \\ -\dfrac{\partial^2 w}{\partial y^2} \\ -2\dfrac{\partial^2 w}{\partial x \partial y} \end{Bmatrix}. \tag{6-125}$$

where the bending moments per unit length (similarly to M_l) arise in the faces (*Fig. 6-22*), and w is the displacement of the plate perpendicular to its plane. The elements of the vector on the right-hand side of the equation are the curvatures of the plate. The material law for the shear deformation is:

$$\left\{ \begin{array}{c} \widetilde{V}_x \\ \widetilde{V}_y \end{array} \right\} = \left[\begin{array}{cc} \widetilde{S}_{11} & \\ & \widetilde{S}_{22} \end{array} \right] \left\{ \begin{array}{c} \dfrac{\partial w}{\partial x} - \chi_x \\ \dfrac{\partial w}{\partial y} - \chi_y \end{array} \right\}. \tag{6-126}$$

where $\widetilde{V}_x$ and $\widetilde{V}_y$ are the shear forces on the faces normal to x and y, respectively (*Fig. 6-22*). The elements of the vector on the right-hand side of the equation are the shear strains, γ_x, γ_y in the x, z and y, z planes, respectively.

The zero elements in the above stiffness matrices show that the axes of orthotropy of the plate coincide with the x and y co-ordinate axes.

We consider a rectangular plate simply supported along all four edges and subjected to uniformly distributed compressive loads, $\widetilde{N}_x$ and $\widetilde{N}_y$ in the x and y directions respectively (*Fig. 6-20*). The critical load parameter (Eqs (6-116), (6-118), (6-119) can be calculated from the following equation [KOLLÁR, 1990]:

$$\lambda_{cr} = \dfrac{\det \begin{bmatrix} F_{33} & F_{34} & F_{35} \\ F_{34} & F_{44} & F_{45} \\ F_{35} & F_{45} & F_{55} \end{bmatrix}}{\det \begin{bmatrix} F_{44} & F_{45} \\ F_{45} & F_{55} \end{bmatrix}}. \tag{6-127}$$

The elements of the matrices are as follows:

$$F_{33} = \widetilde{D}_{11} \dfrac{\pi^4}{l_x^4} + 2 \left(\widetilde{D}_{12} + 2\widetilde{D}_{66} \right) \dfrac{\pi^4}{l_x^2 l_y^2} + \widetilde{D}_{22} \dfrac{\pi^4}{l_y^4}, \tag{6-128}$$

$$F_{34} = -\widetilde{D}_{0,11} \dfrac{\pi^3}{l_x^3} - \left(\widetilde{D}_{0,12} + 2\widetilde{D}_{0,66} \right) \dfrac{\pi^3}{l_x l_y^2}, \tag{6-129}$$

$$F_{35} = -\widetilde{D}_{0,22} \dfrac{\pi^3}{l_y^3} - \left(\widetilde{D}_{0,12} + 2\widetilde{D}_{0,66} \right) \dfrac{\pi^3}{l_x^2 l_y}, \tag{6-130}$$

$$F_{44} = \widetilde{D}_{0,11} \dfrac{\pi^2}{l_x^2} + \widetilde{D}_{0,66} \dfrac{\pi^2}{l_y^2} + \widetilde{S}_{11}, \qquad F_{45} = \left(\widetilde{D}_{0,12} + \widetilde{D}_{0,66} \right) \dfrac{\pi^2}{l_x l_y}, \tag{6-131}$$

$$F_{55} = \widetilde{D}_{0,22} \dfrac{\pi^2}{l_y^2} + \widetilde{D}_{0,66} \dfrac{\pi^2}{l_x^2} + \widetilde{S}_{22}, \tag{6-132}$$

where

$$\widetilde{D}_{11} = \widetilde{D}_{0,11} + \widetilde{D}_{l,11}, \qquad \widetilde{D}_{22} = \widetilde{D}_{0,22} + \widetilde{D}_{l,22}, \tag{6-133}$$

$$\widetilde{D}_{12} = \widetilde{D}_{0,12} + \widetilde{D}_{l,12}, \qquad \widetilde{D}_{66} = \widetilde{D}_{0,66} + \widetilde{D}_{l,66}. \tag{6-134}$$

Note that the buckling load Eq. (6-127) depends on the buckling lengths l_x, l_y. We have to calculate various values of the buckling lengths (or of the number of the half buckling waves) to find the smallest buckling load.

If the shear deformation is negligible ($\tilde{S}_{11}$, $\tilde{S}_{22} \to \infty$), Eq. (6-127) simplifies to the buckling of a classical orthotropic plate:

$$\lambda_{cr} = \tilde{D}_{11}\frac{\pi^4}{l_x^4} + \tilde{D}_{22}\frac{\pi^4}{l_y^4} + 2\left(\tilde{D}_{12} + 2\tilde{D}_{66}\right)\frac{\pi^4}{l_x^2 l_y^2}. \tag{6-135}$$

If the sandwich plate is isotropic, we have

$$\tilde{D}_{0,11} = \tilde{D}_{0,22} = \tilde{D}_0, \qquad \tilde{D}_{0,12} = v_f\tilde{D}_0, \qquad \tilde{D}_{0,66} = \left(1 - v_f\right)\tilde{D}_0\big/2,$$

$$\tilde{D}_{l,11} = \tilde{D}_{l,22} = \tilde{D}_l \qquad \tilde{D}_{l,12} = v_f\tilde{D}_{l0}, \qquad \tilde{D}_{l,66} = \left(1 - v_f\right)\tilde{D}_l\big/2,$$

$$\tilde{S}_{11} = \tilde{S}_{22} = \tilde{S},$$

and Eq. (6-127) simplifies to Eq. (6-115).

6.4.9 Orthotropic shallow sandwich shell

If the midplane of a sandwich panel is curved, the tensile stiffnesses also play a role in the buckling analysis. An orthotropic shell is considered here, the principal curvatures of which lie in the x and y directions. The radii of curvatures in the x,z, and y,z planes are denoted by R_x and R_y. (Assuming a shallow shell, the curvatures are calculated as: $1/R_x = -\partial^2 z/\partial x^2$, $1/R_y = -\partial^2 z/\partial y^2$, where z is the function of the middle surface.) The shell is orthotropic with the axes of orthotropy x and y. The bending and shear stiffnesses were introduced in Eqs (6-124) through (6-126), and the tensile stiffnesses are defined by the following equation:

$$\left\{\begin{array}{c}\widetilde{N_x} \\ \widetilde{N_y} \\ \widetilde{N_{xy}}\end{array}\right\} = \left[\begin{array}{ccc}\tilde{A}_{11} & \tilde{A}_{12} & \\ \tilde{A}_{12} & \tilde{A}_{22} & \\ & & \tilde{A}_{66}\end{array}\right]\left\{\begin{array}{c}\epsilon_x \\ \epsilon_x \\ \gamma_{xy}\end{array}\right\}, \tag{6-136}$$

where on the left-hand side there are the force resultants (i.e. membrane forces, *Fig. 6-22*) and on the right-hand side the normal and shear strains of the middle surface.

We consider a shell subjected to loads which result in membrane forces $\widetilde{N}_x$ and $\widetilde{N}_y$ in the x and y directions respectively. Using the shallow shell approximations, closed form solution was derived for the calculation of the critical load parameter Eqs (6-116), (6-118), (6-119) [KOLLÁR, 1990]:

$$\lambda_{cr} = \frac{\det\left[\begin{array}{ccccc}F_{11} & F_{12} & F_{13} & & \\ F_{12} & F_{22} & F_{23} & & \\ F_{13} & F_{23} & \overline{F}_{33} & F_{34} & F_{35} \\ & & F_{34} & F_{44} & F_{45} \\ & & F_{35} & F_{45} & F_{55}\end{array}\right]}{\det\left[\begin{array}{cc}F_{11} & F_{12} \\ F_{12} & F_{22}\end{array}\right]\det\left[\begin{array}{cc}F_{44} & F_{45} \\ F_{45} & F_{55}\end{array}\right]}, \tag{6-137}$$

where the elements of the matrices are given by Eqs (6-128) through (6-132) and by the following equations:

$$F_{11} = \tilde{A}_{11}\frac{\pi^2}{l_x^2} + \tilde{A}_{66}\frac{\pi^2}{l_y^2}, \qquad F_{22} = \tilde{A}_{22}\frac{\pi^2}{l_y^2} + \tilde{A}_{66}\frac{\pi^2}{l_x^2}, \tag{6-138}$$

$$\overline{F}_{33} = F_{33} + \frac{\tilde{A}_{11}}{R_x^2} + 2\frac{\tilde{A}_{12}}{R_x R_y} + \frac{\tilde{A}_{22}}{R_y^2}, \qquad F_{12} = \left(\tilde{A}_{12} + \tilde{A}_{66}\right)\frac{\pi^2}{l_x l_y}, \qquad (6\text{-}139)$$

$$F_{13} = -\tilde{A}_{11}\frac{\pi}{l_x}\frac{1}{R_x} - \tilde{A}_{12}\frac{\pi}{l_x}\frac{1}{R_y}, \quad F_{23} = -\tilde{A}_{22}\frac{\pi}{l_y}\frac{1}{R_y} - \tilde{A}_{12}\frac{\pi}{l_y}\frac{1}{R_x}. \qquad (6\text{-}140)$$

We have to emphasize again that the buckling load depends on the buckling lengths, l_x, l_y. We have to substitute various values of the buckling lengths to find the smallest buckling load.

Readers interested in the buckling of generally anisotropic shells can find analytical results in [KOLLÁR, 1990, 1994]

6.4.10 Multi-layered sandwich cantilever beam

A sandwich cantilever is considered which consists of $n + 1$ thin faces and n thick core layers (*Fig. 6-23*). The bending and tensile stiffnesses of the thin layers are denoted by EI and EA, respectively, while the shear stiffnesses of the antiplane cores are denoted by S_1. The cores are considered transversely incompressible.

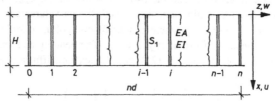

Fig. 6-23 Multi-layered sandwich cantilever beam

The thin layers are subjected to equal concentrated compressive loads on the top, the resultant of which is denoted by N. The derivation of the critical load, N_{cr} was given by [KOLLÁR, 1986a]; here only the results are presented. The critical load is:

$$N_{cr} = N_l + D\sum_{i=1}^{n}\frac{(n+1-i)i}{2}\frac{\sinh(n+1-i)\vartheta + \sinh i\vartheta}{\sinh(n+1)\vartheta}, \qquad (6\text{-}141)$$

where ϑ is defined by the following equation:

$$2\cosh\vartheta = 2 + \frac{D}{S_1}. \qquad (6\text{-}142)$$

and

$$N_l = \frac{(n+1)EI\pi^2}{4H^2}, \qquad S = nS_1, \qquad D = \frac{EAd^2\pi^2}{4H^2}, \qquad (6\text{-}143)$$

By setting $n = 1$, we obtain a three-layered sandwich beam, and Eq. (6-141) becomes identical to Eq. (6-63).

6.5 APPROXIMATE EXPRESSIONS FOR THE CALCULATION OF THE BUCKLING LOAD

In Section 6.2, 6.3, and 6.4 the calculation of the critical load of different models were presented. For a concentrated force on the top, simple formulas were derived, however for distributed loads, in most of the cases, closed form solutions are not available. For these cases we can use the approximate expressions, based for example on the theorems presented in Chapter 2. An approximate expression is applicable in the practice if it is accurate enough or if it gives a conservative estimate, i.e. it provides a lower bound for the critical load.

6.5.1 Parallel and serial connections of beams (Föppl-Papkovich's and Southwell's theorem)

In this section we present a visual treatment of the presented models which enables us to derive simple approximate formulas for the buckling load.

In the following, beams undergoing bending deformation only (Section 6.4.1) will be referred to as B-beams (*Fig. 6-24a*). A B-beam, with bending stiffness D_k, will be denoted by R_{Dk}. Similarly, beams undergoing shear deformation only (Section 6.4.2) will be referred to as S-beams (*Fig. 6-24b*). An S-beam, with shear stiffness S_k, will be denoted by R_{Sk}. The B- and S-beams are referred to as beams with pure deformations.

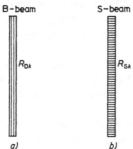

Fig. 6-24 Beam having bending deformations only *a*), and shear deformations only *b*)

We define the parallel and serial connection of beams. The connected structure will be referred to as resultant beam.

Two (vertical) beams, denoted by R_1 and R_2, are connected *parallel*, if the horizontal displacements of the individual beams (w_1, w_2) are identical to each other at every cross section:

$$w = w_1 = w_2. \tag{6-144}$$

We denote the resultant beam (obtained by connecting the two beams) by R, and the parallel connection by the symbol $\|$:

$$R = R_1 \parallel R_2. \tag{6-145}$$

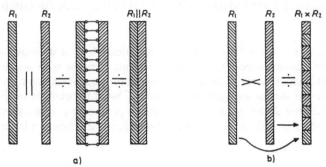

Fig. 6-25 Parallel *a*), and serial *b*) connections of beams

The parallel connection can be visualized by assuming that incompressible hinged bars connect the cross sections of the two beams (*Fig. 6-25a*).

Connection of two (vertical) beams, denoted by R_1 and R_2, is *serial*, if the displacements (w) of the resultant beam are equal to the sum of the displacements (w_1, w_2) of the individual beams at every cross section:

$$w = w_1 + w_2. \tag{6-146}$$

We denote the serial connection by the symbol $\times$:

$$R = R_1 \times R_2. \tag{6-147}$$

The serial connection can be visualized by assuming that the displacements of the resultant bar occur in two steps: first only R_1 deforms, then only R_2. (Note that this approach can be difficult to apply for cases where the boundary conditions contain the sum of the displacements.)

Another visualization of the serial connection is the following. The stiffnesses of the beams R_1 and R_2 are reduced by half, then we cut both beams by vertical planes into short sections. The $R = R_1 \times R_2$ bar is obtained by connecting these sections alternately as illustrated in *Fig. 6-25b*.

Connecting two B-beams we obtain a B-beam, while connecting two S-beams results in an S-beam. The stiffness of the resultant bar, in the case of parallel connection, is equal to the sum of the stiffnesses of the components, while in the case of serial connection the inverse of the stiffness of the resultant bar is equal to the sum of the inverses of the stiffnesses of the components. For example for

$$R_{S3} = R_{S1} \parallel R_{S2}, \qquad S_3 = S_1 + S_2, \tag{6-148}$$

and for

$$R_{D3} = R_{D1} \times R_{D2}, \qquad D_3 = \left(D_1^{-1} + D_2^{-1} \right)^{-1}. \tag{6-149}$$

Using the connection of beams with pure deformations we can obtain the beam models presented in the previous sections (*Fig. 6-26*).

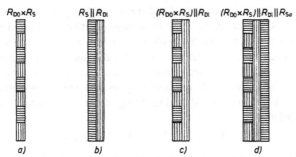

Fig. 6-26 Models of *a*) a Timoshenko-beam, *b*) a Csonka-beam, *c*) a sandwich beam with thick faces, and *d*) a sandwich on elastic foundation

We obtain a *Timoshenko-beam* (sandwich beam with thin faces) by the serial connection of an S- and B-beam:

$$R_{D0} \times R_S, \tag{6-150}$$

we obtain a *Csonka-beam* by the parallel connection of a B- and an S-beam:

$$R_{Dl} \parallel R_S, \tag{6-151}$$

we obtain a *sandwich beam with thick faces* by the parallel connection of a B-beam and a Timoshenko-beam:

$$(R_{D0} \times R_S) \parallel R_{Dl}, \tag{6-152}$$

and we obtain a *sandwich beam with thick faces on an elastic foundation which restrains the rotations* by the parallel connection of an S-beam and a sandwich with thick faces:

$$(R_{D0} \times R_S) \parallel R_{Dl} \parallel R_{Sa}. \tag{6-153}$$

We introduced the concept of the serial and parallel connections because it gives an advantageous tool to estimate the critical load of the resultant beams.

Southwell's theorem
The critical loads of beams R_1 and R_2 are N_1 and N_2, respectively. The critical load of the resultant beam R, obtained by parallel connection, $R = R_1 \parallel R_2$, can be estimated as:

$$\overline{N}_{cr} = N_1 + N_2. \tag{6-154}$$

Föppl-Papkovich's theorem
The critical loads of beams R_1 and R_2 are N_1 and N_2, respectively. The critical load of the resultant beam R, obtained by serial connection, $R = R_1 \times R_2$, can be estimated as:

$$\overline{N}_{cr} = \left(N_1^{-1} + N_2^{-1}\right)^{-1}. \tag{6-155}$$

The above approximations are in most of the cases conservative, i.e. they give a lower bound for the critical load.

For the most important practical case, i.e. for a *cantilever* beam, the above approximations are always conservative, and, as it was shown, for concentrated forces on the top, they result in the accurate buckling load, $\overline{N}_{cr} = N_{cr}$.

It was shown by HEGEDŰS and KOLLÁR [1988c] that the approximation is not conservative only if there is an uncertainty in the boundary conditions of the component beams. This will be illustrated with an example, with the so-called *Plantema* paradox.

In the following we present three examples for the application of *Southwell*'s and *Föppl-Papkovich*'s theorem. First, we consider a *sandwich beam with thick faces on an elastic foundation which restrains the rotations* (*Fig. 6-26d*) fixed at the bottom and free at the top subjected to a uniformly distributed load (*Fig. 6-6e*). The exact critical load can be approximated as follows:

$$N_{cr} \geq \overline{N}_{cr} = \left(N_0^{-1} + S^{-1}\right)^{-1} + N_l + S_a, \tag{6-156}$$

where

$$N_0 = 7.837 \frac{D_0}{H^2}, \qquad N_l = 7.837 \frac{D_l}{H^2}. \tag{6-157}$$

Second, a *Timoshenko-beam* (beam with both shear and bending deformations) is considered built-in at the bottom and free at the top, subjected to a distributed load resulting in a normal force $N(x/H)^n$. The exact critical load was calculated in Section 6.2 and presented by solid lines in *Fig. 6-7*. (The solution for uniformly distributed load belongs to the parameter $n = 1$.) The approximate solution is obtained by *Föppl-Papkovich*'s theorem:

$$\overline{N}_{cr} = \left[\left(m\frac{D_0}{H^2}\right)^{-1} + S^{-1}\right]^{-1}, \tag{6-158}$$

where the term in the parenthesis is the buckling load of cantilever undergoing bending deformation only, and m is given by TIMOSHENKO and GERE [1961] as follows:

n	1	2	3	4	5	6
m	7.837	16.10	27.26	41.30	58.24	78.07

The results obtained from Eq. (6-158) are presented in *Fig. 6-7* by dashed lines. Eq. (6-158) is always a conservative estimates.

The last example is the *Plantema paradox* [PLANTEMA, 1961; , see also Chapter 2]. We consider a *Timoshenko-beam* built-in at the bottom and hinged at the top, subjected to a concentrated load on the top (*Fig. 6-6c*). It was found that the following approximation based on *Föppl-Papkovich*'s theorem is *not a conservative estimate:*

$$N_{cr} < \overline{N}_{cr} = \left[\left(0.699\frac{D_0}{H^2}\right)^{-1} + S^{-1}\right]^{-1}, \tag{6-159}$$

where the first term in the parenthesis is the buckling load of a B-beam built-in at the bottom and hinged at the top, while S is the buckling load of an S-beam. (The exact critical load, N_{cr}, was given in Section 6.2). To solve this puzzle let us consider the displacements of the component beams, i.e. the bending displacements (w_D) and the shear displacements (w_S) of the Timoshenko-beam. The displacements are presented in *Fig. 6-27*. It could be seen that at the top, where the total displacement ($w_D + w_S$) of the beam, is zero, the component beams has nonzero displacements. Consequently, if zero displacement is taken into account at the top of the component beams, then the displacements are more restrained than in reality.

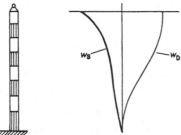

Fig. 6-27 Components of the displacements of a Timoshenko-beam built-in at the bottom and hinged at the top (Plantema paradox)

A conservative estimate can be obtained by assuming no constraint on the top, which yields:

$$\overline{N}_{cr} = \left[\left(D_0/H^2 \right)^{-1} + S^{-1} \right]^{-1}.$$

6.5.2 Cantilever beams on elastic foundation which restrains the rotation

The critical load of the *cantilever beam* $R = R_1 \parallel R_{Sa}$, where R_{Sa} is a beam with shear deformation only, can be estimated by *Southwell*'s theorem

$$N_{cr} \geq \overline{N}_{cr} = N_1 + S_a, \tag{6-160}$$

where $\overline{N}_{cr}$, N_1, S_a are the critical loads of the resultant beam R and the component beams R_1, R_{Sa}, respectively. A better approximation can be obtained by using the theorem presented in Section 6.4.4 The approximate critical load is denoted by $\overline{N}_{cr}$. As it was stated, the buckling load of the beam subjected to a normal force $\overline{N}_{cr}\beta$ and supported by a beam with shear stiffness S_a, is equal to the buckling load of an *unsupported beam* subjected to a normal force $\overline{N}_{cr}\beta - S_a$. It is a conservative estimate if the top sections of the cantilever, where $\overline{N}_{cr}\beta - S_a$ is negative (i.e. tension), is neglected, and a cantilever beam with a reduced length is considered (*Fig. 6-28*). The reduced length is denoted by H'.

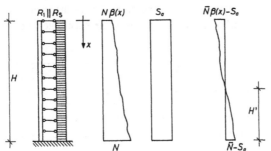

Fig. 6-28 Buckling of a cantilever beam supported by a beam with shear deformation only

For example, if a *Csonka*-beam (Section 6.4.4) subjected to a uniformly distributed load is considered, we obtain

$$\overline{N}_{cr} - S_a = 7.837 \frac{D_l}{(H')^2},\tag{6-161}$$

where H' is calculated from similar triangles, $H' = H\left(\overline{N}_{cr} - S_a\right)/\overline{N}_{cr}$. By introducing this expression into Eq. (6-161), we obtain the following equation for $\overline{N}_{cr}$:

$$\frac{S_a H^2}{7.837 D_l} = \frac{\overline{N}_{cr} H^2}{7.837 D_l} - \left(\frac{\overline{N}_{cr} H^2}{7.837 D_l}\right)^{\frac{2}{3}}.\tag{6-162}$$

(The error in *Southwell*'s expression (6-160) can be more than 60%, while that of Eq. (6-162) is less than 25% [ZALKA, in KOLLÁR, 1991].)

6.5.3 Multi-layered sandwich beam

In Section 6.4.10 the exact buckling load of a multi-layered sandwich cantilever was presented. The axial displacements of the axes of the thin layers are hyperbolically proportional to the distance from the axis of the beam [KOLLÁR, 1986a].

A very simple approximation can be obtained if we assume that the axial displacements of the axes of the thin layers are linearly proportional to the distance from the axis of the beam [KOLLÁR, 1986a]. The approximate critical load is:

$$\overline{N}_{cr} = \left(N_0^{-1} + S^{-1}\right)^{-1} + N_l,\tag{6-163}$$

where

$$N_0 = \frac{\pi^2 D_0}{4H^2}, \qquad D_0 = EAd^2 \frac{n\,(n+1)\,(n+2)}{12},\tag{6-164}$$

and S and N_l are given by Eq. (6-143). Eq. (6-163) is the buckling load of a *sandwich beam with thick faces*, where the local bending stiffness (D_l) is the sum of the bending stiffnesses of the thin layers, the shear stiffness (S) is the sum of the shear stiffnesses of the cores, while the global bending stiffness (D_0) is obtained from the second moment of areas (times E) of the thin layers, i.e. from the parallel axes theorem.

The approximated critical load was compared to the exact one, and the errors are presented below.

n	1	2	3	4	5	10	20	30
$\left(\overline{N}_{cr}/N_{cr} - 1\right)100\,[\%]$	0.00	0.00	0.99	2.04	2.99	6.10	8.66	9.74

In this table the maximum error (as the function of various stiffnesses and the geometry) are presented. Note that the approximation is not conservative, however the results are reasonably accurate.

6.6 SOME APPLICATIONS OF THE SANDWICH THEORY IN STRUCTURAL ENGINEERING

In this section, structures of regular built-up are dealt with, which can be analysed using the methods developed for the analysis of sandwich beams.

6.6.1 Discrete structures with regular built-up

Regular built-up means that the structure can be considered as a chain-like or net-like linkage of uniform structural elements. If the number of the elements is sufficiently large, the stair-like diagrams which characterize the states of stress and strain are close to analogous smooth diagrams of continuous structures. This fact suggests the idea of replacement continua: replacement bars in the case of chain-like linkages and a replacement plates in the case of net-like ones. In this section only replacement bars are dealt with.

Replacement continua can be efficient tools in the analysis of discrete structures with regular built-up. Adequacy of a replacement continuum means that in analogous conditions, the important characteristics of the discrete structure and the continuum can be regarded approximately identical. From a theoretical point of view, we may also require that if the number of structural elements tends to infinity, the behaviour of the discrete structure approximates that of the continuum beyond any limit.

The exact analysis of discrete structures needs a discrete modelling. In the case of regular structures exact analytical solutions of the discrete models can be obtained by using the difference calculus.

6.6.2 Exact analysis of discrete structures using the theory of difference equations

The theory of linear difference equations is a well-explored classical discipline of mathematics, [JORDAN, 1947; BRAND, 1966]; however, its methods and results are not well known among engineers. The theory of ordinary difference equations with *constant coefficients* has been most explored. Monographic treatments of their technical applications including stability analysis of trusses can be found in [BLEICH and MELAN, 1927] and [RATZERS-DORFER, 1936]. In [KÁRMÁN and BIOT, 1940] also the buckling of Vierendeel-columns (or multi-storey single-bay frames) is also presented. KOLLÁR and HEGEDŰS [1985] used partial difference equations in the analysis of space grids and also present a general method to reduce the solution of a system of linear difference equations to that of a single difference equation.

In the following, some fundamentals of the difference calculus are presented.

Functions which are interpreted at a series of discrete values of the independent variable are called functions of discrete variable. Since the set of values of interpretation can always be mapped to the set of integer values, a function of discrete variable can always be interpreted as a function of the serial number i:

$$y_i \equiv y(i). \tag{6-165}$$

In most practical problems, the values of interpretation of $y(i)$ form an algebraic series with the difference h. Functions of this kind are called functions of integer variable. They can be interpreted as the series of values of a common function $y(x/h)$ at values of x/h which are elements of an algebraic series with the difference 1. For practical reasons, also fractional values of x/h, that is, fractional values of the discrete variable i with unit difference, are often used.

Let the elementary shifting operator E be introduced as:

$$E[y(i)] = y(i + 1). \tag{6-166}$$

E is a linear operator with the properties:

$$E^n[y(i)] = E^{n-1}[y(i + 1)] = ... = y(i + n), \tag{6-167}$$

$$E^n[Ay_1(i) + By_2(i)] = Ay_1(i + n) + By_2(i + n). \tag{6-168}$$

Also negative and fractional powers of E can be interpreted, like e.g.

$$E^{-n}\{E^n[y(i)]\} = y(i), \qquad E^{-1/2}[y(i)] = y\left(i - \frac{1}{2}\right). \tag{6-169}$$

We define the *central difference operator*

$$\delta[y(i)] = \left(E^{1/2} - E^{-1/2}\right)[y(i)] = y\left(i + \frac{1}{2}\right) - y\left(i - \frac{1}{2}\right) \tag{6-170}$$

and the *central average operator*

$$\mu[y(i)] = \frac{1}{2}\left(E^{1/2} + E^{-1/2}\right)[y(i)] = \frac{1}{2}\left[y\left(i + \frac{1}{2}\right) - y\left(i - \frac{1}{2}\right)\right], \tag{6-171}$$

which have an important role in establishing the connection between the difference and differential equations. Their higher powers can be formally introduced on the basis of Eqs (6-167), (6-168), e.g.

$$\delta^2[y(i)] = \left(E^{1/2} - E^{-1/2}\right)^2[y(i)] = \left(E^{-1} - 2 + E\right)[y(i)], \tag{6-172}$$

$$\mu^2[y(i)] = \left(\frac{1}{2}\right)^2\left(E^{1/2} + E^{-1/2}\right)^2[y(i)] = \frac{1}{4}\left(E^{-1} + 2 + E\right)[y(i)]. \tag{6-173}$$

Odd powers of δ and μ shift the points of interpretation from integer values of i to fractionals with the fraction 0.5, or from these fractionals to integer values of i.

A homogeneous ordinary linear *difference equation* with constant coefficients of the order n can be expressed in a general form as

$$E^v\left(a_n E^n + a_{n-1}E^{n-1} + ... + a_0\right)y(i) = 0, \tag{6-174}$$

where E^v has no role in the structure of the solution.

The general form of the solution of Eq. (6-174) is [JORDAN, 1947]

$$y = \sum_{k=1}^{m}\left\{(\lambda_k)^i \sum_{r=1}^{l_k} A_{k,r}i^{r-1}\right\}; \tag{6-175}$$

where λ_k, $k = 1, 2, ... \, m \leq n$ are the (real, imaginary or complex) roots of the *characteristic equation*

$$a_n\lambda^n + a_{n-1}\lambda^{n-1} + ... + a_0 = 0, \tag{6-176}$$

and l_k is the multiplicity of λ_k, hence

$$\sum_{k=1}^{m} l_k = n. \tag{6-177}$$

Many problems lead to the second order difference equation

$$\left(\delta^2 + \alpha\right) y = 0 \tag{6-178}$$

with various values of α.

If $\alpha < 0$, then the general solution of Eq. (6-178) is

$$y = A \cosh \vartheta i + B \sinh \vartheta i, \quad \text{where} \quad \vartheta = \operatorname{arcosh}\left(1 - \frac{\alpha}{2}\right), \tag{6-179a}$$

if $\alpha = 0$, then

$$y = A + Bi, \tag{6-179b}$$

if $0 < \alpha < 4$, then

$$y = A \cos \vartheta i + B \sin \vartheta i, \quad \text{where} \quad \vartheta = \arccos\left(1 - \frac{\alpha}{2}\right), \tag{6-179c}$$

if $\alpha = 4$, then

$$y = (-1)^i (A + Bi), \tag{6-179d}$$

finally, if $\alpha > 4$, then

$$y = (-1)^i (A \cosh \vartheta i + B \sinh \vartheta i), \quad \text{where} \quad \vartheta = \operatorname{arcosh}\left(\frac{\alpha}{2} - 1\right). \tag{6-179e}$$

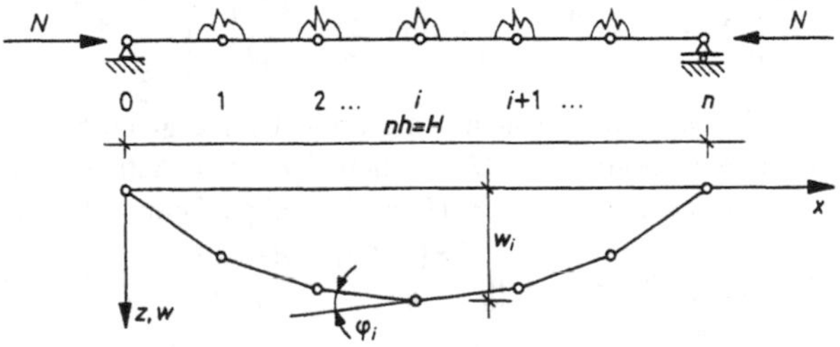

Fig. 6-29 Buckling of a chain of rigid bars

A simple example illustrating the utilization of these solutions is as follows. Let a hinged-hinged beam (*Fig. 6-29*) of the length H be composed of rigid beam elements of equal lengths $h = H/n$ in such a way that the adjacent elements are connected to each other by elastic hinges of equal stiffness

$$C = \frac{M(i)}{\varphi(i)}, \tag{6-180}$$

where $M(i)$ is the moment which develops in the ith hinge if the relative rotation of the connecting bar elements is $\varphi(i)$. Let our task be to determine the critical value of N acting on the beam.

First, the connection between relative rotations $\varphi(i)$ and nodal deflections $w(i)$ is established:

$$\varphi(i) = \frac{1}{h}(-w_{i+1} + 2w_i - w_{i-1}) = -\frac{1}{h}\delta^2 [w(i)]. \tag{6-181}$$

The moment $C\varphi\,(i)$ acting at the ith hinge has to be equal to the moment caused by the eccentricity of the axial load

$$-\frac{C}{h}\delta^2\,[w\,(i)] = Nw\,(i).\qquad\qquad(6\text{-}182a)$$

After rearranging, a homogeneous equation of the structure Eq. (6-178) emerges:

$$\left(\delta^2 + \frac{Nh}{C}\right)w = 0.\qquad\qquad(6\text{-}182b)$$

The boundary conditions are:

$$w\,(0) = 0,\qquad\qquad(6\text{-}183a)$$

$$w\,(n) = 0.\qquad\qquad(6\text{-}183b)$$

The general solution of Eq. (6-182b) is Eq. (6-179c). It follows from (6-183a) that $A = 0$, and from Eq. (6-183b) that

$$\vartheta = \frac{k\pi}{n},\qquad k = 1,2 \ldots (n-1),$$

hence, the equation

$$2\cos\left(\frac{k\pi}{n}\right) = 2 - \frac{Nh}{C}\qquad\qquad(6\text{-}184)$$

must hold. To each k a value of N can be computed from Eq. (6-184), and the minimum of these values is the critical normal force of the beam. N_{cr} belongs to $k = 1$:

$$N_{cr} = \frac{C}{h}2\left(1 - \cos\frac{\pi}{n}\right) = \frac{C}{h}4\sin^2\frac{\pi}{2n} = \frac{Ch\pi^2}{H^2}\left[\frac{\sin\frac{\pi}{2n}}{\frac{\pi}{2n}}\right]^2.\qquad\qquad(6\text{-}185)$$

By increasing the number n of the elements, the critical force given by Eq. (6-185) approaches that of a continuous beam $\pi^2 EI/H^2$, where $EI = Ch$. Actually, by tending with n to infinity in a way that the equation $nh = H$ holds, and performing the limit transitions

$$\lim_{n\to\infty} h = 0,\quad \lim_{n\to\infty} ih = x,\quad \lim_{h\to0}\frac{\delta^2\,[w\,(i)]}{h^2} = \frac{d^2w\,(x)}{dx^2},\quad \lim_{h\to0} Ch = EI\quad(6\text{-}186)$$

on our equations, Eq. (6-182a) can be written as

$$EI\frac{d^2w\,(x)}{dx^2} + Nw\,(x) = 0,\qquad\qquad(6\text{-}187)$$

which is the differential equation of buckling of a continuous beam. In the limit transition, the expression in the brackets of Eq. (6-185) becomes unity and the formula exactly yields *Euler*'s critical force of the continuous bar. This critical force is a very good estimate of that of the bar in *Fig. 6-29* in the case of finite values of n. The difference is about 0.8% if $n = 10$, about 3.2% if $n = 5$, thus, the continuous bar is a fairly adequate replacement continuum of our bar.

6.6.3 Trusses

Let the deformations of the truss shown in *Fig. 6-30* be analysed. The displacements in the x and z directions of the bottom and top joints can be described by integer functions $u_b\,(i)$, $w_b\,(i)$ and $u_t\,(i)$, $w_t\,(i)$ respectively, while the joint loads are given by the components $X_b\,(i)$, $Z_b\,(i)$, and $X_t\,(i)$, $Z_t\,(i)$ in the same way.

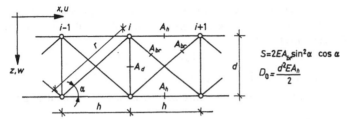

Fig. 6-30 Characteristics of the truss

A substantial simplification can be achieved by decomposing the displacements and forces into symmetrical and antisymmetrical components as follows:

$$u = \frac{1}{2}(u_b + u_t), \qquad \chi = \frac{1}{d}(u_b - u_t),$$

$$w = \frac{1}{2}(w_b + w_t), \qquad \Delta = \frac{1}{2}(w_b - w_t), \qquad (6\text{-}188)$$

$$X = \frac{1}{2}(X_b + X_t), \qquad W = \frac{d}{2}(X_b - X_t),$$

$$Z = \frac{1}{2}(Z_b + Z_t), \qquad V = \frac{1}{2}(Z_b - Z_t). \qquad (6\text{-}189)$$

Using this decomposition, the equilibrium difference equations of the structure take the form as follows [KOLLÁR and HEGEDŰS, 1985]:

$$
\begin{bmatrix}
-2(\alpha_h + \beta_1)\delta^2 & -4\beta_3\mu\delta & & \\
4\beta_3\mu\delta & 2\beta_2\delta^2 + 4(\gamma + 2\beta_2) & 0 & \\
0 & & -2\beta_1\delta^2 & -2d\beta_3\mu\delta \\
& & 2d\beta_3\mu\delta & \frac{d^2}{2}\left[(\beta_1 - \alpha_h)\delta^2 + 4\beta_1\right]
\end{bmatrix}
\begin{bmatrix} u \\ \Delta \\ w \\ \chi \end{bmatrix}
$$

$$
=
\begin{bmatrix} X \\ V \\ Z \\ W \end{bmatrix}
\qquad (6\text{-}190)
$$

where the parameters in the operator matrix are:

$$\alpha_h = \frac{EA_h}{h}, \quad \beta_1 = \frac{EA_{br}}{r}\cos^2\alpha_r, \quad \beta_2 = \frac{EA_{br}}{r}\sin^2\alpha_r, \quad \beta_3 = \sqrt{\beta_1\beta_2}, \gamma = \frac{EA_d}{d}.$$

This difference equation system can also be used for the stability analysis of the truss if second order effects are also taken into account. This can be done in a visual way by assuming a set of fictitious joint loads which replace the effects of bar forces acting in the directions modified by the buckling displacements. If N denotes the normal force acting in the truss, that is, the sum of the chord-bar forces and the horizontal components of forces acting in the inclined bars, the components of this set of fictitious joint loads can be expressed as

$$
dP =
\begin{bmatrix} 0 \\ dV \\ dZ \\ 0 \end{bmatrix}
= -\frac{N}{h}
\begin{bmatrix} 0 \\ \delta^2[\Delta] \\ \delta^2[w] \\ 0 \end{bmatrix}
\qquad (6\text{-}191)
$$

If the boundary conditions attached to Eqs (6-190) can be separately expressed in terms of symmetrical and antisymmetrical displacement components, respectively, the analysis can also be split into separate analyses of symmetrical and antisymmetrical components. The components of the *symmetrical buckling modes*, u and Δ, are the solutions of the homogeneous difference equation system

$$\begin{bmatrix} -2\left(\alpha_h+\beta_1\right)\delta^2 & -4\beta_3\mu\delta \\ 4\beta_3\mu\delta & 2\beta_2\delta^2+4\left(\gamma+2\beta_2\right)+\dfrac{N}{h}\delta^2 \end{bmatrix}\begin{bmatrix} u \\ \Delta \end{bmatrix}=\begin{bmatrix} 0 \\ 0 \end{bmatrix}, \qquad (6\text{-}192)$$

The components of the *antisymmetrical buckling modes*, w and χ, can be determined as the solutions of the equation system

$$\begin{bmatrix} -2\beta_1\delta^2 & -2d\beta_3\mu\delta \\ 2d\beta_3\mu\delta & \dfrac{d^2}{2}\left[\left(\beta_1-\alpha_h\right)\delta^2+4\beta_1\right]+\dfrac{N}{h}\delta^2 \end{bmatrix}\begin{bmatrix} w \\ \chi \end{bmatrix}=\begin{bmatrix} 0 \\ 0 \end{bmatrix}. \qquad (6\text{-}193)$$

Assuming that the boundary conditions permit the above splitting into two components, the analysis of *symmetrical buckling modes* can be reduced to that of a single difference equation by eliminating u from the second equation of (6-192). In this way we obtain:

$$\delta^2\left(\delta^2+\alpha\right)[\Delta]=0 \qquad (6\text{-}194)$$

where

$$\alpha=4\frac{2\beta_2\alpha_h+\gamma\left(\alpha_h+\beta_1\right)}{2\beta_2\alpha_h+\left(\alpha_h+\beta_1\right)}.$$

Solutions of Eq. (6-194) are, in the case of $\alpha=0$:

$$\Delta=A+Bi+Ci^2+Di^3,$$

and in the cases $\alpha\neq0$:

$$\Delta=A+Bi+y, \qquad (6\text{-}195)$$

where y is the solution of Eq. (6-178). Since α is positive, the solution is given by Eqs (6-179c–e). If the value of n is sufficiently high, for any boundary conditions oscillatory buckling modes as shown in *Fig. 6-31* develop and the critical force can be calculated using the value $\alpha=4$. This yields the critical normal force

$$N_{\mathrm{cr}}=\frac{EA_dh}{d}, \qquad (6\text{-}196)$$

which is in all practical cases substantially higher than the force which causes individual (local) buckling of the bars. For this reason, symmetric buckling mode of trusses has no practical importance.

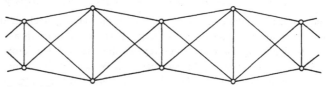

Fig. 6-31 Symmetrical buckling mode

The analysis of *antisymmetric buckling* can also be reduced to that of a single difference equation by eliminating w from Eq. (6-193). In this way the following homogeneous difference equation emerges:

$$\left\{\frac{d^2\alpha_h h}{2}\delta^4+N\left[h^2\mu^2\delta^2-\frac{d^2\alpha_h}{4\beta_2}\delta^4\right]\right\}\chi=0. \qquad (6\text{-}197)$$

Instead of the lengthy procedure of solving the fourth order eigenvalue problem stated by Eq. (6-197) and by the boundary conditions for w and χ, let us perform a limit transition on our equations, similar to that used in the previous section in Eqs (6-186):

$$\lim_{h\to 0} hi = x, \quad \lim_{h\to 0} \frac{\delta^4 [\chi(i)]}{h^4} = \frac{d^4\chi(x)}{dx^4}, \quad \lim_{h\to 0} \frac{\mu^2\delta^2 [\chi(i)]}{h^2} = \frac{d^2\chi(x)}{dx^2}. \tag{6-198}$$

Introducing the notations

$$D_0 = \frac{d^2 EA_h}{2} \tag{6-199}$$

and

$$S = 2EA_r \sin^2 \alpha_r \cos \alpha_r, \tag{6-200}$$

we obtain the differential equation

$$D_0 \frac{d^4\chi(x)}{dx^4} + N \left[\frac{d^2\chi(x)}{dx^2} - \frac{D_0}{S} \frac{d^4\chi(x)}{dx^4} \right] = 0, \tag{6-201}$$

which is the buckling differential equation of a sandwich beam with thin faces and with bending and shear rigidities D_0 and S respectively. This result shows that the replacement continuum of the truss is a sandwich beam and, consequently, the critical normal force of the truss can be approximated by that of this sandwich beam:

$$N_{cr} = \left[\left(\frac{\pi^2 D_0}{L^2} \right)^{-1} + S^{-1} \right]^{-1} \tag{6-202}$$

where L stands for the buckling half wave-length.

A detailed analysis shows that the critical forces of this replacement continuum approximate those of the real structure with the same accuracy as it was observed in the previous section.

Trusses with networks shown in *Fig. 6-32a–c.* can also be analysed using replacement sandwich bars. Formulas for the effective rigidities can also be found in the figures.

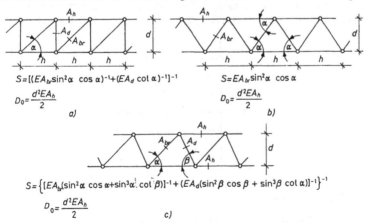

$$S = [(EA_{br}\sin^2\alpha \ \cos \alpha)^{-1} + (EA_d \cot \alpha)^{-1}]^{-1}$$

$$D_0 = \frac{d^2 EA_h}{2}$$

a)

$$S = EA_{br}\sin^2\alpha \ \cos \alpha$$

$$D_0 = \frac{d^2 EA_h}{2}$$

b)

$$S = \left\{ [EA_{br}(\sin^2\alpha \ \cos \alpha + \sin^3\alpha \ \cot \beta)]^{-1} + (EA_d(\sin^2\beta \ \cos \beta + \sin^3\beta \ \cot \alpha)]^{-1} \right\}^{-1}$$

$$D_0 = \frac{d^2 EA_h}{2}$$

c)

Fig. 6-32 Replacement stiffnesses of trusses

The slenderness λ is a traditional measure of the propensity of bars to buckle. Its square is inversely proportional to the buckling load:

$$N_{cr} = A\sigma_{cr} = \frac{A\pi^2 E}{\lambda^2}. \tag{6-203}$$

For trusses, an *effective slenderness* λ_{eff} can be derived on the basis of the formula for the critical load of sandwich beams with thin faces.

Assuming $A = 2A_h$, (with A_h as the area of one chord,) the square of λ_{eff} can be obtained by setting equal the right-hand sides of Eqs (6-203) and (6-202). In this way we can express λ_{eff}^2 as

$$\lambda_{eff}^2 = 4\frac{L^2}{d^2} + \frac{2A_h E}{S}. \tag{6-204}$$

Note that this effective slenderness does not depend on the individual slenderness of the chordbars because local buckling does not interact with global buckling, see also Section 3.2.

6.6.4 Laced (Vierendeel) column

Laced columns (*Fig. 6-33*) develop buckling deformations which can be split into two parts: one with changes in length of the chord bars and another without it. The former part can be considered as flexural deformation, the latter as shear deformation of sandwich bars. The stability analysis of the laced columns has to take into account both types of deformations. It has been made on the basis of discrete models by RATZERSDORFER [1936], KÁRMÁN and BIOT [1940], and on the basis of various replacement bar models by TIMOSHENKO and GERE [1961], ASZTALOS [1972], ZALKA [1976, 1977, 1979], ZALKA and ARMER [1992], KOLLÁR [1986*b*].

First the discrete analysis of a laced column subjected to a vertical force N (see in *Fig. 6-33*) is summarized here. The detailed analysis is presented in [HEGEDŰS and KOLLÁR, 1999].

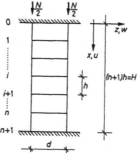

Fig. 6-33 Laced (Vierendeel) column

In deriving the differential equation of the laced column the differential equation of each continuous column section must be solved. The boundary conditions of these differential equations are the yet unknown nodal displacements.

Assuming antisymmetrical buckling, the laced column can be characterized by the following displacements of the nodes (*Fig. 6-34*):

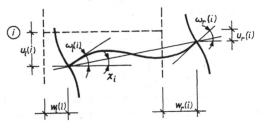

Fig. 6-34 Displacements of the nodes

The horizontal displacements:

$$w_l(i) = w_r(i) = w(i); \qquad (6\text{-}205)$$

the vertical displacements:

$$u_l(i), \qquad u_r(i); \qquad (6\text{-}206)$$

and the rotations:

$$\omega_l(i) = \omega_r(i) = \omega(i). \qquad (6\text{-}207)$$

By introducing the rotations of the connecting lines of the nodes

$$\chi(i) = \frac{1}{d}[u_l(i) - u_r(i)], \qquad (6\text{-}208)$$

the following difference equation can be derived [HEGEDŰS and KOLLÁR, 1999]:

$$\{\delta^4 + x_1\delta^2 + x_2\}[\chi(i)] = 0, \qquad (6\text{-}209)$$

where

$$x_1 = \frac{hEI_b}{dEI_c}\frac{6\sin\alpha}{\alpha} - \frac{24hEI_b}{EA_cd^3} + 2(1 - \cos\alpha), \qquad (6\text{-}210)$$

$$x_2 = -\frac{48hEI_b}{EA_cd^3}(1 - \cos\alpha), \qquad (6\text{-}211)$$

I_c and A_c are the moment of inertia and the area of the vertical bars, respectively, I_b is the moment of inertia of the horizontal bars, and

$$\alpha = h\sqrt{\frac{N}{2EI_c}},$$

where N is the sum of the vertical forces (*Fig. 6-33*). The analytical solution of this equation is presented in [HEGEDŰS and KOLLÁR, 1999].

To derive a replacement continuum, let us perform the limit transition

$$\lim_{h\to 0} hi = x, \quad \lim_{h\to 0}\frac{\delta^2[\chi(i)]}{h^2} = \frac{d^2\chi(x)}{dx^2}, \quad \lim_{h\to 0}\frac{\delta^4[\chi(i)]}{h^4} = \frac{d^4\chi(x)}{dx^4} \qquad (6\text{-}212)$$

on Eq. (6-209). It yields the differential equation:

$$\frac{d^4\chi(x)}{dx^4} + \frac{x_1}{h^2}\frac{d^2\chi(x)}{dx^2} + \frac{x_2}{h^4}\chi(x) = 0. \qquad (6\text{-}213)$$

By introducing the effective rigidities

$$D_0 = \frac{EA_cd^2}{2}, \qquad (6\text{-}214)$$

and

$$S_b = \frac{12EI_b}{hd},$$ (6-215)

x_1, and x_2 can be written in the form

$$x_1 = -\frac{h^2 S_b}{2EI_c}\frac{\sin\alpha}{\alpha} - \frac{h^2 S_b}{D_0} + 2(1 - \cos\alpha),$$ (6-216)

$$x_2 = \frac{h^2 S_b}{D_0} 2(1 - \cos\alpha).$$ (6-217)

The properties of the fictitious bar described by the buckling differential equation (6-213) can be interpreted from some aspects; however, its behaviour is too complex to be used in practical applications. A considerable simplification can be achieved by assuming α sufficiently small for replacing its trigonometric functions by their *Taylor* expansions:

$$\sin\alpha \approx \alpha, \quad \cos\alpha \approx 1 - \frac{\alpha^2}{2}.$$ (6-218)

In this way N explicitly appears in the differential equation (6-213) which can be assumed as that of a sandwich bar with thick faces:

$$\frac{D_0(2EI_c)}{S_b}\frac{d^4\chi(x)}{dx^4} - (D_0 + D_l)\frac{d^2\chi(x)}{dx^2} + N\left[\frac{D_0}{S_b}\frac{d^2\chi(x)}{dx^2} - \chi(x)\right] = 0,$$ (6-219)

where S_b stands for the shear rigidity, and the local bending rigidity is

$$D_l = 2EI_c.$$ (6-220)

We found that the replacement continuum of the laced column can be the sandwich beam having the rigidities defined by Eqs (214), (215), and (220). This continuum also fulfils the mathematical condition for the adequacy because the differences between the discrete and continuous models vanish when the number of the elements tends to infinity. However, the applied limit transition eliminates the effect of sway from the differential equation, because the distance between the horizontal beams tends to zero. Hence, in the case of a finite n, the replacement continuum yields good results only if n is large and the effect of sway can be, in fact, neglected.

Let the approximations Eq. (6-218) be modified by assuming that

$$\sin\alpha \approx \alpha - \frac{\alpha^3}{\pi^2}, \quad \cos\alpha \approx 1 - \frac{\alpha^2}{2}.$$ (6-221)

By so doing, Eq. (6-213) results in

$$\frac{D_0 D_l}{S}\frac{S_c}{S_c + S_b}\frac{d^4\chi(x)}{dx^4} - (D_0 + D_l)\frac{d^2\chi(x)}{dx^2} + N\left[\frac{D_0}{S}\frac{d^2\chi(x)}{dx^2} - \chi(x)\right] = 0,$$ (6-222)

where

$$S_c = 2\frac{\pi EI_c}{h^2} \qquad S = \left(S_b^{-1} + S_c^{-1}\right)^{-1}.$$ (6-223)

This differential equation slightly differs from that of a sandwich beam with thick faces, thus, the formula for the critical load also differs from that of a sandwich beam. For a cantilever beam the expression

$$N_{cr} = \frac{1 + N_l\left(N_0^{-1} + S_b^{-1}\right)}{N_0^{-1} + S_b^{-1} + S_c^{-1}}$$ (6-224)

can be derived by introducing

$$\chi(x) = \cos \frac{\pi x}{2H} \tag{6-225}$$

into Eq. (6-222), where

$$N_0 = \frac{\pi^2 D_0}{4H^2}, \quad \text{and} \quad N_l = \frac{\pi^2 D_l}{4H^2} \tag{6-226}$$

are the *Euler*-loads of cantilever bars with the bending stiffnesses D_0 and D_l, respectively.

If $D_l \ll D_0$ holds, Eq. (6-222) can be approximated by

$$\frac{D_0 D_l}{S} \frac{S_c}{S_c + S_t} \frac{d^4\chi(x)}{dx^4} - \left(D_0 + D_l \frac{S_c}{S_c + S_t} \right) \frac{d^2\chi(x)}{dx^2} + N \left[\frac{D_0}{S} \frac{d^2\chi(x)}{dx^2} - \chi(x) \right] = 0, \tag{6-227}$$

which is the differential equation of a sandwich beam with thick faces having the local bending stiffness

$$\tilde{D}_l = 2EI_c \frac{S_c}{S_c + S_b}. \tag{6-228}$$

Introducing the notation

$$\widetilde{N}_l = \frac{\pi^2 \tilde{D}_l}{4H^2} \tag{6-229}$$

Eq. (6-227) yields a critical load

$$N_{cr} = \left(N_0^{-1} + S^{-l} \right)^{-l} + \widetilde{N}_l, \tag{6-230}$$

which is lower than Eq. (6-224).

For practical use a replacement bar belonging to Eq. (6-227) is suggested, thus, a sandwich bar with thick faces having the rigidities

$$D_0 = \frac{EA_c d^2}{2}, \quad \tilde{D}_l = 2EI_c \frac{S_c}{S_c + S_b}, \tag{6-231}$$

$$S = \left(S_b^{-1} + S_c^{-1} \right)^{-1} \tag{6-232}$$

where

$$S_b = \frac{12EI_b}{hd}, \quad S_c = \frac{2EI_c\pi^2}{h^2}. \tag{6-233}$$

The same replacement bar can also be used in the stability analysis of laced columns loaded at intermediate levels. In this case N is a function of x.

In the following we present some formulas which were derived by various authors for the critical loads of single-bay frames.

TIMOSHENKO and GERE [1961] modelled the frame by a sandwich beam with thin faces, with the bending and shear stiffnesses:

$$D = D_0 + 2EI_c, \quad \tilde{S} = \left(\frac{1}{S_b} + \frac{1}{\tilde{S}_c} \right)^{-1}, \tag{6-234}$$

where S_b is defined by Eq. (6-233), and

$$\tilde{S}_c = \frac{24EI_c}{h^2}. \tag{6-235}$$

The critical load can be calculated as

$$\overline{N}_{cr} = \left[(N_0 + N_l)^{-1} + \tilde{S}^{-1} \right]^{-1}, \tag{6-236}$$

where N_0, N_l are the same as defined by Eqs (6-226). They recommend to decrease the shear stiffness depending on the normal force ([TIMOSHENKO and GERE, 1961], p. 160. Eq. 2-65), which may lead to an error up to 50% [KOLLÁR, 1986b].

ASZTALOS [1972] derived the formula

$$\overline{N}_{cr} = \left[\left(N_0^{-1} + \tilde{S}^{-1}\right)^{-1} + N_l\right]^{-1}, \tag{6-237}$$

which is very similar to Eq. (6-230), the essential difference being the calculation of the shear stiffness. This formula, in the case of infinitely rigid beams, may result in a non conservative estimate, the error of which is 21.6%.

If the frame has at least four storeys, the error of formula (6-230) is less than 8%, and the approximation is conservative (*Fig. 6-35*), see also [HEGEDŰS and KOLLÁR, 1999].

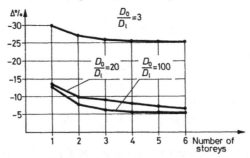

Fig. 6-35 Error in the approximate critical force of laced column (Eq. (6-230))

We have to mention that some authors suggested different replacement continua for the calculation of frames subjected to wind load [CSONKA, 1962].

Practical calculations use the *buckling length* l_0 and the *slenderness* λ to characterize the stability properties of compressed bars. For laced columns, an *effective slenderness* λ_{eff} can be introduced on the basis of Eq. (6-203) as follows.

Let L be the buckling length which belongs to the global buckling of the structure. (In the case of a cantilever, $L = 2H$). Let the following notations be introduced:

$$\lambda_0^2 = L^2 \frac{2EA_c}{D_0}, \qquad \lambda_l^2 = L^2 \frac{2EA_c}{\tilde{D}_l}, \tag{6-238a}$$

$$\lambda_{S0}^2 = h^2 \frac{EA_c}{EI_c}, \qquad \lambda_{Sg}^2 = dh \frac{\pi^2 EA_c}{6EI_b}, \tag{6-238b}$$

where D_0 and $\tilde{D}_l$ are the global and the local bending stiffnesses of the replacement sandwich bar defined by Eqs (6-231).

By introducing Eq. (6-230) and Eqs (6-238) into Eq. (6-203), the critical axial stress can be expressed as

$$\sigma_{cr} = \pi^2 E \left[\frac{1}{\lambda_0^2 + \lambda_{S0}^2 + \lambda_{Sg}^2} + \frac{1}{\lambda_l^2}\right] = \frac{\pi^2 E}{\lambda_{eff}^2} \tag{6-239}$$

The square of the effective slenderness of the laced column can be expressed as the reciprocal of the sum in the bracket of the above equation:

$$\lambda_{eff}^2 = \left[\frac{1}{\lambda_0^2 + \lambda_{S0}^2 + \lambda_{Sg}^2} + \frac{1}{\lambda_l^2} \right]^{-1}. \tag{6-240}$$

The *effective buckling length* l_0 of a bar having the bending stiffness D and the critical force N_{cr} is defined by the equation

$$N_{cr} = \frac{\pi^2 D}{l_0^2}. \tag{6-241}$$

For deriving an *effective buckling length for the vertical bar sections* of the laced column, let us rearrange Eq. (6-230) as follows:

$$N_{cr} = \pi^2 (2EI_c) \left[\left(\frac{2EI_c}{D_0} L^2 + h^2 + \frac{\pi^2}{12} \frac{2EI_c}{EI_b} dh \right)^{-1} + \frac{\tilde{D}_l}{2EI_c} \frac{1}{L^2} \right]. \tag{6-242}$$

The square of the effective buckling length of the column section can be expressed as the reciprocal of the sum in the square bracket of the above equation. Omitting the last term in the square brackets, the approximate formula for the square of the effective buckling length emerges as:

$$l_0^2 = \left[\frac{2EI_c}{D_0} L^2 \right] + \left[h^2 + \frac{\pi^2}{12} \frac{2EI_c}{EI_b} dh \right]. \tag{6-243}$$

6.6.5 Frames and shear walls

The methods presented in the previous section can simply be extended to the analysis of multi-storey, multi-bay frames and shear walls. That is, both types of structures can be replaced by sandwich bars with thick faces and the global analysis of frames braced by shear walls can be performed as that of connected sandwich bars.

Multi-bay frames
Multi-bay frames and shear walls with series of openings are often analysed using multi-layer sandwich beam replacement continua [BECK, 1956, 1959; ROSMAN, 1965, 1968; SZERÉMI, 1975, 1984; LIGETI, 1974]. Stiff layers (faces) of the sandwich represent the columns of the frame or the solid vertical strips of the wall, soft layers (cores) represent the effect of horizontal elements which connect the columns or wall strips. Though the global critical normal force of these structures is fairly high, their stability analysis cannot be omitted because the dimensioning of the column sections needs design values which can only be assumed on the basis of this analysis.

If the soft layers of the multi-layer sandwiches can be assumed incompressible in the horizontal direction, then their buckling modes cannot essentially differ from those of three layer sandwiches (see Section 6.5.3). Since in the case of soft layers representing the effect of the horizontal elements of multi-bay frames, this assumption can always be used, a simplified though sufficiently accurate stability analysis of multi-bay frames can be

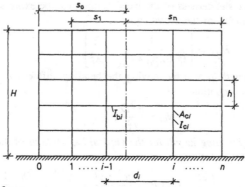

Fig. 6-36 Multi-bay frame

performed using three layer sandwich replacement beams [HEGEDÜS and KOLLÁR, 1987].
The rigidities of the replacement sandwich beam can be assumed, using the notations of
Fig. 6-36, as

$$D_0 = \sum_{i=0}^{n} EA_{ci} s_i^2, \qquad (6\text{-}244)$$

$$D_l = \left(\sum_{i=0}^{n} EI_{ci} \right) \frac{S_c}{S_c + S_b}, \qquad (6\text{-}245)$$

$$S = \left(S_c^{-1} + S_b^{-1} \right)^{-1}, \qquad (6\text{-}246)$$

where A_{ci} denotes the cross sectional area of the columns,

$$S_c = \sum_{i=0}^{n} \frac{\pi^2 EI_{ci}}{h^2}, \qquad S_b = \sum_{i=1}^{n} \frac{12 EI_{bi}}{d_i h}, \qquad (6\text{-}247)$$

and s_i are the distances of the centres of the columns from the global centre of gravity of
their cross sections.

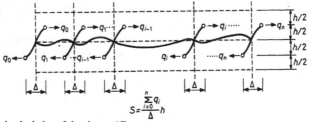

Fig. 6-37 Refined calculation of the shear stiffness

This simplified model shows the structure somewhat stiffer than reality. The main reason
for this is that it assumes equal horizontal displacements of all columns along their whole
length (although this assumption is only valid in the case of so called *proportional frames*
[CSONKA, 1956; SZERÉMI, 1978, 1984], that is, frames with specific rigidities, whose joints
perform uniform rotations by levels under the action of horizontal loads). If the lengths and

rigidities of the horizontal beam sections or the rigidities of the columns strongly differ from each other, a more realistic effective shear stiffness can be obtained by assuming a displacement pattern shown in *Fig. 6-37*, and using reduced column stiffnesses in the calculation with the reduction factor $\pi^2/12$.

For finding the *effective buckling lengths* of the column sections, let us analyse the formula (6-243). It clearly shows the separate effects of the global bending stiffness and the shear stiffness. The same separation can be assumed in our case. If $\bar{l}_0$ denotes the buckling length of a column section determined by the methods presented in Section 5.2, the buckling length which also takes into account the global buckling of the structure can be calculated as

$$l_0^2 = \frac{\sum\limits_{i=0}^{n} EI_{ci}}{D_0} L^2 + \bar{l}_0^2, \tag{6-248}$$

where D_0 is the global bending rigidity and L is the global buckling length of the replacement sandwich bar. For sandwich cantilever beams loaded on the top, $L = 2H$, for cantilever beams subjected to uniformly distributed axial load $L \approx H\sqrt{\pi^2/7.837} \approx 1.2H$.

Shear walls

As discrete models of coupled shear walls (walls with series of openings) multi-storey frames can be used whose vertical and horizontal beams represent the vertical solid wall strips and the straps connecting them, respectively. This discrete model has to be different from common frames, because the shear deformation of the strips must not be neglected, and sections of the horizontal strips which coincide with the vertical wall strips have to be assumed undeformable. These differences make the discrete stability analysis cumbersome. Nonetheless, replacement continua defined by Eqs (6-244)–(6-247) in the previous section can be used, provided S_b is taken according to the following expression [ZALKA, in KOLLÁR, 1991].

$$S_b = \sum_{i=1}^{n} \frac{12EI_{bi}}{\dfrac{l_i^2 h}{d_i^2}\left[I + 12\dfrac{1.2E}{l_i^2 G}\dfrac{I_{bi}}{A_{bi}}\right]}, \tag{6-249}$$

where l_i is the distance between the inner faces of the columns adjacent to the ith beam, d_i is the distance between the axes of the columns, G is the shear modulus, I_{bi} and A_{bi} are the moment of inertia and the area of the ith beam, respectively.

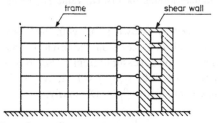

Fig. 6-38 Frame braced by a shear wall

Shear walls without openings are modelled, as a rule, with simple *Euler-Bernoulli* beams. Nevertheless, less slender walls may develop considerable shear deformation, hence it may expedient to model them with *Timoshenko*-beams (i.e. sandwiches with thin faces).

Plane frames braced by shear walls

The stability analysis of multi-bay frames braced by coupled shear walls (*Fig. 6-38*) can be performed on replacement sandwich beams elastically restrained against rotation. Both the coupled shear wall and the frame have to be substituted for sandwiches with thick faces having the rigidities D_{0w}, S_w, D_{lw}, and D_{0f}, S_f, D_{lf}, respectively. Since the relation $D_{0w} \ll D_{0f}$ practically always holds, the replacement beam of the frame can be regarded as a *Csonka*-beam with $D_{0f} = \infty$, and its contribution can be taken into account as an elastic restraint against rotation with the spring constant S_f acting on the sandwich beam with the global rigidities D_{0w}, S_w, and the increased local bending rigidity $D_l = D_{lw} + D_{lf}$.

In some cases, simpler replacement continua can also be used. ROSMAN [1974] proposed to use a bar with pure shear deformations for replacing the frame, and another with pure bending deformations for the wall. In this case, the replacement continuum of the coupled structures is a single *Csonka*-beam.

6.6.6 Combined torsional and in-plane buckling of multi-storey buildings

The structure of a k-storey building with identical storeys, of the total height H consists of k slabs, n supporting columns and a set of m shear walls with arbitrary arrangement (*Fig. 6-39a*). If k is not a small number ($k > 3$) and the ground plan is not assembled of narrow rectangles, we do not commit a substantial error by neglecting the plate-action of the walls, assuming that the slabs are infinitely rigid in their planes and their loads transferred to the columns are 'smeared' out along the height. In this way the stability analysis reduces to that of the buckling of the coupled system of n columns and m sandwich beams which replace the elements of the stiffening system.

If the shear walls are parallel to each other, this model is the same as used by SZERÉMI [1978]. ROSMAN [1980] analysed the problem of general arrangement of the walls assuming simple *Euler-Bernoulli* beams. GOSCHY [1970] used *Timoshenko*-beams in the same problem, HEGEDŰS and KOLLÁR [1987] proposed sandwich bars with thick faces as replacement continua for the elements of the shear wall system. The concise outline below follows the last cited paper in determining the critical value N_{cr} of the vertical load acting at the top of the structure.

Let the shear walls be replaced by sandwich bars with thick faces with the rigidities $D_{0i}, S_i, D_{li}, i = 1, 2, \ldots, m$. Their distances from the origin r_i and the angles ε_i which determine their directions in ground plan measured from z (*Fig. 6-39a*) are also given. Coordinates of the columns in ground plan be $y_j, z_j, j = 1, 2, \ldots, n$. The total normal load is denoted by N, and the jth column carries $\beta_j N$. Obviously,

$$\sum_{j=1}^{n} \beta_j = 1, \qquad (6\text{-}250)$$

and the sums

$$y_N = \sum_{j=1}^{n} y_j \beta_j, \qquad (6\text{-}251a)$$

$$z_N = \sum_{j=1}^{n} z_j \beta_j \qquad (6\text{-}251b)$$

determine the co-ordinates y, z of N.

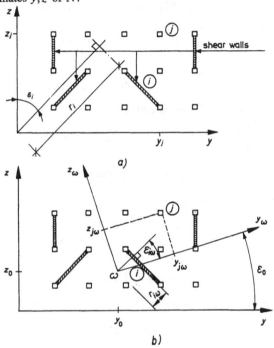

a)

b)

Fig. 6-39 Plan of the stiffening system of a building

Having introduced the individual critical forces of the replacement sandwich bars

$$N_{\text{cr},i} = \left[\left(\frac{\pi^2 D_{0i}}{4H^2} \right)^{-1} + S_i^{-1} \right]^{-1} + \frac{\pi^2 D_{1i}}{4H^2}, \qquad (6\text{-}252)$$

it is expedient to shift the origin of the co-ordinate system by y_0 and z_0 to point ω which is the shear center of the stiffening system and to rotate the axes by the angle ε_0 in directions y_ω, z_ω which are the principal directions of the system. This shifting is defined by the equations

$$y_0 = \frac{\left(\sum_{i=1}^{m} y_i N_{\text{cr},i} \right)}{N_0}, \qquad z_0 = \frac{\left(\sum_{i=1}^{m} z_i N_{\text{cr},i} \right)}{N_0}, \qquad (6\text{-}253)$$

where

$$N_0 = \sum_{i=1}^{m} N_{\text{cr},i}, \tag{6-254}$$

and the angle of rotation by

$$\tan 2\varepsilon_0 = \frac{\sum_{i=1}^{m} N_{\text{cr},i} \, (y_i - y_0) \, (z_i - z_0)}{\sum_{i=1}^{m} N_{\text{cr},i} \left[(y_i - y_0)^2 - (z_i - z_0)^2 \right]}. \tag{6-255}$$

The transformed co-ordinates are

$$y_{i\omega} = r_{i\omega} \cos \varepsilon_{i\omega}, \qquad z_{i\omega} = r_{i\omega} \sin \varepsilon_{i\omega}, \tag{6-256}$$

where

$$r_{i\omega} = \sqrt{(y_i - y_0)^2 + (z_i - z_0)^2} \quad \text{and} \quad \varepsilon_{i\omega} = \arctan \left(\frac{y_i - y_0}{z_i - z_0} \right) - \varepsilon_0. \tag{6-257}$$

In this co-ordinate system the following equations hold:

$$\sum_{i=1}^{m} N_{\text{cr},i} r_{i\omega} \cos \varepsilon_{i\omega} = 0, \qquad \sum_{i=1}^{m} N_{\text{cr},i} r_{i\omega} \sin \varepsilon_{i\omega} = 0, \tag{6-258}$$

$$\sum_{i=1}^{m} N_{\text{cr},i} r_{i\omega} \cos \varepsilon_{i\omega} \sin \varepsilon_{i\omega} = 0. \tag{6-259}$$

The new co-ordinates of the columns are:

$$y_{j\omega} = \sqrt{(y_j - y_0)^2 + (z_j - z_0)^2} \cos \varepsilon_{j\omega}, \tag{6-260}$$

$$z_{j\omega} = \sqrt{(y_j - y_0)^2 + (z_j - z_0)^2} \sin \varepsilon_{j\omega}, \tag{6-261}$$

where

$$\varepsilon_{j\omega} = \arctan \left(\frac{y_j - y_0}{z_j - z_0} \right) - \varepsilon_0. \tag{6-262}$$

Defining the square of the polar radius of inertia in the transformed co-ordinate system as

$$e_\omega^2 = \sum_{j=1}^{m} \beta_j \left(y_{j\omega}^2 + z_{j\omega}^2 \right), \tag{6-263}$$

and the values

$$N_y = \sum_{i=1}^{m} N_{\text{cr},i} \, r_{i\omega} \cos^2 \varepsilon_{i\omega}, \tag{6-264}$$

$$N_z = \sum_{i=1}^{m} N_{\text{cr},i} \, r_{i\omega} \sin^2 \varepsilon_{i\omega}, \tag{6-265}$$

$$N_\omega = \frac{1}{e_\omega^2} \sum_{i=1}^{m} N_{\text{cr},i} r_{i\omega}^2, \tag{6-266}$$

the critical value N_{cr} of the vertical load is obtained as the lowest root of the third degree algebraic equation

$$\det \begin{bmatrix} N_{\text{cr}} - N_y & 0 & -N_{\text{cr}} y N_\omega \\ 0 & N_{\text{cr}} - N_z & N_{\text{cr}} z N_\omega \\ -N_{\text{cr}} y N_\omega & N_{\text{cr}} z N_\omega & (N_{\text{cr}} - N_\omega) e_\omega^2 \end{bmatrix} = 0, \tag{6-267}$$

where

$$y_{N\omega} = \sum_{j=1}^{m} y_{j\omega}\beta_j,$$ (6-268)

$$z_{N\omega} = \sum_{j=1}^{n} z_{j\omega}\beta_j.$$ (6-269)

Eq. (6-26) is formally the same as presented in [WLASSOW, 1965] Band 2(3.2) and in [TIMOSHENKO and GERE 1961] for the combined *Euler*-type and torsional buckling load of thin walled bars. Actually, critical forces N_y, N_z, N_ω can also be interpreted as critical forces belonging to 'pure' *Euler*-type and torsional buckling of the structure, and in the cases of symmetric arrangements of the elements Eq. (6-267) splits into three independent equations defining these critical forces. However, this analogy is not perfect, because WLASSOW [1965], and TIMOSHENKO and GERE [1961] assume only flexural deformations while in Eqs (6-264) through (6-266) shear deformations are also taken into account.

   If sandwiches with thin faces are assumed in Eqs (6-252), (6-264) through (6-266) and (6-267), then the outcoming results become identical with those of [GOSCHY, 1970].

7

Bracing of building structures against buckling

Lajos KOLLÁR and Károly ZALKA

7.1 BASIC PRINCIPLES

The vertical load bearing structure of a building under vertical loads is stable in itself only in some special cases. The lateral stiffness is normally provided by a dedicated structural element or a system of structural elements. Such a structural element or system will be referred to as a *bracing element* (or *bracing core* in its most general form) or a *bracing system* consisting of shear walls and/or cores. Two questions are usually raised regarding the bracing system:

- What is the necessary stiffness of the bracing system to provide the building with adequate stability?
- How the vertical structural elements (supported by the bracing system) should be sized?

These questions will be answered in this chapter. It is assumed that the horizontal structural elements connecting the elements of the bracing system and the other load bearing elements do not develop deformation in their plane, i.e. they only develop translations and rotations but do not change their shape. The floor slabs (considered stiff in their plane) of multi-storey buildings satisfy this assumption.

It is also assumed that the bracing system is not directly loaded, i.e. it is under a small amount of load only and in the elastic state when buckling occurs.

Sections 7.2 and 7.3 deal with buildings whose lateral stiffness is provided by a single core, giving design guidelines of general validity regarding the necessary stiffness and strength of the bracing core. The more general case of buildings braced by a *system* of shear walls and cores is discussed in Section 7.4. Section 7.5 treats the stability analysis of the columns of the building.

7.2 THE NECESSARY STIFFNESS OF THE BRACING CORE

For the sake of simplicity, it is assumed in this section that the product of inertia of the cross section of the core is zero in the co-ordinate system x, y which are therefore the principal axes.

A building braced by a core can develop three basic types of buckling: it can develop flexural (*Euler*-type) buckling in the two principal planes and pure torsional buckling. These deformations can also combine resulting in flexural-torsional buckling. Accordingly, the relevant lateral and torsional stiffnesses of the bracing core should be determined first.

7.2.1 The bending stiffness of the bracing core

Consider the structure in *Fig. 7-1*. The vertical load is carried by the column on the left-hand side. The column has pinned joints on each floor level and therefore it has no lateral stiffness at all. The lateral stiffness of the structure is provided by the column on fixed support on the right-hand side representing the bracing core.

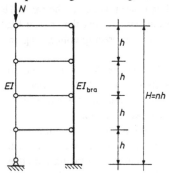

Fig. 7-1 Bracing a column having pinned joints on storey levels

Assume first that the structure has an infinite number of storeys ($n \to \infty$), i.e. the beams are densely placed. In this case the deformation curves of the two (loaded and unloaded) columns are identical. According to the energy method, the external work of the concentrated force acting on the column on the left-hand side is the same as if it acted on the bracing column. The internal work is provided by the bracing column anyway, so the critical load of the structure for $n \to \infty$ is identical to that of the case when the load acts on the bracing column.

When the number of storeys is reduced, then the buckling shape of the column on the left-hand side assumes a polygon shape and becomes more and more different from the smooth curve of the bracing column. Consequently, the top point of the unloaded column undergoes a smaller downward shift than that of the unloaded bracing column. It follows that a smaller critical load is obtained by using the top vertical translation of the bracing column than by using the top translation of the unloaded column. The calculation of the critical load of the structure of finite number of storeys assuming that the load acts on the bracing column therefore leads to a conservative estimate. The magnitude of error of such a calculation is insignificant for a four-storey structure ($n = 4$), 5.7% for a two-storey structure and 21.5% for a single-storey structure [KOVÁCS and FABER, 1963]. (This latter case is investigated in detail in Section 7.5.)

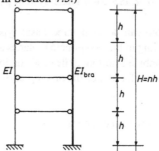

Fig. 7-2 Bracing a continuous column

The above-mentioned error practically vanishes when the unloaded column is continuous over the height of the structure (*Fig. 7-2*) as the vertical (downward) translations of the top of the loaded and unloaded columns are practically the same. However, the loaded column also develops internal work in this case so that the critical load of the structure shown in *Fig. 7-2* is also greater than that of the bracing column, again resulting in a conservative estimate.

The important conclusion that can be drawn from the above is this. If the bending stiffness of the bracing core is chosen in such a way that its elastic critical load (P_{cr}) is identical to the load on the braced column, then the exact critical load of the whole system is equal to or greater than P_{cr} [DULÁCSKA, 1966].

When this calculation is carried out, the *elastic* critical load of the bracing core has to be considered, without taking into account the magnitude of the stresses that the load P_{cr} might cause in the bracing core. Actually, the bracing core does not have to carry the total load of the building; the above criterion is only needed to ensure that the bracing core has the necessary *bending stiffness*.

It is also the elastic critical load which has to be considered when the safety of the whole structure against buckling is investigated. (The effects of plastic deformations are briefly discussed in Section 7.4.4.)

Based on the above principle, other load cases can also be investigated. If, for example, the structure is subjected to a vertical load of the same magnitude on each storey level, then the load can be considered as uniformly distributed over the height and the bracing core can be analysed by producing the overall critical load of the structure [TIMOSHENKO and GERE, 1961] as shown in *Fig. 7-3*:

$$(nN_1)_{cr} = (pH)_{cr} = \frac{7.84EI_{bra}}{H^2} \tag{7-1}$$

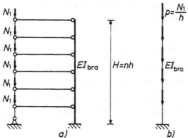

Fig. 7-3 Bracing a column subjected to distributed load over the height. *a*) The structure, *b*) principle for the stability analysis

Flexural (*Euler*-type) buckling in the two perpendicular principal planes do not combine and therefore they can be treated independently. It follows that the bracing core has to have the same EI_{bra} bending stiffness in the two directions. Interaction only occurs when the principal directions vary over the height (which only happens very rarely in practical cases), or when combined flexural-torsional buckling develops (see also Sections 7.2.3 and 7.4).

7.2.2 The torsional stiffness of the bracing core

The torsion of the building around the vertical axis is prevented by the torsional stiffness of the bracing core. This phenomenon will be investigated first by using a simple model

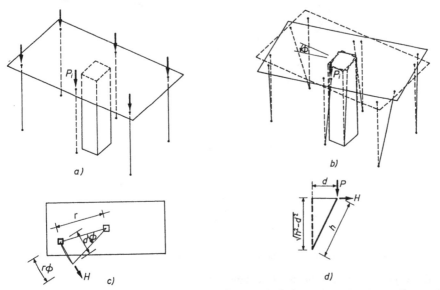

Fig. 7-4 Bracing the building against pure torsional buckling. *a*) Before torsion, *b*) after torsion, *c*), *d*) exact geometrical relationships

[KOLLÁR, 1977], then, based on a more detailed analysis, design formulae and guidelines
are presented for practical application.

A simple model for pure torsional buckling

Consider the single-storey structure shown in *Fig. 7-4a*. The structure is braced by a single
unloaded bracing core. The vertical load of the floor slab is transmitted by the columns
which carry their load share of P_i. The columns have pinned joints at both ends. Assume
for the time being that the core only has *Saint-Venant* torsional stiffness GI_t while its
warping stiffness is zero ($EI_\omega = 0$).

When the floor slab is rotated around the shear centre of the bracing core (T) by
angle φ (*Fig. 7-4b*), then the columns and their compressive forces P_i assume an inclined
position. (The phenomenon is similar to the torsional buckling of thin-walled columns with
open cross section where the originally vertical compressive stresses also become inclined
as they cannot leave the plane of the wall sections.)

Moment equilibrium around the vertical axis is expressed as follows (*Fig. 7-5*).

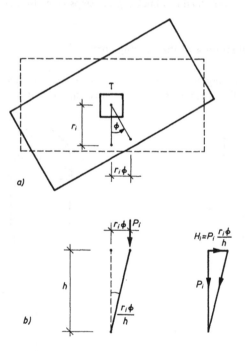

Fig. 7-5 Calculation of the torsional moment. *a*) Plan, *b*) elevation

The angle of inclination of the ith column is proportional to its distance from the shear
centre and assumes the value $r_i \Phi/h$ (*Fig. 7-5b*). Its load has to be decomposed into two
components: a horizontal one and another one in the axis of the column. The horizontal

component assumes the value (*Fig. 7-5b*)

$$H_i = P_i \frac{r_i \Phi}{h} \qquad (7\text{-}2)$$

and its moment arm with respect to the shear centre is r_i. The total (external) torsional moment on the floor slab is therefore

$$M_t^e = \sum_i P_i \frac{r_i^2 \Phi}{h}. \qquad (7\text{-}3a)$$

The balancing (internal) torsional moment originates from the torsional stiffness of the bracing core:

$$M_t^i = GI_t \frac{\Phi}{h}. \qquad (7\text{-}3b)$$

After making Eqs. (7-3a) and (7-3b) equal, expressing P_i as

$$P_i = \alpha_i P_1 \qquad (7\text{-}4)$$

where the arbitrarily chosen 'basic force' P_1 and multiplier α_i are introduced, the critical value of this force P_l is obtained as

$$P_{1cr} = \frac{GI_t}{\sum_i \alpha_i r_i^2} \qquad (7\text{-}5)$$

The total critical load is needed for practical applications. It is assumed that the floor slab is subjected to a uniformly distributed load of intensity q. Force P_i on the ith column is proportional to area ΔA_i supported by the column, according to

$$P_i = q \Delta A_i \qquad (7\text{-}6)$$

After substituting ΔA_i for α_i and q for P_i in Eq. (7-4), formula (7-5) can be rearranged:

$$q_{cr} = \frac{GI_t}{\sum_i \Delta A_i r_i^2} \qquad (7\text{-}7)$$

The denominator represents the polar moment of inertia of the area $\left(I_p \right)$ of the floor slab with respect to the shear centre of the bracing core, provided that ΔA_i is small enough for its polar moment of inertia with respect to its own centroid to be neglected compared to that of the floor slab. (This approximation is normally conservative, except when there are relatively few columns and some of them are located along the perimeter of the floor slab.) After multiplying both sides of Eq. (7-7) by the total floor area A, the overall critical load is obtained as

$$Q_{cr}^{tors} = q_{cr} A = \frac{GI_t}{I_p / A} = \frac{GI_t}{i_p^2} \qquad (7\text{-}8)$$

where

$$i_p = \sqrt{\frac{I_p}{A}} \qquad (7\text{-}9)$$

is the radius of gyration of the plan area of the building with respect to the shear centre of the bracing core.

Loads of arbitrary arrangements can also be handled applying the above procedure. After equating formulae (7-3a) and (7-3b) and multiplying both sides by $\sum P_i = Q$, formula

(7-8) is obtained but with

$$i_p^2 = \frac{\sum\limits_i P_i r_i^2}{\sum\limits_i P_i} \qquad (7\text{-}10)$$

When the floor slab is subjected to a distributed load of variable intensity $q(x,y)$, the radius of gyration is obtained using the second moment of area 'weighted' by the loads:

$$i_p^2 = \frac{\int\limits_{(A)} q(x,y)\left[(x - x_0)^2 + (y - y_0)^2\right] dA}{\int\limits_{(A)} q(x,y)dA} \qquad (7\text{-}11)$$

where x_0 and y_0 are the co-ordinates of the shear centre in the x,y co-ordinate system, i.e. $(x - x_0)^2 + (y - y_0)^2 = r^2$.

Detailed torsional analysis of a building braced by a single core

The above guidelines can be used for the torsional buckling analysis of multi-storey buildings under normal load of variable distribution. In addition to the *Saint-Venant* torsional stiffness GI_t, the core may have warping stiffness EI_ω as well. It is assumed that the columns of the building are placed densely enough to be considered 'longitudinal fibres' linking the floor slabs and transmitting the vertical load. It is also assumed that there are enough floor slabs to make the columns deform as if they were 'longitudinal fibres' which belong to the cross section of the bracing core.

As only pure torsional buckling is considered here, it is assumed that the shear centre of the bracing core and the centroid of the plan of the building coincide (*Fig. 7-6a*). The building is subjected to a concentrated load of P on top of the building and a uniformly distributed load of intensity p over the height (*Fig. 7-6b*).

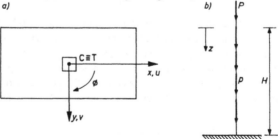

Fig. 7-6 Bracing the building against pure torsional buckling, subjected to distributed load over the height. *a*) Plan, *b*) elevation, *c*) stresses due to twisting of the fibres, *d*) stresses due to bending of the fibres

The equilibrium of the bracing core is expressed by the governing differential equation [TIMOSHENKO and GERE, 1961]

$$EI_\omega \Phi'''' - GI_t\Phi'' = \overline{m}_z \qquad (7\text{-}12)$$

where Φ is the rotation of the cross section and m_z is the torque per unit length. Assuming geometrically perfect structures, the torque is due to the horizontal components of the originally vertical normal stresses, caused by the twisting and bending of the longitudinal fibres. The magnitude of these stresses is

$$\sigma = -\frac{P + pz}{A} \qquad (7\text{-}13)$$

if the load is considered positive and the compressive stresses are considered negative.

Consider the elementary section of the longitudinal fibre of length dz. Because of its twisting (*Fig. 7-6c*), increment $d\sigma$ of stress σ results in a horizontal component. Its magnitude per unit length is obtained by multiplying $\sigma' = d\sigma/dz$ by the angle of twist which equals $u' = du/dz$ and $v' = dv/dz$ in directions x and y, if u and v denote the translations of the points of the elementary section in directions x and y, respectively.

The bending of the elementary section (*Fig. 7-6d*) also results in a horizontal component. Because of curvature u'' and v'' of the elementary section, stress σ results in resultant $\sigma u''$ and $\sigma v''$ in directions x and y. The total horizontal components are $\bar{q}_x = \left(\sigma' u' + \sigma u''\right)$ and $\bar{q}_y = \left(\sigma' v' + \sigma v''\right)$.

Taking the moments of these forces with respect to shear centre T and producing their integral over area A leads to torque $\bar{m}_z$:

$$\bar{m}_z = -\int\limits_{(A)} \left[\left(\sigma' u' + \sigma u''\right) y - \left(\sigma' v' + \sigma v''\right) x\right] dA \qquad (7\text{-}14)$$

The translations of the point of co-ordinates x, y of the surface can be expressed by the angle of rotation of the cross section Φ. As pure torsion is considered, the building only rotates about shear centre T of the bracing core and – as the origin of the co-ordinate system coincides with T – the translations are

$$u = -y\Phi \quad \text{and} \quad v = x\Phi \qquad (7\text{-}15)$$

After substituting for u, v and σ in Eq. (7-14), carrying out the integration and making use of formula (7-9), the equation

$$\bar{m}_z = -(P + pz)i_p^2 \Phi'' - pi_p^2 \Phi' \qquad (7\text{-}16)$$

is obtained which leads to the governing differential equation of pure torsional buckling as

$$EI_\omega \Phi'''' + \left[(P + pz)i_p^2 - GI_t\right]\Phi'' + pi_p^2 \Phi' = 0 \qquad (7\text{-}17)$$

Differential equation (7-17) was obtained by examining the equilibrium of the elementary section of length dz of the bar; this derivation method is called the 'differential approach'. (Another possibility is the 'integral approach' when stresses acting on the whole bar are examined. This latter approach leads to a differential equation of lower order.)

Consider first the load case when the structure is subjected to a concentrated force on top ($p = 0$). Eq. (7-17) simplifies to

$$EI_\omega \Phi'''' + \left(Pi_p^2 - GI_t\right)\Phi'' = 0 \qquad (7\text{-}18)$$

The solution to this differential equation is readily available [TIMOSHENKO and GERE, 1961]. After making use of *Timoshenko*'s solution with the relevant boundary conditions, the critical load is obtained as:

$$P_{cr} = \frac{1}{i_p^2}\left(GI_t + \frac{\pi^2 EI_\omega}{4H^2}\right) \qquad (7\text{-}19)$$

where H is the height of the building.

When the bracing core has no *Saint-Venant* stiffness, i.e. $GI_t = 0$, formula (7-19) simplifies to

$$P_{cr} = \frac{\pi^2 EI_\omega}{4H^2 i_p^2} \tag{7-20}$$

where the term $4H^2$ in the denominator is the square of the 'bending' buckling length $2H$.

When the bracing core has no warping stiffness, i.e. $EI_\omega = 0$, then the original differential equation (7-18) has to be used which assumes the form

$$\left(Pi_p^2 - GI_t\right)\Phi'' = 0 \tag{7-21}$$

If the first term (in brackets) equals zero, i.e. if

$$P_{cr} = \frac{GI_t}{i_p^2} \tag{7-22}$$

holds, then $\Phi(z)$ is an arbitrary function and the critical load does not depend on the height of the building.

The fact that the function of the angle of rotation is of arbitrary nature means that the building does not have a defined buckling shape, i.e. any function $\Phi(z)$ can represent the buckling shape. This phenomenon is similar to the one seen at the buckling analysis of bars having both bending and shear stiffnesses (see Chapter 6). It also follows that when the part critical loads associated with the EI_ω and GI_t stiffnesses are added up using the *Southwell* theorem, the summation produces the exact solution, i.e. formula (7-19), since the undefined buckling shape which belongs to GI_t 'conforms' to the defined buckling shape which belongs to EI_ω.

Consider now the uniformly distributed load case ($P = 0$). Differential equation (7-17) assumes the form

$$EI_\omega \Phi'''' + \left(pzi_p^2 - GI_t\right)\Phi'' + pi_p^2\Phi' = 0 \tag{7-23}$$

Assume first that the core has no *Saint-Venant* stiffness ($GI_t = 0$). Eq. (7-23) simplifies to

$$EI_\omega \Phi''' + pzi_p^2\Phi' = 0 \tag{7-24}$$

This differential equation is analogous to the differential equation characterizing buckling by bending [TIMOSHENKO and GERE, 1961] so that the solution is also analogous:

$$p_{cr}H \equiv Q_{cr} = \frac{7.84 EI_\omega}{H^2 i_p^2} \tag{7-25}$$

If the core has no warping stiffness ($EI_\omega = 0$), then Eq. (7-23) simplifies to

$$\left(pzi_p^2 - GI_t\right)\Phi' = 0. \tag{7-26}$$

It is not necessary to solve this first-order differential equation of variable coefficients in the formal way as the following simple physical considerations yield the solution.

The left-hand side of Eq. (7-26) represents a product which can only assume zero value if one of its terms vanishes. As coefficient z in the brackets varies over the height, the term in brackets can equal zero at one point where Φ' can be of arbitrary value. Function Φ' has to assume zero value everywhere else. It follows that the function of rotation $\Phi'(z)$ assumes the shape shown in *Fig. 7-7a*: the core rotates at one point only and the rest of the core does not develop deformation. Accordingly, the formula for the critical load is

$$p_{cr} = \frac{GI_t}{zi_p^2}. \tag{7-27a}$$

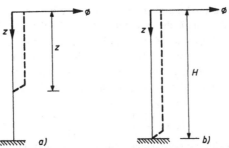

Fig. 7-7 Buckling shape of the structure subjected to distributed load over the height. *a*) In general, *b*) buckling shape for the smallest load

The value of the critical load is the smallest when the value of z is the greatest, i.e. when $z = H$. Consequently, function $\Phi'(z)$ characterizing the rotation of the core assumes the shape shown in *Fig. 7-7b* and the critical load is

$$p_{cr}H \equiv Q_{cr} = \frac{GI_t}{i_p^2} \tag{7-27b}$$

The value of the critical load has turned out to be the same as with the load case when the core is subjected to a concentrated force at the top (formula (7-22)). This result can also be interpreted in a different way. The structure develops torsional buckling at the cross section where the compressive force is maximum along 'an infinitely short buckling length'. This can happen because (for a constant compressive load) the buckling shape is undefined, likewise the buckling length; the latter can therefore become as short as possible. It follows that, as far as torsional buckling is concerned, it is irrelevant whether the compressive force acts at a single point or over a longer section.

The analogy with the buckling of columns with both bending and shear stiffnesses also exists for the distributed load case (Chapter 6). If the two part critical loads are added up using the *Southwell* theorem, then, contrary to the concentrated load case, the resulting formula

$$p_{cr}H = Q_{cr} = \frac{1}{i_p^2}\left(GI_t + \frac{7.84EI_\omega}{H^2}\right) \tag{7-28}$$

is not exact because the buckling shape shown in *Fig. 7-7b* differs from the 'bending' type buckling shape which belongs to the case when the *Saint-Venant* torsional stiffness is zero ($GI_t = 0$). Formula (7-28) is conservative (see also Section 7.4 where the exact solution is presented.)

Finally, it may be useful to note that when area A equals the area of the cross section of the bracing core, then the formulae presented above also represent the solution of the pure torsional buckling problem of thin-walled columns.

7.2.3 Generalization of the results. Spatial behaviour

The evaluation of formula (7-28) leads to the following conclusions.

a) The necessary torsional (and warping) stiffness of the bracing core is defined by the total load on the building, as is the case with flexural (*Euler*-type) buckling.

b) As the plan length and breadth of the building increase, the critical load decreases (in proportion to the square of the plan sizes). This can be easily demonstrated for rectangular plans. If the shear centre of the bracing core and the centroid of the plan coincide, the square of the radius of gyration assumes the form:

$$i_p^2 = \frac{\dfrac{ab^3}{12} + \dfrac{ba^3}{12}}{ab} = \frac{a^2 + b^2}{12} \tag{7-29}$$

This result is totally different from that obtained for flexural (*Euler*-type) buckling where there is no relation whatsoever between the plan area of the building and the critical load.

c) The greater the eccentricity of the shear centre of the bracing core in relation to the centroid of the plan of the building, the smaller the critical load. This is clearly seen by examining the formula for the square of the radius of gyration which, with the eccentricity of the shear centre denoted by t (*Fig. 7-9b*), assumes the form

$$i_p^2 = \frac{\dfrac{ab^3}{12} + \dfrac{ba^3}{12} + t^2 ab}{ab} = \frac{a^2 + b^2}{12} + t^2 \tag{7-30}$$

This can also be interpreted in a different way: the more eccentric the bracing core is with respect to the centroid of the plan of the building, the less effective the core becomes against pure torsional buckling.

d) As with *Euler*-type buckling, the critical load should be calculated using elastic theory as (most of) the load of the building does not act on the bracing core.

The critical load $\left(p_{cr}H \equiv Q_{cr}^{\Phi}\right)$ given by formula (7-28) represents the critical load for pure torsional buckling and can only be readily used if pure torsional buckling does not combine with flexural (*Euler*-type) buckling. To examine the problem of combination in more detail, consider the phenomena using an ordinary column.

According to the theory of torsional buckling of thin-walled columns of open cross section [TIMOSHENKO and GERE, 1961], if the shear centre and the centroid of the cross section do not coincide, then pure torsional buckling combines with the flexural buckling which is perpendicular to the principal axis in which direction the shear direction is eccentric to the centroid. This can easily be demonstrated using common sense. Consider for example the column with an L-section (*Fig. 7-8a*) which is subjected to a centrally applied compressive load. Assume that the column develops pure flexural buckling in direction y (*Fig. 7-8b*). Because of the bending of the longitudinal fibres, the resulting axial compressive stresses develop forces $\overline{q}_y$ in direction y (*Fig. 7-8b,c*) whose resultant acts in the line of action of the external compressive load, i.e. in the centroid of the cross section and therefore exerts a torsional moment around the shear centre. (See also Eq. (7-12)). It follows that flexural buckling cannot develop in itself and the column also has to undergo torsion.

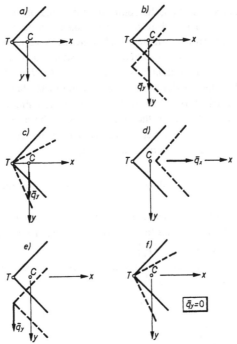

Fig. 7-8 Combination of buckling modes. *a*) Cross section of the column, *b*) compressive force in the centroid; buckling in direction *y*, *c*) compressive force in the centroid; torsional buckling, *d*) compressive force in the centroid; buckling in direction *x*, *e*) compressive force in the shear centre; buckling in direction *y*, *f*) compressive force in the shear centre; torsional buckling

Assume now that the column develops pure torsional buckling about the shear centre axis (*Fig. 7-8c*). Because of the bending of the centroidal axis, the external compressive load develops forces $\bar{q}_y$ again in direction *y*, resulting in bending in direction *y*, i.e. flexural buckling. It follows that pure torsional buckling cannot develop in itself either and it has to combine with flexural buckling in direction *y*.

However, buckling in direction *x* does not combine with pure torsional buckling because, on the one hand, pure torsional buckling (*Fig. 7-8c*) does not develop forces $\bar{q}_x$ in direction *x* which would in turn cause bending in direction *x*, and, on the other hand, the resultant of forces $\bar{q}_x$ acts at the centroid (*Fig. 7-8d*) and passes through the shear centre and therefore does not cause torsion.

In the case of buildings braced by a core, the role of the centroid is taken by the centroid of the (loaded) floor slab(s). The above reasoning therefore has to be modified accordingly. The difference between the buckling of an ordinary column and that of a building is that the load does not act on the cross section of the column in a uniformly distributed manner but it acts outside the 'column' (i.e. the core), distributed over the area of the plan of the building. It follows that the role of the geometrical characteristics of the cross section as the loaded area is taken by the geometrical characteristics of the area of the plan of the building: instead of the centroid and the radius of gyration with respect to the shear centre of the cross section, the centroid of the area of the plan and the radius of gyration of the area of the plan with respect to the shear centre of the bracing core have to be considered for uniformly distributed load. (However, as far as the *stiffness*-related characteristics of the cross section are concerned, the situation is unchanged and the shear centre, the principal axes and the moments of inertia (I_t, I_ω, I_x, I_y) have to be calculated using the geometrical characteristics of the cross section of the bracing core.)

The geometrical principal axes ξ, η of the layout of the building do not play any role in the above considerations. However, they might be useful because it is easier to calculate the polar moment of inertia with respect to centroid C in the co-ordinate system whose co-ordinate axes coincide with ξ and η, as the formula

$$I_{p,C} = I_\xi + I_\eta \tag{7-31}$$

known in the strength of materials shows than in any other co-ordinate system with skew axes (*Fig. 7-9c*).

If the load is not *uniformly* distributed over the plan, then, instead of the centroid of the plan, the centroid of the loads has to be used and i_p^2 also has to be calculated using the loads, according to the formula $\sum \left(r_i^2 P_i \right) \big/ \sum P_i$.

The combination of the flexural and torsional buckling modes of the building braced by a core can now be characterized as follows.

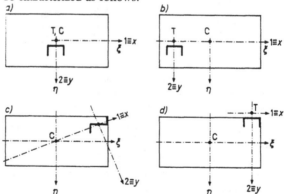

Fig. 7-9 Relative location of the centroid of mass C with respect to the principal axes of the core

The three buckling modes do not combine when the centroid of the loads (C) and the shear centre of the core (T) coincide (*Fig. 7-9a*). This corresponds to the case shown in *Fig. 7-8* when the external compressive force acts at the shear centre and not at the centroid. In this case – apart from the fact that the cross section is under eccentric compression in direction x – pure torsional buckling and flexural buckling in direction y do not combine because forces $\overline{q}_y$ originating from flexural buckling in direction y pass through the shear centre (*Fig. 7-8e*) and do not cause torsion; in the case of pure torsion the shear centre axis remains straight and therefore forces $\overline{q}_y$ do not even develop (*Fig. 7-8f*).

If the centroid of the load (C) does not coincide with the shear centre of the core (T) but lies on one of the principal axes of the core passing through T (e.g. on axis x: *Fig. 7-9b,c*), then pure torsional buckling combines with the *Euler*-type buckling perpendicular to axis x and the combination reduces the value of the critical load. A conservative estimate of the combined critical load can be obtained applying the *Föppl-Papkovich* theorem (discussed in Chapter 2) to the part (pure torsional) critical load Q_{cr}^{tors} and the part critical load Q_{cr}^x (which represents flexural buckling in direction y):

$$\frac{1}{Q_{cr}^{comb}} = \frac{1}{Q_{cr}^{tors}} + \frac{1}{Q_{cr}^x} \tag{7-32a}$$

If the centroid of the load does not lie on either of the principal axes passing through the shear centre of the core (T) (*Fig. 7-9d*), then pure torsional buckling combines with *Euler*-type buckling perpendicular to both axes x and y. The *Föppl-Papkovich* formula again can be used to produce a conservative estimate:

$$\frac{1}{Q_{cr}^{comb}} = \frac{1}{Q_{cr}^{tors}} + \frac{1}{Q_{cr}^x} + \frac{1}{Q_{cr}^y} \tag{7-32b}$$

The above *Föppl-Papkovich* formulae are simple to use and are always conservative. However, their use in some cases may lead to considerably uneconomical structural solutions as the error of the formulae can be as much as 67%. The more sophisticated – and slightly more complicated – exact method for taking into account the combination of the part critical loads is given in Section 7.4.2.

Section 7.4 also discusses the case when – instead of a single core – the building is braced by a system of shear walls and cores.

Finally, another important aspect of buckling is discussed here briefly. It is important to establish the nature of the *post-critical behaviour* as it shows whether or not the structure is sensitive to initial imperfections.

As far as *Euler*-type buckling is concerned (*Figs 7-1, 7-2, 7-3*), the post-critical behaviour of the structure is identical to that of ordinary columns, i.e. it is considered constant.

On the other hand, the behaviour of a structure braced against pure torsional buckling is totally different from that of an ordinary column developing pure torsional buckling. According to Section 1.5, an ordinary thin-walled column develops stiffening post-critical behaviour because the load can transfer to the internal stiffer parts of the cross section and also the walls of the thin-walled cross section can benefit from membrane action. However, the load of the structure shown in *Fig. 7-4* cannot be transferred and there are no elements which can develop membrane actions. The post-critical behaviour of the structure has to be investigated in another way. For the sake of simplicity, the investigation concentrates on a single column.

The horizontal translation of the top of the column during the rotation of the structure in *Fig. 7-4a* is shown in *Fig. 7-4c*. The translation assumes the value:

$$d = r2\sin\frac{\Phi}{2} \tag{7-33a}$$

According to the inclined position of the column, the horizontal component of force P assumes the value

$$H = P\frac{d}{\sqrt{h^2 - d^2}} \tag{7-33b}$$

The (external) torsional moment due to force H (*Fig. 7-4c*) is

$$M_e = Hr \cos \frac{\Phi}{2} \tag{7-33c}$$

Substituting for H and d and after some rearrangement, the torsional moment is obtained as

$$M_e = \frac{2 \sin \frac{\Phi}{2} \cos \frac{\Phi}{2} r^2 P}{\sqrt{h^2 - 4r^2 \sin^2 \frac{\Phi}{2}}} \cong \frac{r^2 P}{h} \left[\Phi + \left(\frac{r^2}{h^2} - \frac{1}{3} \right) \frac{\Phi^3}{2} \right]. \tag{7-33d}$$

The internal (resisting) torsional moment of the core is directly proportional to the angle of rotation:

$$M_i = GI_t \frac{\Phi}{h} \tag{7-33e}$$

Making formulae (7-33d) and (7-33e) equal results in

$$\frac{r^2}{h^2} = \frac{1}{3}$$

It follows that if

$$\frac{h}{r} < \sqrt{3}$$

holds, then the external torsional moment increases faster than the internal torsional moment and the structure develops unstiffening post-critical behaviour. If, however

$$\frac{h}{r} > \sqrt{3}$$

holds, then the external torsional moment increases slower than the internal torsional moment and the post-critical behaviour is that of stiffening.

According to this result obtained from this investigation of the post-critical behaviour of one column, the conclusion can be drawn that for multi-storey buildings with more columns the post-critical behaviour depends on the value of the ratio h/r which is calculated by producing the averages of h and r. The post-critical behaviour tends to be stiffening for great h/r values and unstiffening for small h/r values.

7.3 THE NECESSARY STRENGTH OF THE BRACING CORE

The external loads on the building (wind, seismic force, forces due to misalignment) have to be transmitted by the bracing core to the foundation. Well-known bending effects are not investigated here; only two torsional phenomena often neglected in structural analysis are discussed briefly in this section.

The columns of the structures are often constructed in such an inclined way that, because of their inclined position, they develop torsional moments about the shear centre of the bracing core.

Another effect originates from the wind even when, contrary to common sense, the building has a doubly symmetrical rectangular plan. Measurements in wind tunnels demonstrate that the resultant of the wind load exerts a torsional moment to the centre of the rectangular plan (C) when it has an angle with the axes of symmetry [FLACHSBART, 1932]. It assumes the maximum value of $0.1pa^2$ per unit height for narrow rectangular cross sections when the angle is $32°$ (*Fig. 7-10*), where p is the wind pressure. (If the core is eccentrically located or the layout is not rectangular, the wind or the seismic force develops additional torsional moments.)

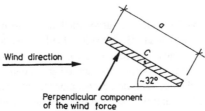

Wind direction

Perpendicular component
of the wind force

Fig. 7-10 Torsional moment due to inclined wind

The value of the bending and torsional moments originating from horizontal loads has to be increased to take into consideration the fact that the deformations caused by the horizontal loads further increase due to axial forces. The phenomenon is basically identical to the increase of moments with *Euler* buckling which is shown in *Fig. 7-11*.

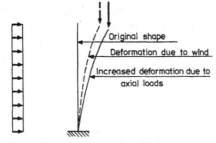

Original shape

Deformation due to wind

Increased deformation due to
axial loads

Fig. 7-11 Increased deformations and moments due to the axial forces

If it is assumed that the deflection shape of the core due to the horizontal load and the buckling shape (the first eigenfunction) are similar, then the amplification factor of *Euler* buckling is obtained as

$$\Psi = \frac{1}{1 - \dfrac{Q}{Q_{cr}}} \qquad (7\text{-}34)$$

where Q is the total load of the building and Q_{cr} is the elastic critical load of the core [TIMOSHENKO and GERE, 1961, see also Eq. (1-14)]. The value of the bending moment (originally M_{ben}) is calculated from

$$M = \Psi M_{ben}. \qquad (7\text{-}35)$$

In order to limit the value of amplification factor Ψ and to ensure adequate safety against buckling, Q_{cr} should be at least four times as great as Q. Some Code of Practice provisions, however, set more rigorous limits. Stability criteria in BS5950, for example, involve a combination of a drift and lateral load criteria which lead to a ratio of $Q_{cr}/Q = 10$ [MACLEOD and ZALKA, 1996]. These conditions also give an indication how to choose the value of Q_{cr}/Q which can be considered as a global safety factor.

The above reasoning is also valid for pure torsional buckling when Q_{cr} in formula (7-34) should represent the pure torsional critical load of the core and formula (7-35) should refer to torsional moments.

If flexural buckling and pure torsional buckling combine, then the guidelines given below should be followed.

Combined flexural-torsional buckling is associated with one or two flexural critical loads and a pure torsional critical load. Each critical load is characterized by a different buckling shape (eigenfunction) where the (one or two) flexural and the torsional deformation components are in a different proportion to each other. Theoretically, the deformation originating from the horizontal forces should be decomposed into components corresponding to the relevant eigenfunctions and then each should be multiplied by the corresponding amplification factor (Ψ) obtained by using the relevant Q_{cr}^{comb}.

Instead, the following simple procedure can also be used.

One or two amplification factor(s) (Ψ) obtained by using greater Q_{cr}^{comb} will definitely be smaller than amplification factor Ψ obtained using the smallest Q_{cr}^{comb}. It will therefore result in a conservative estimate if the total deformation due to the horizontal load, i.e. both the total bending and the torsional moments are multiplied by amplification factor Ψ (formula (7-34)) which is calculated using the smallest Q_{cr}^{comb} of the combined flexural-torsional buckling. In doing so, it is no longer relevant to investigate how much the ratio of the flexural and torsional deformations resulting from the horizontal load differs from the ratio of the combined flexural and torsional deformation components calculated using the smallest Q_{cr}^{comb}. The analysis becomes very simple as formulae (7-32) always result in the smallest Q_{cr}^{comb}.

7.4 BRACING SYSTEM OF SHEAR WALLS AND CORES

In this section the more general case will be discussed, i.e. when the building is braced by a system of shear walls and cores. Even the exact stress analysis of such structures is complicated, partly because of the interaction between the horizontal and vertical systems and among the elements of the horizontal and vertical systems themselves and partly because of the great number of elements to be involved in the analysis. The finite element method has proved to be efficient for the stress analysis [ZIENKIEWITZ, PAREKH and TEPLY, 1971]. As far as the stability analysis of large structures is concerned, however, an additional problem emerges as the stiffness matrix of the structure is often ill-conditioned. Even sophisticated FE procedures may have difficulties in determining the critical load. The LUSAS 'User Manual' [1995] warns that the solution is not without its problems and convergence problems might emerge in the iterative procedure. The most effective approximate solutions rely on the theory of thin-walled structures [PODOLSKI, 1970; GLÜCK and GELLÉRT, 1972; DANAY, GELLÉRT and GLÜCK, 1975; STAFFORD SMITH and COULL, 1991]. According to investigations into the behaviour of large structures, under certain circumstances, the overall stability analysis of buildings can be traced back to the torsional buckling analysis of a single cantilever of thin-walled cross section, subjected to uniformly distributed vertical load. The biggest challenge through the procedure is the solution of the eigenvalue problem of the differential equation of variable coefficients of pure torsional buckling [BARTA, 1967; YARIMCI, 1972]. Instead of solving this problem, the authors of different approximate procedures often introduce different simplifying assumptions concerning the torsional or warping characteristics of the bracing system [AKESSON, 1980; ROSMAN, 1980].

Based on the continuum method and using the equivalent column approach, the exact solution is given in this section for the calculation of the critical load of pure torsional buckling. The exact interaction of the flexural and pure torsional critical loads is taken into account using a simple algebraic equation.

Only the formulae for practical application are given here; the theoretical background and the derivation of the governing equations are available elsewhere [ZALKA and ARMER, 1992; ZALKA, 1994].

Only regular structures are considered. Structures are defined regular, if:

a) the floor slabs are at a regular distance from each other, in other words, if the storey height is constant;

b) the arrangement and the cross sections of the vertical load bearing and bracing elements do not vary over the height of the building.

It is also assumed that the bracing elements are not pierced by openings.

7.4.1 The equivalent column

The stability analysis is based on the equivalent column which represents the whole bracing system. It is obtained by combining all the bracing elements to form a single column and has a thin-walled cross section. The first step is to find the location of the shear centre of the equivalent column which coincides with the shear centre of the bracing system (O).

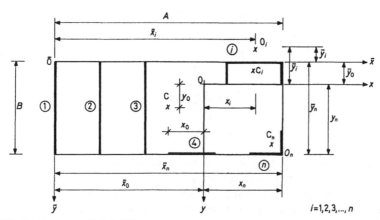

Fig. 7-12 Bracing system of the building

For rectangular building layouts, the calculation is carried out in the co-ordinate system $\bar{x}, \bar{y}$ whose origin lies in the upper left corner of the plan of the building and whose axes are aligned with the sides of the building (*Fig. 7-12*). (For layouts of general arrangement, a convenient location of co-ordinate system $\bar{x}, \bar{y}$ is chosen in such a way that the calculation is the simplest.) The co-ordinates are obtained by making use of the geometrical and stiffness characteristics of the bracing elements [BECK and SCHÄFER, 1969]

$$\bar{x}_0 = \frac{I_{xy}\left(\sum I_{y,i}\bar{y}_i - \sum I_{xy,i}\bar{x}_i\right) - I_y\left(\sum I_{xy,i}\bar{y}_i - \sum I_{x,i}\bar{x}_i\right)}{I_x I_y - I_{xy}^2} \qquad (7\text{-}36a)$$

$$\bar{y}_0 = \frac{I_x\left(\sum I_{y,i}\bar{y}_i - \sum I_{xy,i}\bar{x}_i\right) - I_{xy}\sum I_{xy,i}\bar{y}_i - \sum I_{x,i}\bar{x}_i}{I_x I_y - I_{xy}^2} \qquad (7\text{-}36b)$$

where $I_{x,i}$, $I_{y,i}$ and $I_{xy,i}$ represent the moments of inertia and the product of inertia of the ith element of the bracing system with respect to its local centroidal co-ordinate axes which are parallel to axes $\bar{x}$ and $\bar{y}$. Quantities

$$I_x = \sum_1^n I_{x,i}, \quad I_y = \sum_1^n I_{y,i},$$

and

$$I_{xy} = \sum_1^n I_{xy,i}$$

are the sums of the second moments of area and the product of inertia of the elements of the bracing system, where $i = 1, ..., n$ and n is the number of bracing elements.

Knowing the location of the shear centre, the equivalent column can now be created by incorporating the individual bracing elements into a single column. The stiffnesses of the equivalent column are special sums of those of the individual bracing elements, computed by the following principles. For the bending and pure torsional stiffnesses, the corresponding values of the individual bracing elements have to be added up:

$$EI_x = E\sum_1^n I_{x,i} \qquad (7\text{-}37a)$$

$$EI_y = E\sum_1^n I_{y,i} \qquad (7\text{-}37b)$$

$$EI_{xy} = E\sum_1^n I_{xy,i} \qquad (7\text{-}37c)$$

$$GI_t = G\sum_1^n I_{t,i} \qquad (7\text{-}37d)$$

where E is the modulus of elasticity, G is the modulus of elasticity in shear and $I_{t,i}$ is the *Saint-Venant* torsion constant of the ith bracing element. The establishment of formulae $(7\text{-}37a, ..., d)$ follows from the fact that the stiffness of the floor slabs of the building is infinitely great in their plane and infinitely small perpendicular to their plane. The infinitely great in-plane stiffness of the floor slabs makes the bracing elements work together in the horizontal direction, resulting in identical deflections. However, the infinitely small stiffness of the floor slabs perpendicular to their plane cannot make them work together in the vertical direction (since no vertical shear forces are transmitted) so that the individual bracing elements cannot act as the longitudinal fibres of a bar and do not develop axial deformation from the point of view of the whole system. It follows that, as far as bending is concerned, the floor slabs act like pinned bars linking the cores.

The warping stiffness of the equivalent column is obtained from the following formula [BRANDT, SCHÄFER and REEH, 1975]:

$$EI_\omega = E \left(\sum_1^n I_{\omega,i} + \sum_1^n I_{x,i} x_i^2 + \sum_1^n I_{y,i} y_i^2 - 2 \sum_1^n I_{xy,i} x_i y_i \right) \tag{7-37e}$$

Coordinates x_i and y_i in this formula are the co-ordinates of the shear centre of the ith bracing element in the co-ordinate system x, y, whose origin coincides with the shear centre of the bracing system and whose x and y axes are parallel with the corresponding sides of the ground plan of the building with a rectangular layout (*Fig. 7-12*).

Formula (7-37e) needs some explanation. The first term in the formula shows that all the bracing elements may develop warping deformation. The remaining terms in the formula are responsible for the fact that the equivalent column as a whole can also develop warping since the floor slabs (the 'cross sections' of the equivalent column) are infinitely flexible perpendicular to their plane and do not remain plane. They also show that, in addition to their own warping stiffnesses, the individual bracing elements also contribute to the overall warping stiffness of the system by utilizing their bending stiffnesses and their 'torsion arm' about the shear centre of the bracing system.

In many practical cases the product of inertia of the bracing elements (and of the bracing system) is zero. In such cases axes x and y are also the principal axes of the bracing system and the establishment of the equivalent column is complete. However, I_{xy} may be different from zero (e.g. for Z and L shaped bracing elements) and in such cases it is necessary to calculate the bending stiffnesses with respect to the principal axes as the stability analysis (later in the section) has to be carried out in the co-ordinate system whose horizontal co-ordinate axes point to the direction of the principal axes. The angle of principal axis X with axis x (*Fig. 7-12*) is given by

$$\alpha = \frac{1}{2} \arctan \frac{2I_{xy}}{I_y - I_x} \tag{7-38}$$

Principal axis Y is perpendicular to axis X. The moments of inertia with respect to the principal axes are obtained from

$$I_X = I_x \cos^2 \alpha + I_y \sin^2 \alpha - I_{xy} \sin 2\alpha \tag{7-39a}$$

and

$$I_Y = I_x \sin^2 \alpha + I_y \cos^2 \alpha + I_{xy} \sin 2\alpha \tag{7-39b}$$

Finally, the bending stiffnesses with respect to the principal axes are given by EI_X and EI_Y (and $EI_{XY} = 0$ as the product of inertia vanishes in the co-ordinate system whose axes are the principal axes).

In addition to the above bending and torsional stiffnesses, the radius of gyration of the floor area is also needed. As demonstrated in Section 7.2.3, the radius of gyration is defined by the size and shape of the plan of the building. The radius of gyration is obtained from

$$i_p^2 = \frac{A^2 + B^2}{12} + t^2 \tag{7-40}$$

for rectangular plans, where A and B are the plan length and breadth of the building and $t = \left(x_c^2 + y_c^2\right)^{1/2}$ is the distance between the shear centre (O) and the centre of the vertical load (C) (*Fig. 7-12*).

Formulae for other than rectangular plans are given in Section 7.2.3 where the accuracy of the formulae is also discussed.

7.4.2 Uniformly distributed load over the height

The combined flexural-torsional buckling of the equivalent column with the above stiffness characteristics and subjected to uniformly distributed axial load [VLASOV, 1961; ZALKA and ARMER, 1992] is defined by the system of governing differential equations:

$$u'''' + \left(\frac{N}{EI_X}\left(u' - y_c\varphi'\right)\right)' = 0 \tag{7-41a}$$

$$u'''' + \left(\frac{N}{EI_Y}\left(v' + x_c\varphi'\right)\right)' = 0 \tag{7-41b}$$

$$EI_\omega\varphi'''' - GI_t\varphi'' + \left(N\frac{I_0}{A}\varphi'\right)' - \left(N\left(x_c v' - y_c u'\right)\right) = 0 \tag{7-41c}$$

where
$N = qz$ is the total axial load at z,
A is the plan area
$I_0 = I_x + I_y + A\left(x_c^2 + y_c^2\right)$ is the polar moment of inertia of the plan
                                   area with respect to the shear centre.

The derivation of Eq. (7-41c) (with $x_c = y_c = 0$) is also given in Section 7.2.2.

The boundary conditions in the co-ordinate system whose origin is fixed to the top cross section of the equivalent column are as follows.

The translations and rotation are zero at the top of the column:

$$u(0) = v(0) = \varphi(0) = 0 \tag{7-42a}$$

The slopes of deflection as well as warping are zero at the bottom of the column:

$$u'(H) = v'(H) = \varphi'(H) = 0 \tag{7-42b}$$

where H is the height of the building.

The bending moments as well as the warping stresses are zero at the top of the column:

$$u''(0) = v''(0) = \varphi''(0) = 0 \tag{7-42c}$$

The shear forces as well as the torsional moment equal zero at the top of the column:

$$u'''(0) = v'''(0) = GI_t\varphi'(0) - EI_\omega\varphi'''(0) = 0 \tag{7-42d}$$

Equations (7-41a,b,c) are the three simultaneous differential equations for buckling by bending and torsion and can be used for the determination of the critical load. The angle of rotation φ appears in all three equations showing that in the general case torsion about the shear centre and bending in both principal axes of inertia occur simultaneously (see also Section 7.2.3). In the case of columns with thin-walled open cross section, the phenomenon of torsional-flexural buckling is very important because the critical load of pure torsional buckling can be considerably smaller than those of flexural buckling and the critical load of the combined torsional-flexural buckling is even smaller.

For the actual calculation of the combined critical load, the flexural and pure torsional critical loads are needed. These are defined by the case when the shear centre of the bracing system and the centroid of the vertical load coincide, i.e. by Eqs (7-41a,b,c) with $x_c = y_c = 0$. Equations (7-41a,b,c) simplify considerably and at the same time become independent from each other. They can be solved one by one and the solutions represent the two flexural critical loads

$$N_{cr,X} = qH = \frac{7.84EI_X}{H^2},$$ (7-43a)

$$N_{cr,Y} = qH = \frac{7.84EI_Y}{H^2},$$ (7-43b)

and the pure torsional critical load

$$N_{cr,\varphi} = qH = \alpha N_{cr,\varphi}^{\omega}.$$ (7-44)

In formulae (7-43a,b), EI_X and EI_Y are the bending stiffnesses of the bracing system with respect to the principal axes.

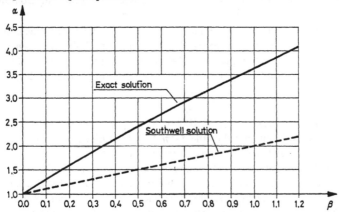

Fig. 7-13 Critical load parameter α as a function of β ($0 \leq \beta \leq 1.2$)

In formula (7-44)

$$N_{cr,\varphi}^{\omega} = \frac{7.84EI_{\omega}}{i_p^2 H^2}$$ (7-45)

is the warping part of the torsional critical load. Values for critical load parameter α are given in *Table 7-1* and in *Fig. 7-13* as a function of stiffness parameter β

$$\beta = \frac{N_{cr,\varphi}^t}{N_{cr,\varphi}^{\omega}}$$ (7-46)

where

$$N_{cr,\varphi}^t = \frac{GI_t}{i_p^2}$$ (7-47)

is the pure torsional part of the torsional critical load.

Bracing of buildings

Table 7-1 Critical load parameter α

β	α	β	α	β	α
0.000	1.000	2	5.624	60	82.3
0.001	1.003	3	7.427	70	94.4
0.01	1.030	4	9.100	80	106.4
0.1	1.295	5	10.70	90	118.4
0.2	1.578	6	12.24	100	130.2
0.3	1.856	7	13.75	120	153.8
0.4	2.123	8	15.23	140	177.1
0.5	2.382	9	16.68	160	200.3
0.6	2.633	10	18.12	180	223.3
0.7	2.878	20	31.82	200	246.2
0.8	3.116	30	44.86	500	581.8
0.9	3.349	40	57.54	1000	1115
1.0	3.576	50	69.99	10 000	10 000

Fig. 7-13 also shows the approximate solution of pure torsional buckling obtained by using the *Southwell* theorem (see also formula (7-28) in Section 7.2.2). The *Southwell* solution is always conservative and the maximum error is 47%.

Formula (7-44) cannot be used in the special case when $EI_\omega = 0$, i.e. when the column only has torsional stiffness. The denominator of the fraction in Eq. (7-46) is zero and the value of the fraction tends to infinity. Equation (7-44) assumes the form $0 \cdot \infty$ and collapses. In this particular case the critical load is obtained by making use of formula (7-28):

$$N_{cr,\varphi} = \frac{GI_t}{i_p^2} \tag{7-48}$$

Table 7-1 has values for the range $0 \le \beta \le 10000$. When β is greater than 10000, i.e. when the torsional behaviour is dominated by the *Saint-Venant* stiffness, then again formula (7-48) can be used for the calculation of the torsional critical load.

The critical loads for flexural and pure torsional buckling (formulae (7-43*a,b*) and (7-44)) can directly be used for the stability analysis when the centroid of the load and the shear centre of the bracing system coincide. The three buckling modes do not combine and the smallest one of the three critical loads is also the critical load of the building.

However, the above symmetry condition (i.e. identical shear centre and centroid) is not fulfilled in many practical cases so that the three buckling modes interact and the building develops a combined flexural-torsional buckling. It has been demonstrated that the exact effect of the interaction can be taken into account in a very simple way [ZALKA, 1994] and the combined critical load (N_{cr}) is obtained as the smallest root of the cubic equation

$$N^3 + \frac{N_{cr,X}\tau_X^2 + N_{cr,X}\tau_Y^2 - N_{cr,\varphi} - N_{cr,X} - N_{cr,Y}}{1 - \tau_X^2 - \tau_Y^2}N^2 + \tag{7-49}$$

$$+ \frac{N_{cr,X}N_{cr,Y} + N_{cr,\varphi}N_{cr,X} + N_{cr,\varphi}N_{cr,Y}}{1 - \tau_X^2 - \tau_Y^2}N - \frac{N_{cr,\varphi}N_{cr,X}N_{cr,Y}}{1 - \tau_X^2 - \tau_Y^2} = 0$$

where

$$\tau_X = \frac{x_c}{i_p} \quad \text{and} \quad \tau_Y = \frac{y_c}{i_p} \tag{7-50}$$

characterize the eccentricity of the load. The flexural critical loads $N_{cr,X}$ and $N_{cr,Y}$ and the pure torsional critical load $N_{cr,\varphi}$ in Eq. (7-49) are defined by formulae (7-43a,b) and (7-44).

7.4.3 Concentrated load at top floor level

Cubic equation (7-49) can also be used to take into account the exact effect of the inter-action of the flexural and pure torsional critical loads when the building is subjected to concentrated load on top floor level [TIMOSHENKO and GERE, 1961] (e.g. a water tank or a panorama restaurant). The concentrated critical loads

$$P_{cr,X} = \frac{\pi^2 EI_X}{4H^2} \tag{7-51a}$$

$$P_{cr,Y} = \frac{\pi^2 EI_Y}{4H^2} \tag{7-51b}$$

$$P_{cr,\varphi} = \frac{1}{i_p^2}\left(GI_t + \frac{\pi^2 EI_\omega}{4H^2}\right) \tag{7-52}$$

have to be substituted for N_{crX}, N_{crY} and $N_{cr,\varphi}$ and the smallest root of the cubic equation yields the combined critical load P_{cr}.

7.4.4 Supplementary remarks

The effect of the distribution of the floor load over the height
The smallest root of cubic equation (7-49) represents the exact solution for the combined flexural-torsional buckling of a column which is subjected to uniformly distributed vertical load over the height. With multi-storey buildings, however, the situation is somewhat different as the actual vertical load consists of concentrated forces at floor levels, each representing the load of one storey. The uniformly distributed load of the equivalent column is obtained by distributing these forces *downwards* along the centroidal axis. In this way, a model, representing a more favourable situation than the original one, is created and the procedure leads to an overestimated critical load.

This approximation may be significant for low-rise buildings as the distribution of the load over the height leaves a concentrated force on the top of the building which is not 'accounted for' and whose effect, compared to the effect of the distributed load, is more significant than in the case of taller buildings. However, the unconservative effect of this approximation can be eliminated by applying *Dunkerley*'s additive theorem [ZALKA and ARMER, 1992]. According to the additive formula, the critical load which takes into account the effect of the distribution of the vertical load is obtained from

$$N_{cr}^D = \left(\frac{1}{2n\,P_{cr}} + \frac{1}{N_{cr}}\right)^{-1} \tag{7-53}$$

where P_{cr} is the concentrated critical load, N_{cr} is the critical load for the uniformly distributed load case and n is the number of storeys.

Shear centre - centre of torsion

When the equivalent column is established, it is assumed that the location of the shear centre only depends on the geometrical characteristics of the cross section. This – often tacit – assumption [BORNSCHEUER, 1952; SZMODITS, 1975] is correct for open cross sections. For closed cross sections, however, this assumption does not hold. This follows from the fact that the above assumption is equivalent to neglecting the shear deformation resulting from bending. According to investigations, the error resulting from the above approximation rarely exceeds 5% [PEARSON, 1956]. This error is reduced when the bracing system of the building has cores with both open and closed cross sections. No error is made, of course, when there are only cores with open cross sections.

In taking into account shear deformations as well, the location of the shear centre also depends on the load and may vary along the axis of the bar [STÜSSI, 1965]. Depending on the load distribution over the height of the structure, the location of the shear centre can vary considerably on the floor levels. This phenomenon may play an important role in the dynamic analysis of tall buildings [KÓKAI, 1983].

The shear centre also varies over the height of the structure when the arrangement of the shear walls and cores on the floor levels are different. Some of the shear walls and cores may not continue to the top of the structure as they are not needed for the support system or some of them are left out on ground floor level as the architect needs more space there. Such structures cannot be considered regular and their stability analysis is not covered here.

Pierced bracing elements

The bracing system of buildings often consists of pierced elements: coupled shear walls and cores with openings. It is difficult to take into account the weakening effect of the openings. The state of stress of the structure becomes complex and the problem can only be defined by a large system of differential equations. This is why approximate methods of different accuracy are used in structural engineering practice.

The simplest and less accurate method is obtained when the effect of the beams linking the walls is neglected altogether. The application of this method is justified when the structure is subjected to great loads (e.g. earthquake) which would probably make the beams fail anyway. The great advantage of this method is simplicity and the fact that it is conservative, i.e. it results in a lower bound to the critical load. The state of stress of the structure is simple and the effects of the individual stiffness characteristics are seen clearly which makes it possible to strengthen the structure where it is most needed.

Columns with thin-walled open cross sections mainly resist torque by making use of their warping stiffness. This phenomenon can be traced back to the bending of the elements of the cross section [VLASOV, 1961; ZALKA, 1980]. It follows that, as an approximation, pierced shear walls and cores can be replaced by solid shear walls and cores with equivalent thicknesses. The calculation of the equivalent thickness can be based on the requirement of identical maximum deflection or identical average deflection. Well-established methods are available for this calculation [ROSMAN, 1965; IRWIN, 1984; TARANATH, 1988]. In this way the effect of the axial deformation of the walls, which may be considerable for tall buildings, can also be taken into account [DANAY, GELLÉRT and GLÜCK, 1975]. Another possible approach may be the requirement of identical critical load.

A simple approximate method is available for the spatial stability analysis of a system of coupled shear walls, subjected to concentrated forces at top floor level. After assuming that torsional buckling is dominated by warping and therefore neglecting the *Saint-Venant* torsional stiffness, HEGEDŰS and KOLLÁR [1987] established the differential equations of such systems and showed that, after producing the critical loads for buckling in the principal planes and for pure torsional buckling, the combined critical load is obtained from Eq. (7-49).

Plastic effects

If the load on the core, i.e. on the bracing walls, (Q_{bra}), is no longer negligible, compared to the total load of the building (Q), then safety against plastic buckling also has to be investigated. To this end, the value of the 'plastic limit' of the core is needed. For

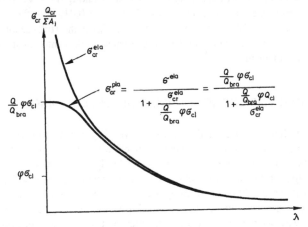

Fig. 7-14 Plastic critical stress

reinforced concrete shear walls and cores it is obtained by multiplying the buckling coefficient φ and the limit of the strength of concrete σ_{cl}. The buckling coefficient φ can be taken from the Building Codes.

Since the core is under Q_{bra} only, which is only part of the total load Q, but buckling is caused by the total load Q, $\varphi\sigma_{cl}$ should be increased by Q/Q_{bra}. Using this modification and applying the *Föppl-Papkovich* formula, a good approximation for the plastic critical stress is obtained as

$$\frac{1}{\sigma_{cr}^{pla}} = \frac{1}{\sigma_{cr}^{ela}} + \frac{1}{\dfrac{Q}{Q_{bra}}\varphi\sigma_{cl}} \tag{7-54}$$

which leads to the formula given in *Fig. 7-14.*

The elastic critical stress $\left(\sigma_{cr}^{ela}\right)$ is obtained by dividing Q_{cr}^{ela} – calculated earlier – by the sum of the areas of the bracing elements $\left(\sum A_i\right)$.

Finally, to obtain the plastic critical load, the elastic critical load should be reduced according to the ratio of the two critical stresses:

$$Q_{cr}^{pla} = Q_{cr}^{ela}\frac{\sigma_{cr}^{pla}}{\sigma_{cr}^{ela}} = \frac{Q_{cr}^{ela}}{1 + \dfrac{\sigma_{cr}^{ela}}{\dfrac{Q}{Q_{bra}}\varphi\sigma_{cl}}} \tag{7-55}$$

The minimum level of safety against plastic buckling should be 2.0. (Safety regarding Q_{cr}^{ela} is again required to be at least 4.0 - see also Section 7.3.)

Recommendations for reinforced concrete bracing elements
When the bending and torsional stiffnesses of reinforced concrete bracing elements are established for the calculation of the critical load, the following guidelines should be considered. The value of the modulus of elasticity is obtained by taking the lower bound of the modulus which belongs to the nonpermanent loads E_c, i.e. $E = 0.8E_c$. For the value of the second moment of area, $I = 0.7I_{uc}$ is recommended where I_{uc} is the (full) second moment of area of the uncracked cross section and I (approximately) takes into consideration the effects of cracks in the elements.

7.5 STABILITY ANALYSIS OF THE COLUMNS OF THE BUILDING

The columns of a building can lose stability in two ways: they can develop sway buckling (when the joints develop translations) or nonsway buckling. Of the critical loads of the two types the smaller one is of practical importance. Clearly, the columns must have the necessary stiffness against nonsway buckling. As for sway buckling, engineering common sense suggests that if the conditions regarding bracing given in Section 7.2 are fulfilled, then it is also ensured that the columns do not develop sway buckling, or more precisely, the critical load of sway buckling is always greater than that of non-sway buckling. In this section the validity of this statement will be investigated.

Consider a column with three types of support, braced by a core (*Figs. 7-15, 7-16* and *7-17*).

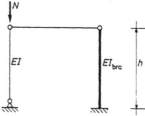

Fig. 7-15 Column with pinned supports braced by the core

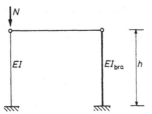

Fig. 7-16 Column with fixed lower and pinned upper supports braced by the core

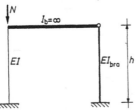

Fig. 7-17 Column with fixed supports braced by the core

The stiffness of the core is chosen to be equal to the load of the column in all three cases, i.e.

$$EI_{bra} = \frac{4h^2}{\pi^2} N \tag{7-56}$$

7.5.1 Sway critical loads

Calculate the exact sway critical load of each column braced by a core having the stiffness defined by formula (7-56).

From the point of view of the column, the bracing core represents a horizontal support in the form of a spring with stiffness c.

a) Column with two pinned ends (Fig. 7-15).
The sway critical load can be formulated in a simple way. Assuming a horizontal translation of δ at the top of the column, moment equilibrium with respect to the lower pinned support results in

$$N\delta = (c\delta)h$$

from which the critical load is

$$N_{cr} = ch \tag{7-57}$$

showing that the critical load increases in direct proportion to c. The limit to the critical load is the non-sway *Euler* critical load:

$$N_{cr}^{fix} = N_E = \frac{\pi^2 EI}{h^2} \tag{7-58}$$

The different critical loads will be referred to this critical load which will be used to produce nondimensional quantities.

As a function of spring constant c, two intersecting lines in co-ordinate system $(N_{cr}/N_E, c/N_E)$ show the critical loads which belong to the two types of buckling (*Fig. 7-18*). Of the two critical loads, the smallest one is of relevance (marked by a thick line).

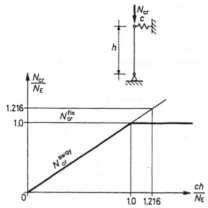

Fig. 7-18 Behaviour of the column in *Fig. 7-15*

b) Column with one fixed and one pinned ends (Fig. 7-16)

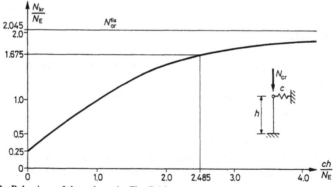

Fig. 7-19 Behaviour of the column in *Fig. 7-16*

The sway critical load is obtained using *Timoshenko*'s solution (Eq. (i) on page 72 in [TIMOSHENKO and GERE, 1961]), taking into consideration that the column is half of the one shown in *Fig. 7-21c*, assuming that the latter develops symmetric buckling (shown using a broken line). Applying the notation used here to transcendental equation (i) of the solution of the problem and leaving out the term $\sin 2u$ which would represent the asymmetric buckling of the column in *Fig. 7-21c*, the equation

$$- \sin 2u_1 + 8u_1 \cos 2u_1 \left(\frac{1}{4} - \frac{N_{cr}^{sway}}{4ch} \right) = 0 \qquad (7\text{-}59a)$$

is obtained (see also [PFLÜGER, 1950], p. 259), where

$$u_1 = \frac{h}{2}\sqrt{\frac{N_{cr}^{sway}}{EI}}.$$ (7-59b)

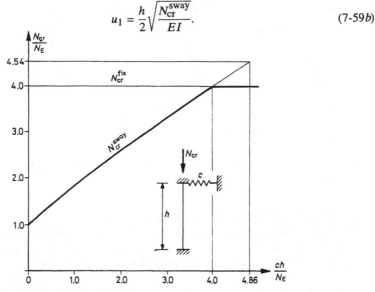

Fig. 7-20 Behaviour of the column in *Fig. 7-17*

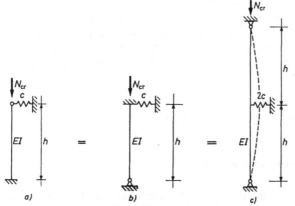

Fig. 7-21 Tracing back a column with fixed lower and pinned upper ends to a column with two pinned ends

The solution of Eq. (7-59a) for a number of corresponding values of N_{cr}^{sway} and c results in the diagram in *Fig. 7-19*. The co-ordinate axes again represent ratios N_{cr}/N_E and ch/N_E, where N_E is the nonsway critical load of the column of height h and having pinned ends as defined by formula (7-58).

The diagram also shows the nonsway critical load of the column:

$$N_{cr}^{fix} = \frac{\pi^2 EI}{(0.699h)^2} = 2.045\frac{\pi^2 EI}{h^2} = 2.045N_E$$ (7-60)

which represents an asymptote to N_{cr}^{sway}.

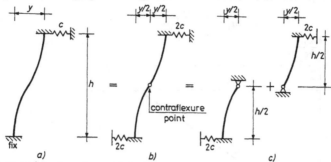

Fig. 7-22   Tracing back a column with fixed lower and upper ends to a column with fixed lower and pinned upper ends

c) Elastically supported column with two fixed ends (Fig. 7-17)
The calculation of the critical load can be traced back to the previous case. *Fig. 7-22* shows that the column with one fixed end and one supported by a spring of stiffness c (*Fig. 7-22a*) is equivalent to the column supported by a spring of stiffness $2c$ at both ends (*Fig. 7-22b*), as the relative translation y of the two ends develops the same force in the spring $\left(yc = \frac{y}{2}2c\right)$. According to *Fig. 7-22c*, the latter column can be decomposed into two columns, each with $2c$ and $h/2$, instead of c and h.

It follows that the diagram in *Fig. 7-19* can be used but the scales have to be converted since, although the product $ch = 2c\frac{h}{2}$ remains the same, N_E is reduced to one quarter. The diagram in *Fig. 7-20*, which in fact is identical to the starting section of the diagram in *Fig. 7-19*, reflects the necessary alterations. The horizontal line representing the nonsway critical load of the column

$$N_{cr}^{fix} = 4\frac{\pi^2 EI}{h^2} = 4N_E \tag{7-61}$$

limits the increasing value of the sway critical load.

7.5.2 Sway versus non-sway critical loads

The above results make it possible to calculate the sway critical load of columns braced by a core of stiffness EI_{bra} (formula (7-56)) and to compare it to the nonsway critical load.

According to formula (7-56), the stiffness of the spring representing the core is

$$c = \frac{3EI_{bra}}{h^3} = \frac{12}{\pi^2}\frac{N}{h} = 1.216\frac{N}{h} \tag{7-62}$$

Check the nonsway stability of the braced column in all three cases.

Case a) (Fig. 7-15):

$$N_{cr}^{fix} = N_E = N \tag{7-63}$$

The stiffness of the spring representing the core is obtained from formula (7-62). With $N = N_E$ it assumes the value

$$c = 1.216\frac{N_E}{h} \tag{7-64}$$

and the sway critical load of the column (*Fig. 7-18*) is

$$N_{cr}^{sway} = 1.216N_E > N_{cr}^{fix} = N_E \tag{7-65}$$

N_{cr}^{sway} is greater than N_{cr}^{fix} by 17.7%.

Case b) (Fig. 7-16):

$$N_{cr}^{fix} = 2.045N_E = N \tag{7-66}$$

with which

$$c = 1.216\frac{2.045N_E}{h} = 2.485\frac{N_E}{h} \tag{7-67}$$

and the sway critical load (*Fig. 7-19*) is

$$N_{cr}^{sway} = 1.675N_E < N_{cr}^{fix} = 2.045N_E \tag{7-68}$$

N_{cr}^{sway} is now 18% smaller than N_{cr}^{fix}.

Case c) (Fig. 7-17):

$$N_{cr}^{fix} = 4N_E = N \tag{7-69}$$

$$c = 1.216\frac{4N_E}{h} = 4.86\frac{N_E}{h} \tag{7-70}$$

and the sway critical load (*Fig. 7-20*):

$$N_{cr}^{sway} = 4.54N_E > N_{cr}^{fix} = 4N_E \tag{7-71}$$

N_{cr}^{sway} is 17.7% greater than N_{cr}^{fix}.

7.5.3 Conclusions

The above results show that the stability analysis is conservative for columns with two pinned ends and for columns with two fixed ends and unconservative for columns with pinned upper and fixed lower ends as in the latter case the sway critical load of the column is 18% smaller than the nonsway critical load.

The following conclusions can be drawn.

If the elastic support of the column is such that the horizontal line representing the nonsway buckling of the column *intersects* the diagram representing the sway critical load (*Figs 7-18* and *7-20*), then by choosing the stiffness of the core as described, it is ensured that the sway critical load is greater than the nonsway one because the spring representing the bracing effect of the core is stronger than the spring which would be needed to produce a sway critical load identical to that of the nonsway buckling.

If, however, the diagram of the sway critical critical load *asymptotically* approaches the horizontal line representing the nonsway critical load (*Fig. 7-19*), then no spring stiffness of finite magnitude exists that would ensure a critical load as great as that of the nonsway buckling.

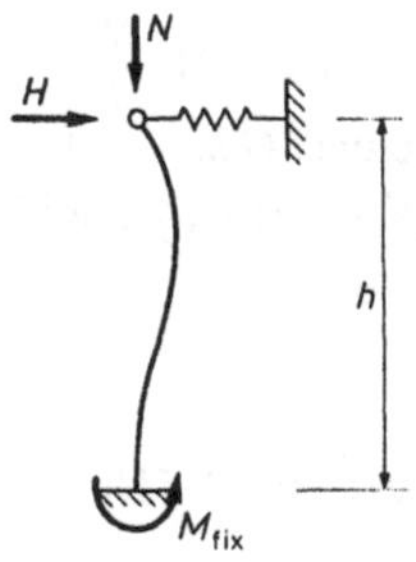

Fig. 7-23 Forces on a column with fixed lower and pinned upper ends

The physical difference between the two phenomena is as follows. In the case of nonsway buckling shown in *Figs 7-18* and *7-20, no horizontal force* develops which would try to push the core. On the other hand, the nonsway buckling of the column in *Fig. 7-19* is accompanied by horizontal force *H* (*Fig. 7-23*) because it is needed to ensure moment equilibrium. This horizontal force *H* causes the translation of the core unless it has an infinitely great stiffness. It follows therefore that a core with finite stiffness cannot prevent the translation of the upper end of the column and, consequently, the magnitude of the critical load can never be as great as that of nonsway buckling.

The greater the difference between the fixing effects of the supports at the two ends of the column, the greater the magnitude of this horizontal force. No horizontal force develops when the fixing effects are identical (*Figs 7-18* and *7-20*) because the two fixing moments are of the same magnitude and are in equilibrium. The horizontal force assumes maximum value when the difference between the fixing effects at the two ends is maximum, i.e. when one end is totally fixed and the other end is pinned (*Fig. 7-23*). The 18% error which belongs to the structure shown in *Fig. 7-16* (and in *Fig. 7-19*) is therefore a maximum.

This unconservative error can always be eliminated if the nonsway critical load of columns with different fixing stiffnesses is calculated in such a way that the fixing stiffness of the 'weaker end' is taken into account at the 'stronger end' as well. If then the critical load is calculated using the smaller fixing stiffness at both ends, then the sway critical load is always greater than the nonsway critical load and the latter can always be used for the analysis.

Consider for example the structure which would result in the greatest error (*Fig. 7-16*). If both ends are assumed to be pinned, then the case shown in *Fig. 7-15* is obtained where, as demonstrated earlier, the nonsway critical load is 17.7% smaller than that of the sway one.

It should also be mentioned here that the sway and nonsway buckling shapes are orthogonal to each other and therefore they do not combine. There is no need therefore to use *Dunkerley's* theorem (see Chapter 2) and it is simply the smaller critical load which has to be used for the analysis. The two buckling shapes do not interact in the post-critical range either because the situation here is not identical to the one investigated in Sections 3.2-3.4, i.e. it is not a part of the structure that develops local buckling reducing the total stiffness of the structure but two different eigenfunctions of the same structure are considered here.

In addition to cantilever type bracing cores with a fixed lower support, the above conclusions are also valid for bracing structures with any other type of support. The effect of the flexible support of the bracing system is not discussed here; it is only pointed out that detailed information on soil-superstructure interaction and the effect of the performance of soil on the stability of the building is available elsewhere [LIPTÁK, 1973a, b; Tall Buildings...., 1980; STAFFORD SMITH and COULL, 1991]. The critical load of the bracing structure has to be calculated according to the type of support and it has to be ensured that it is greater (by a sufficient margin) than the total load of the building.

8

Buckling of arches and rings

Lajos KOLLÁR

The stability problems of bars with plane curved axes are rather manifold, and can be systematized as follows:
- the *axis* can be circular or can have any other shape;
- the *load* causes central compression before buckling, but during buckling it can exhibit three kinds of behaviour: it can maintain its original direction, it can always remain perpendicular to the (deformed) bar axis (hydrostatic pressure), or it can always pass through a given point ('central load');
- the *boundary conditions* of the bar can be manifold: the arch can be independent (its ends can be clamped or hinged; there can be also a hinge at their middle cross section); or the arch can be connected to other structural elements (arches with hangers or struts);
- their *cross section* can be full (closed) or thin-walled (open); in the latter case the cross section may also deform in its own plane;
- the *loss of stability* may occur either in the plane of the arch (by bifurcation: steep arches; by limit point or snapping through: flat arches), or perpendicularly to the plane of the arch (by bifurcation); the post-critical behaviour can also be manifold;
- depending on what we *neglect in the computation*: with the buckling phenomena in the plane of the arch we can consider from the bending, compression or shear deformations either only the first, or also the second, or all three; we can consider or neglect the horizontal displacements of the points of the axis of the arch.

We thus have to systematize the treatment according to these points of view.

In this chapter we only treat bars whose axis is originally a plane curve which is merely 'slightly' curved, i.e. the dimensions of the cross section are much smaller than its radius of curvature. We assume, as a rule, that the structure is elastic.

We will not deal with the local buckling of the elements of the thin-walled cross sections, since this belongs to the theory of plate or shell buckling. (In special cases, as with the torsional buckling of arches with open cross section, the local buckling may coincide with the global buckling of the arch, see in [KOLLÁR, 1973].)

8.1 BUCKLING OF BARS WITH CURVED AXIS (ARCHES) IN THEIR OWN PLANE

In the following – if not said the contrary – we will deal with the buckling of centrally compressed rings and arches, i.e. we will not consider any initial imperfection. This means, as a rule, bifurcation of equilibrium, or, in the case of flat arches, snapping through.

As is well known, to treat bifurcation problems by the equilibrium method we need the first powers of the displacement componenets, while for the energy method we need their second powers. In the latter case it it thus important to choose appropriate approximate geometric relations, see in detail [BATOZ, 1981].

Except for the case of very flat arches (Section 8.1.3) we can always assume that the compression of the arch axis prior to buckling is negligible, so that we can take the original (unloaded) axis as a basis. This also means that we can consider the axis inextensional also during buckling.

8.1.1 Buckling of rings and arches with circular axis

Of all bars with curved axes, it is the circular arc with renders the mathematical treatment most simple. First we deal with the bucking of the closed circular ring under uniform radial load (*Fig. 8-1*), considering three different behaviours of the load during buckling.

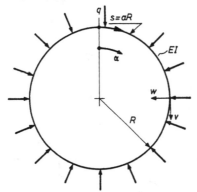

Fig. 8-1 The radially compressed circular ring

Buckling of the closed circular ring
First of all we have to clarify whether the three different kinds of load are conservative or not.

The constant directional load is by definition conservative.

The hydrostatic pressure always remains perpendicular to the surface, so that it is, as a rule, not conservative.

The central load always points to the original centre of the circle (as e.g. the radial cable forces acting on the edge ring of the 'bicycle wheel' type hanging roof), so that it is generally also not conservative.

However, BODNER [1958] showed that if the loads acting on a closed circular ring maintain their magnitude during buckling, then both the hydrostatic and the central loads are conservative. That is, if a hydrostatic pressure p acts on a closed surface, and the change of volume enclosed by this surface, due to p, is ΔV, the work done by these forces is $p \cdot \Delta V$, independently of the path. On the other hand, in the case of a central load the work done uniquely depends on the distance measured from the centre of the circle, again independently of the path.

Consequently, we can investigate the buckling of the ring in all three loading cases by the usual static method, and there is no need to apply the more complicated kinetic method treated in [TIMOSHENKO and GERE, 1961].

The value of the critical load is different in the three cases. To gain a better insight, we investigate the problem in detail on the basis of [SINGER and BABCOCK, 1970]. Let us consider the *rigid-body displacement* (shifting and rotation) of the ring and determine: which kind of load performs work during this displacement.

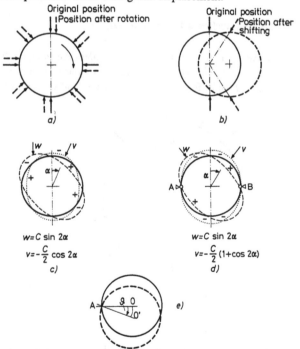

Fig. 8-2 The influence of the rigid-body displacement of the circular ring. a) Rotation of the ring loaded by constant directional forces, b) shifting of the ring loaded by central load, c) buckling deformation of the ring performing 'average zero tangential displacement', d) buckling deformation of the ring supported at two points, e) rigid-body rotation of the ring around point A

Simple engineering consideration shows that the hydrostatic pressure does not perform any work during rigid-body shifting of the ring, but the two other kinds of load do. The constant directional load performs positive work during rotation of the ring (*Fig. 8-2a*), while the central load performs negative work during shifting of the ring (*Fig. 8-2b*).

Let us now investigate the question: in which cases does the ring perform a rigid-body displacement during buckling?

The axis of the ring remains inextensional during buckling. The elongation of a bar with curved axis is composed of two parts. The first part is the derivative of the tangential displacement v (*Fig. 8-1*) with respect to the arc length:

$$\varepsilon^I = \frac{dv}{ds}, \tag{8-1a}$$

and the second part is caused by the fact that the radial displacement w carries the arc element from the circle with the radius R onto a circle with a different radius (*Fig. 8-1*):

$$\varepsilon^{II} = -\frac{w}{R}. \tag{8-1b}$$

The total elongation will thus be, using the relation $s = \alpha R$ and denoting the differentiation with respect to α by $'$:

$$\varepsilon^I + \varepsilon^{II} = \frac{1}{R}\left(v' - w\right). \tag{8-2}$$

Hence the condition of inextensionality is:

$$w = v'. \tag{8-3}$$

We assume the buckling deformation in the form of a sine series, of which we take, for the case of simplicity, only the first term:

$$w = C \sin 2\alpha. \tag{8-4}$$

The corresponding tangential displacement can be obtained from (8-3):

$$v = -\frac{C}{2}\cos 2\alpha + D, \tag{8-5}$$

where D is an integration constant, unknown for the time being, representing the rigid-body rotation of the ring. (These simple deformation functions do not contain the rigid-body shifting.)

In *Fig. 8-2c* we depicted the radial and tangential displacements according to Eqs (8-4) and (8-5) for the case $D = 0$. It can be seen that if we require that the ring should perform no rigid-body displacement, we must not support any point of the ring at the same time in both directions. If, however, we support two points of the ring both in radial and tangential directions (as in the case of two-hinged arches), then in Eq. (8-5) D cannot be equal to zero, so that the ring has to rotate as a rigid body. Thus if we support e.g. the points $\alpha = \pi/2$ and $\alpha = 3\pi/2$ (this corresponds to an arc with a central angle $180°$), the value of D becomes $(-C/2)$, see *Fig. 8-2d*.

Figs 8-2c,d also show that if we clamp one cross section of the ring (i.e. we prevent not only its shifting in two directions, but also its rotation), then we force the ring to shift as a rigid body. Let us clamp e.g. its cross section $\alpha = 3\pi/2$, denoted by A. Since the deformation sketched in *Fig. 8-2d* entails the rotation of this cross section (the diagram of w does not join tangentially the original position of the ring), this rotation has to be removed by rotating the ring around point A by an angle θ, which entails the shifting of the centre point O to O'. That is, the ring – in addition to its rotation – also shifts as a rigid body (*Fig. 8-2e*).

The shifting is described in the general case by the deformation functions

$$w = C_1 \sin \alpha + C_2 \cos \alpha, \tag{8-6a}$$

$$v = -C_1 \cos \alpha + C_2 \sin \alpha. \tag{8-6b}$$

On the basis of what has been said so far, we can state the following on the critical values of the three kinds of loading.

The critical value of the *hydrostatic* pressure is, independently of the support conditions:

$$q_{cr}^h = 3.0 \frac{EI}{R^3}, \tag{8-7}$$

and the critical value of the *central* load is, provided we support the ring in such a way that it does not shift as a rigid body:

$$q_{cr}^c = 4.5 \frac{EI}{R^3} \tag{8-8}$$

[CHWALLA and KOLLBRUNNER, 1938; BORESI, 1955; TIMOSHENKO and GERE, 1961; WEMPNER and KESTI, 1962]. If we support the ring in such a way that it has to shift as a rigid body, then – according to the explanation made in connection with *Fig. 8-2b* – the critical load increases due to the negative work performed. This happens if we clamp one cross section of the ring [WEMPNER and KESTI, 1962; SCHMIDT, 1981].

The critical value of the *constant directional* load is, provided the ring cannot rotate as a rigid body:

$$q_{cr}^d = 4.0 \frac{EI}{R^3} \tag{8-9a}$$

[PEARSON, 1956]. This kind of support is realized e.g. with tubes embedded in the soil or with barrel vaults resting on diaphragms.

If we prevent the displacements in two directions of two cross sections of the arch (as with two-hinged arches), then we force the ring to rotate as a rigid body as shown in *Fig. 8-2c*, which causes – due to the positive work done by the load, as explained in connection with *Fig. 8-2a* – the critical value of Eq. (8-9a) to decrease. Thus the critical load of an arch with the central angle of e.g. 180° is

$$q_{cr}^d = 3.2712 \frac{EI}{R^3} \tag{8-9b}$$

[CHWALLA and KOLLBRUNNER, 1938]; the improved value given here is taken from [SCHMIDT, 1980a]. It should be remarked that in this case the displacement functions (8-4) and (8-5) do not represent the exact solution, but more complicated functions are needed.

If we do not prevent the rigid-body rotation of the ring at all, then the value of q_{cr}^d decreases to zero: the ring 'buckles' (i.e. rotates) under the smallest load.

Let us shortly deal with the buckling of rings *elastically supported* in the radial direction, which can be considered as the model of edge rings of cable net roofs: the elastic support represents (approximately) the fact that the cable forces loading the ring change due to the radial displacement of the ring. The solution of this problem can be found in [BRUSH and ALMROTH, 1975] for *hydrostatic* loading:

$$q_{cr}^h = \left(n^2 - 1 \right) \frac{EI}{R^3} + \frac{k_r R}{n^2 - 1}, \tag{8-10a}$$

where k_r [kN/m^2] is the constant of the elastic support, and n is the number of (full) waves, i.e. $n \geq 2$. It should be remarked that in the case of elastic support it is not always the shape with $n = 2$ which yields the minimum load, but we have to look for the value of n yielding q_{cr}^{min}.

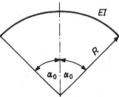

Fig. 8-3 Notations

For *central* load we find a solution in [DYM, 1977], which contains no rigid-body shifting:

$$q_{cr}^c = \frac{\left(n^2 - 1\right)^2}{n^2 - 2} \frac{EI}{R^3} + \frac{k_r R}{n^2 - 1}. \tag{8-10b}$$

We can easily compute the critical value of *constant directional* load using the paper of DYM [1977]. We assume for the radial and tangential displacements the functions

$$w = C \sin n\alpha \tag{8-11a}$$

and

$$v = -\frac{C}{n} \cos n\alpha + D. \tag{8-11b}$$

Assuming that the ring does not perform a rigid-body rotation, then from the last equation we can omit D, also appearing in (8-5). Using the relations of [DYM, 1977], the expression of the strain energy has to be completed by the term

$$R \int_{-\pi}^{+\pi} k_r w^2 d\alpha \tag{8-12}$$

representing the strain energy of the elastic support. Thus we obtain the following expression for the critical load:

$$q_{cr}^d = n^2 \frac{EI}{R^3} + \frac{n^2 k_r R}{\left(n^2 - 1\right)^2}. \tag{8-13}$$

All these formulas consider the elastic support as 'constant directional'.

Buckling of arches with circular axis under uniform radial load
The notations used with arches having a circular axis are to be seen in *Fig. 8-3*.

Two-hinged arches
As mentioned in Section 8.1.1, the solution of the complete ring under central and hydrostatic loads can be used for two-hinged arches too (*Fig. 8-4a*), if we chose the buckling length in such a way that two half waves develop along the arch axis. CHWALLA and KOLLBRUNNER [1938] gave numerical results for some values of the central angle $2\alpha_0$. Closed

formulas are to be found for hydrostatic pressure in [TIMOSHENKO and GERE, 1961], and for central load in [DYM, 1977]:

Hydrostatic pressure:

$$q_{cr}^{h} = \left[\left(\frac{n\pi}{\alpha_0} \right)^2 - 1 \right] \frac{EI}{R^3}, \tag{8-14a}$$

Central load:

$$q_{cr}^{c} = \frac{\left[\left(\frac{n\pi}{\alpha_0} \right)^2 - 1 \right]^2}{\left(\frac{n\pi}{\alpha_0} \right)^2 - 2} \frac{EI}{R^3}. \tag{8-14b}$$

These two formulas are exact.

In the case of constant directional load the constant D appearing in the expression of the tangential displacement (8-12b) has to be determined in such a way that at the supports $v = 0$ should be fulfilled ($D = C/n$). By so doing DYM obtained the following formula.

Constant directional load:

$$q_{cr}^{d} = \frac{\left(\frac{n\pi}{\alpha_0} \right)^2 \left[\left(\frac{n\pi}{\alpha_0} \right)^2 - 1 \right]^2}{\left[\left(\frac{n\pi}{\alpha_0} \right)^2 - 1 \right]^2 + 2} \frac{EI}{R^3}, \tag{8-14c}$$

which is a very good approximation.

In the above formulas n is the number of *full* waves developing on the arch. The minimum critical load is always given by $n = 1$ (a buckling shape with two half waves).

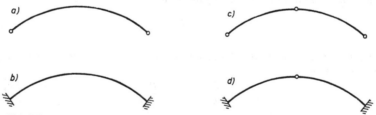

Fig. 8-4 The different kinds of arches. *a*) Two-hinged, *b*) clamped, *c*) three-hinged, *d*) one-hinged

Clamped arches

The buckling shape of clamped arches (*Fig. 8-4b*) is – as that of the two-hinged arches – also antisymmetric.

For *hydrostatic* pressure we find the solution in [TIMOSHENKO and GERE, 1961]. The critical load is given by the formula

$$q_{cr} = \left(k_h^2 - 1 \right) \frac{EI}{R^3} \tag{8-15}$$

where the constant k_h is the smallest root of the characteristic equation

$$k_h \tan \alpha_0 \cdot \cot \left(k_h \alpha_0 \right) = 1. \tag{8-16}$$

Some numerical values are:

$$\alpha_0 \quad = 15° \quad 30° \quad 45° \quad 60° \quad 75° \quad 90°$$
$$k_h^2 - 1 = 294 \quad 73.3 \quad 32.4 \quad 18.1 \quad 11.5 \quad 8.00$$

The solution for *constant directional* load is given by SCHMIDT [1980b]:

$$q_{cr} = k^2 \frac{EI}{R^3}, \tag{8-17}$$

with k as the smallest root of the transcendental equation:

$$(2\alpha_0 + \sin 2\alpha_0)\, k^2 - (2\alpha_0 - \sin 2\alpha_0) = 4k \cos^2 \alpha_0 \cdot \tan (k\alpha_0). \tag{8-18}$$

Some numerical values are:

$$\alpha_0 = \quad 30° \quad 40° \quad 50° \quad 60° \quad 70° \quad 80° \quad 90°$$
$$k^2 = 74.95 \quad 42.70 \quad 27.75 \quad 19.59 \quad 14.62 \quad 11.33 \quad 9.00$$

For *central load* the problem has been solved by WEMPNER and KESTI [1962]. The critical load is again given by Eq. (8-17), and k is the smallest root of the equation:

$$\lambda \tan (\mu\alpha_0) = \mu \tan (\lambda\alpha_0), \tag{8-19}$$

where

$$2\lambda^2 = k^2 + 2 + k\sqrt{k^2 - 4}, \tag{8-20a}$$

$$2\mu^2 = k^2 + 2 - k\sqrt{k^2 - 4}. \tag{8-20b}$$

Some numerical value can be taken from [SCHMIDT, 1981]:

$$\alpha_0 = \quad 15° \quad 30° \quad 45° \quad 60° \quad 70° \quad 80° \quad 90°$$
$$k^2 = 295.94 \quad 75.06 \quad 34.26 \quad 20.11 \quad 15.40 \quad 12.46 \quad 10.60$$

Three-hinged arches

The three-hinged arch (*Fig. 8-4c*) may buckle in two different shapes. One is the antisymmetric shape, the critical load of which is equal to that of the two-hinged arch, since the latter develops zero bending moment in the middle cross section, so that it behaves as if it were a three-hinged arch. The second is the symmetric shape (with three 'half waves'), which contains a relative rotation at the middle hinge, so that the corresponding critical load is lower than that pertaining to the symmetric buckling shape of the two-hinged arch. The symmetric shape yields a lower critical load than the antisymmetric one if the arch is shallow.

The solution has been established for *hydrostatic* pressure only [TIMOSHENKO and GERE, 1961].

The constant k_h appearing in Eq. (8-15), pertaining to the symmetric shape, is the smallest root of the equation

$$k_h^3 \tan \alpha_0 - (k_h^2 - 1)\, k_h\alpha_0 - 2 \tan (k_h\alpha_0/2) = 0 \tag{8-21a}$$

We take the improved numerical values of $(k_h^2 - 1)$ from [SCHMIDT, 1979a]:

$$\alpha_0 \quad = 15° \quad 30° \quad 45° \quad 60° \quad 75° \quad 90°$$
$$k_h^2 - 1 = 108.4 \quad 27.08 \quad 12.02 \quad 6.76 \quad 4.32 \quad 3.00$$

To the antisymmetric buckling shape the value $k_h = \pi/\alpha_0$ belongs, in accordance with Eq. (8-14a). Of the two k_h values the smaller one is the most critical.

In [SCHMIDT, 1979a] we also find a very good approximate formula for $\left(k_h^2 - 1\right)$ pertaining to the symmetric buckling shape:

$$\left(k_h^2 - 1\right) = \frac{24350}{\left(\alpha_0\right)^2}, \tag{8-21b}$$

where α_0 has to be substituted in grades. The formula yields a close approximation in the range $15° \leq \alpha_0 \leq 90°$.

One-hinged arches

The antisymmetric buckling shape and the pertaining critical load of the one-hinged arch (*Fig. 8-4d*) coincides with that of the clamped arch. The critical load pertaining to the symmetric buckling shape (which is most critical in the case of shallow arches) has been determined for *hydrostatic* load [TIMOSHENKO and GERE, 1961]; improved values can be found in [SCHMIDT, 1979a]. Some numerical values of the factor $\left(k_h^2 - 1\right)$ appearing in Eq. (8-15) are:

$$\begin{array}{llllll}
\alpha_0 = & 15° & 30° & 45° & 60° & 75° & 90° \\
\left(k_h^2 - 1\right) = & 160.4 & 40.21 & 17.95 & 10.16 & 6.56 & 4.61
\end{array}$$

The constant k_h is the smallest root of the following equation:

$$\left(k_h^2 - 1\right)\left[k_h^2\alpha_0 \cot\alpha_0 + 2 - 2\sec\left(k_h\alpha_0\right)\right] - k_h^4 + \tag{8-22a}$$

$$\left[\left(k_h^2 - 1\right)\alpha_0 + \cot\alpha_0\right]k_h \tan\left(k_h\alpha_0\right) = 0.$$

In [SCHMIDT, 1979a] the following approximate formula can be found, which yields very good values in the range $15° \leq \alpha_0 \leq 90°$:

$$\left(k_h^2 - 1\right) = 0.162 + \frac{36060}{\left(\alpha_0\right)^2}, \tag{8-22b}$$

where α_0 has to be introduced in grades.

Post-critical behaviour of arches with circular axis

As explained in Chapter 1, it is very important to know whether the post-critical load-bearing capacity of a structure is decreasing or increasing (*Figs. 1-1* to *1-3*), since this decides whether it is imperfection sensitive or not. In the following, we present the results of investigations on this problem. It should be mentioned that since the higher powers of the displacement functions have also to be taken into account, it is expedient to start from the exact relations, see e.g. in [SCHMIDT, 1979b], because by so doing we can avoid the possibility of error in the approximate methods, namely that we neglect quantities which would be necessary to calculate exactly the higher order terms.

Post-critical behaviour of the complete ring

The solution can be found for hydrostatic pressure in [CARRIER, 1947; BUDIANSKY, 1974], for constant directional pressure in [EL NASCHIE, 1975], for central load in [SILLS and BU-DIANSKY, 1978]. The circular ring exhibits an increasing post-critical load-bearing capacity for all three kinds of loading. [In his solution for constant directional load EL NASCHIE assumed a tangential displacement with an average value of zero, i.e. zero rotation, which corresponds to the critical load according to Eq. (8-9a).]

Post-critical behaviour of arches

Two-hinged arches. For *hydrostatic* pressure the solution can be found in [SCHMIDT, 1979c] or in [SCHMIDT, 1979d]. The post-critical load-bearing capacity of the arch is decreasing when $\alpha_0 < 56.3°$ (*Fig. 8-3*), and becomes increasing if $\alpha_0 > 56.3°$.

For *constant directional* load DYM [1973] published a solution, using the approximate equations of SANDERS [1963]. The load-bearing capacity is decreasing if $\alpha_0 < 75°$, and increasing if $\alpha_0 > 75°$.

Clamped arches. Under *hydrostatic* pressure the arch exhibits a decreasing load-bearing capacity when $\alpha_0 < 75°$, and an increasing one if $\alpha_0 > 75°$ [SCHMIDT, 1979e].

For *constant directional* load the (approximate) investigation of DYM [1973] yielded a decreasing load-bearing capacity for all values of α_0. It is, however, possible that a more exact analysis would result in an increasing load-bearing capacity at α_0 values greater than a certain limit value, as is the case with two-hinged arches.

Three-hinged arches. In the literature we found a solution for *hydrostatic* pressure only [SCHMIDT, 1978, 1979f], yielding a decreasing load-bearing capacity for all values of α_0. This result, however, contradicts that for two-hinged arches with $\alpha_0 > 56.3°$: in this domain both kinds of arches buckle antisymmetrically, with the same shape, so that their post-critical behaviours should also be the same. Hence this problem needs further clarification.

8.1.2 Arches with noncircular axes

In this section we deal only with *constant directional*, vertical loads.

Arches with parabolic axis

A uniformly distributed load causes no bending in an arch whose axis is a parabola of the second degree (*Fig. 8-5*). DISCHINGER [1937] determined the critical load of such arches, taking only the vertical displacements w of the points of the axis into account. To derive the differential equation of buckling, it is necessary to consider that the change in slope of the tangent to the arch axis is dw/dx, and the curvature κ is the variation of this quantity along the arch axis, i.e.

$$\kappa = \frac{d}{ds} \frac{dw}{dx}$$

Using the relation $ds = \frac{dx}{\cos\alpha}$ (*Fig. 8-5*) the expression $\kappa = \cos\alpha \frac{d^2w}{dx^2}$ is obtained. Since only the vertical displacements are taken into account, the pressure line of the loads do not change during buckling, so that the bending moment is equal to $-Hw$, where $H = ql^2/8$ is the horizontal projection of the compressive force in the arch.

Introducing all these into the well-known differential equation of the bent bar $EI\kappa = M$, this assumes the following form:

$$Hw + EI\cos\alpha \frac{d^2w}{dx^2} = 0. \tag{8-23}$$

which has constant coefficients if the moment of inertia of the arch varies according to the law: $I = I_0/\cos\alpha$, where I_0 is the moment of inertia at the crown. DISCHINGER thus obtained for the critical value of the horizontal force acting at the supports of a two-hinged arch $\left[H = q l^2 \big/ (8f) \right]$ the expression

$$H_{cr} = \frac{\pi^2 E I_0}{(l/2)^2},$$

(8-24)

which is the critical load of a straight bar, having a constant moment of inertia I_0, in the case of a buckling shape with two half waves. This critical load is independent of the rise f of the arch.

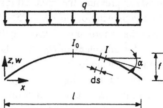

Fig. 8-5 The parabolic arch loaded by uniformly distributed forces

The fact that when assuming a variation of the moment of inertia $I = I_0 / \cos \alpha$ the differential equation of buckling of arches has constant coefficients was the reason why many researchers dealt mainly with such arches. This variation is advantageous, however, only with clamped arches, but not with two-hinged ones. A further, more important deficiency of this approximate theory is that the horizontal displacements of the cross sections of the arch can be neglected only if the arch is flat, but these arches do not buckle antisymmetrically, but snap through symmetrically (Section 8.1.3). Hence there is no domain in which this theory could furnish results accurate enough.

Taking into account also the horizontal displacements of the points of the arch axis, the problem has been solved for several cases. The critical load (or critical horizontal force) has been determined for arches with constant cross section $(I = I_0)$ by LOKSCHIN [1936] and STÜSSI [1935] (see also in [KOLLBRUNNER and MEISTER, 1961]); for arches with a variation of the moment of inertia $I = I_0 / \cos \alpha$ by HILMAN [1930] and DISCHINGER [1939]; for the variation of the moment of inertia $I = I_0 / \cos^3 \alpha$ by LOKSCHIN [1936] and DINNIK [1955]. Of these, the solution of LOKSCHIN and DINNIK refers to a load which rotates together with the cross section on which it acts, so that it has the character of hydrostatic load, but is not perpendicular to the arch, i.e. it maintains the angle between itself and the tangent of the arch axis [RÓZSA, 1964; AUSTIN, 1971]. The other solutions assume constant directional load. According to what has been said in Section 8.1.1, the results of LOKSCHIN and DINNIK yield results inferior to those pertaining to the constant directional load [cf. Eqs (8-7a) and (8-8)].

Most of these numerical values can be found in [TIMOSHENKO and GERE, 1961] and in [KOLLBRUNNER and MEISTER, 1961].

The development of computers made it possible to calculate the critical value of the constant directional load for every case mentioned so far. This has been done by TIETZE [1973]; his results are to be found in [PETERSEN, 1980]. TIETZE replaced the arch by a chord polygon consisting of ten sections on which the loads act at the kink points. His results thus deviate from the exact ones to the unsafe side, since the chord polygon is somewhat shorter than the arch, but this error is negligibly small.

Table 8-1 The factor β of the critical horizontal force [Eq. (8-25)] of parabolic arches

f/l	$I = I_0$ = constant			$I = I_0/\cos\alpha$			$I = I_0/\cos^3\alpha$	
	clamped	two-hinged	three-hinged	clamped	two-hinged	three-hinged	clamped	two-hinged
0	80.8	39.4	(29.8)	80.8	39.4	(29.7)	80.8	39.4
0.1	76.2	36.2	(28.5)	78.4	37.2	(29.4)	83.2	39.0
0.2	64.5	29.1	(24.9)	71.0	31.6	(27.7)	86.2	37.9
0.3	50.1	20.9	20.9	61.1	25.1	25.1	89.0	35.9
0.4	36.7	14.4	14.4	50.8	19.4	19.4	91.0	34.0
0.5	26.3	9.6	9.6	41.6	15.0	15.0	93.2	32.8
0.6	18.9	6.9	6.9	33.8	12.0	12.0	94.7	31.1
0.8	10.1	4.0	4.0	22.9	7.9	7.9	96.3	28.7
1.0	6.0	2.0	2.0	16.2	5.3	5.3	96.7	26.9

The values in parentheses belong to the symmetric buckling shape

In *Table 8-1* we give, on the basis of the computations of TIETZE, the values of the factor β necessary for calculating the critical horizontal force

$$H_{cr} = \beta\frac{EI_0}{l^2},\qquad(8\text{-}25)$$

and in *Fig. 8-6* we plotted them in a diagram. The curves show that the critical force decreases with increasing f/l also in the case of $I = I_0/\cos\alpha$, as contrasted to Eq. (8-24).

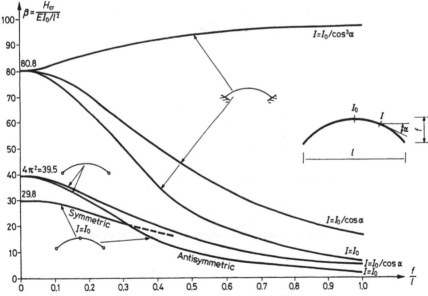

Fig. 8-6 The critical horizontal forces of various parabolic arches according to TIETZE

In the case of flat three-hinged arches (up to about $f/l \leq 0.3$) the symmetric buckling is the most critical; but more high-rise arches buckle antisymmetrically and their critical force is equal to that of the two-hinged arches. The values of the critical force pertaining to the symmetric buckling shape, appearing in *Table 8-1* and *Fig. 8-6*, have been taken for $I = $ const. from STÜSSI [1935], and for $I = I_0/\cos\alpha$ from DISCHINGER [1939].

The initial sections of the curves ($f/l \to 0$) are not valid because very flat arches snap through symmetrically, see in Section 8.1.3.

The critical load can be characterized, instead of the horizontal force H_{cr}, also by the critical load intensity q_{cr}. Using the relation $H = ql^2/(8f)$ and Eq. (8-25), we can express q_{cr} as

$$q_{cr} = \frac{8f}{l}\beta\frac{EI_0}{l^3} = \gamma\frac{EI_0}{l^3},\tag{8-26a}$$

where the notation

$$\gamma = \frac{8f}{l}\beta\tag{8-26b}$$

has been introduced. Plotting γ against f/l, we obtain the curves of *Fig. 8-7* for arches with constant cross section. The numerical values are also given in *Table 8-2*. These values are taken from [TIMOSHENKO and GERE, 1961], hence they are smaller than those computed from the data of *Table 8-1*. We thus wanted, on the one hand, to show the difference between the values of TIETZE and some older results and, on the other hand, to give values for one-hinged arches, not dealt with by TIETZE.

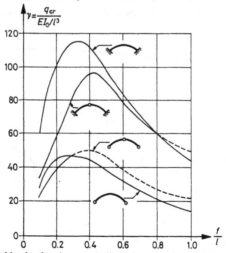

Fig. 8-7 The critical vertical loads of various parabolic arches according to TIMOSHENKO

Table 8-2 The factor γ of the critical load (8-26a) of parabolic arches of constant cross section

f/l	clamped	one-hinged	two-hinged	three-hinged
0.1	60.7	(33.8)	28.5	(22.5)
0.2	101	(59)	45.4	(39.6)
0.3	115	– –	46.5	46.5
0.4	111	(96)	43.9	43.9
0.5	97.4	– –	38.4	38.4
0.6	83.8	(80)	30.5	30.5
0.8	59.1	59.1	20.0	20.0
1.0	43.7	43.7	14.1	14.1

The values in parentheses belong to the symmetric buckling shape

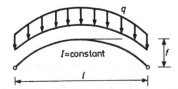

Fig. 8-8 Catenary arch

We can observe that all gamma curves have a peak around $f/l = 0.3$, which means that arches with a constant cross section are 'optimal' in the vicinity of $f/l = 0.3$.

The variation of the moment of inertia according to $1/\cos\alpha$ and $1/\cos^3\alpha$ is advantageous above all for clamped arches. For two- and three-hinged arches the variation according to *Fig. 8-9* is expedient. For these arches there are model test results [DEUTSCH, 1940], and also an approximate method [DIN 4114] which recommends that a replacement constant moment of inertia should be assumed by taking a beam with constant cross section according to *Fig. 8-10a* producing the same deflection under a concentrated load as the original but straightened half arch. AUSTIN [1971] proposes a variant of this method: we should determine the moment of inertia of a bar with constant cross section according to *Fig. 8-10b*, which has the same critical load as the original but straightened half arch. The numerical examples of AUSTIN show that the results of his method always lie on the safe side, while by the substitution according to *Fig. 8-10a* the error is smaller, but sometimes it lies on the unsafe side.

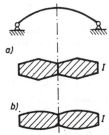

Fig. 8-9 Parabolic arches with variable moment of inertia

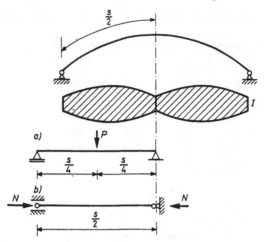

Fig. 8-10 Determining the moment of inertia of the replacement arch with constant cross section

FIROOZBAKSH and FARSHAD [1976] determined the optimum material distribution of the funicular arch loaded by constant directional forces, i.e. the arch with variable cross section which needs the least material for a given critical load. They found that in the case of a circular arch the material consumption can be reduced by 13% if we properly vary the cross section.

In the case of roofing structures the arch is loaded mainly by its own weight (and the roof covering), which can be regarded as uniformly distributed along the arch length (*Fig. 8-8*). The funicular curve of such a load is the catenary (hyperbolic cosine). The critical load of two-hinged arches of this shape with a constant cross section can be written as

$$q_{cr} = \gamma_{cat} \frac{EI}{l^3}. \tag{8-27}$$

The values of the factor γ_{cat} are given in *Table 8-3*, on the basis of [LOKSCHIN, 1936] and [DINNIK, 1955]. (The same values are to be found in [TIMOSHENKO and GERE, 1961].) Since the solution of LOKSCHIN considers a loading of the hydrostatic type, his values most probably deviate somewhat from those valid for constant directional load to the safe side, as explained earlier.

Table 8-3 The factor γ_{cat} of the critical load of catenary arches of constant cross section under a load uniformly distributed along the arch axis [Eq. (8-27)]

f/l	clamped	two-hinged
0.1	59.4	28.4
0.2	96.4	43.2
0.3	112.0	41.9
0.4	92.3	35.4
0.5	80.7	27.4
1.0	27.8	7.06

The optimum rise of the catenary arch is around $f/l = 0.25$. It should be noted that the minimum compressive force at the supports under a given load intensity q arises at $f/l = 0.34$.

8.1.3 Snapping through of flat arches

As already mentioned in Section 8.1.2, very flat arches do not buckle in an antisymmetric shape, but snap through symmetrically. This phenomenon can be investigated in the simplest way if we take a flat arch with constant cross section, having a sine curve axis, on which a load distributed according to a sine curve is acting (*Fig. 8-11*). That is, both the deformation and the bending moment diagram of such an arch has a sine form, and thus closed formulas can be derived.

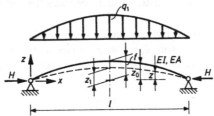

Fig. 8-11 Flat arch with a sine curve axis

The derivation can be found in [FUNG and KAPLAN, 1952; HOFF and BRUCE, 1953; HUANG, 1967]. It results in the critical load intensity causing snapping

$$q_1^{\text{snap}} = \frac{\pi^4 EI f}{l^3} \frac{f}{l} \left[1 + \frac{2}{3\sqrt{3}k} \left(k^2 - 1 \right)^{3/2} \right] \tag{8-28}$$

with the notation

$$k = \frac{f}{2i}, \tag{8-29}$$

where $i = \sqrt{I/A}$ is the radius of gyration of the cross section.

We can also write the rise of the arch z_1 and the horizontal force H, both belonging to the snapping load intensity:

$$z_1^{\text{snap}} = \pm \frac{f}{\sqrt{3}} \sqrt{1 - \frac{1}{k^2}}, \tag{8-30}$$

$$H_{\text{snap}} = \frac{\pi^2 EI}{l^2} \left[1 + \frac{2}{3} \left(k^2 - 1 \right) \right]. \tag{8-31}$$

In *Fig. 8-12* we plotted q_1 against the changing rise z_1 of the arch. The diagram clearly shows the phenomenon of snapping: after reaching q_1^{snap} the load-bearing capacity decreases, and begins to increase again only after reaching a negative value of z_1, i.e. after assuming a shape bowing *under* the horizontal position. The diagram also shows that the snapping flat arch has a decreasing post-buckling load-bearing capacity.

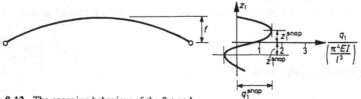

Fig. 8-12 The snapping behaviour of the flat arch

Eqs (8-28) and (8-31) show that the load causing snapping depends not only on the data appearing in the critical load of 'steep' arches (EI, l, f), but also on the ratio $k = f/2i$ (which is, in the case of an I-shaped cross section, approximately equal to the ratio: rise of the arch to height of the cross section).

Let us now determine the rise of the arch which separates the antisymmetric buckling from the symmetric snapping.

Due to the flatness of the arch we can take the expression (8-24) as the critical horizontal force of antisymmetric buckling (with $I_0 = I$ as the moment of inertia assumed constant along the length of the arch). Equating this to (8-31), we obtain the following equation for k:

$$1 + \frac{2}{3}\left(k^2 - 1\right) = 4$$

which yields the value of k as

$$k = \frac{f}{2i} = \frac{\sqrt{22}}{2} = 2.345. \tag{8-32}$$

Thus the arch buckles antisymmetrically if $k > 2.345$, and snaps through symmetrically if $k < 2.345$.

For subsequent comparison we write the (changed) rise of the arch at which it buckles antisymmetrically:

$$z_1^{\text{antisymm}} = \sqrt{f^2 - \frac{4\pi^2 EI}{l^2 EA}\frac{4l^2}{\pi^2}} = f\sqrt{1 - \frac{4}{k^2}}, \tag{8-33}$$

and also the intensity of the load (characterized by its middle value) causing antisymmetric buckling:

$$q_1^{\text{antisymm}} = \frac{\pi^4 EI f}{l^3}\frac{}{l}\left(1 + 3\sqrt{1 - \frac{4}{k^2}}\right). \tag{8-34}$$

We can obtain a clear picture of buckling and snapping if we plot the respective critical load intensities against the ratio $k = f/2i$. It is expedient to make q_1 dimensionless by dividing it by the load intensity $q_{1\text{cr}}^*$ which would cause antisymmetric buckling on the arch with unchanged shape. Since the pertaining critical horizontal force can be assumed as that given by (8-24), again setting $I_0 = I$, we obtain:

$$q_{1\text{cr}}^* = \frac{H_{\text{antisymm}}\pi^2 f}{l^2} = 4\frac{\pi^4 EI f}{l^3}\frac{}{l}. \tag{8-35}$$

The values of $q_{1\text{cr}}/q_{1\text{cr}}^*$ are plotted in *Fig. 8-13*. Comparing also the values z_1 pertaining to snapping and antisymmetric buckling respectively, we can draw the following conclusions.

At large values of k the snapping load q_1^{snap} is greater than than the load $q_1^{antisymm}$ causing buckling, so that this latter form of buckling occurs. However, $q_1^{antisymm}$ is not exactly equal to the critical load q_{1cr}^* of the arch considered as inextensional (which has been computed anyway by neglecting the horizontal displacements, see the explanations to Eq. (8-24)). The load $q_1^{antisymm}$ approaches q_1^* only asymptotically, but, according to Eq. (8-34), at $k = 12$ it is only 1% less, which yields information about the limits of validity of the assumption on the inextensibility of steeper arches, made earlier. We depicted this phenomenon in the domain $k > 2.345$ of *Fig. 8-13* in such a way that the line of antisymmetric buckling bifurcates (horizontally) from the $q_1(w_1)$ diagram of the symmetric deformation, leading to snapping, before its peak. Now $w_1 = f - z_1$ is the deflection of the crown of the arch, considered positive when pointing downwards. At $k = 2.345$ the critical loads of the two kinds of instability are equal to each other, so that here the line of antisymmetric buckling bifurcates from the peak point of that of the symmetric deformation.

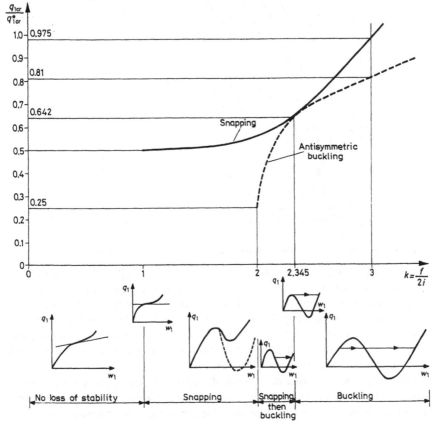

Fig. 8-13 The various kinds of loss of stability of flat and steep arches

In the domain $2 < k < 2.345$ $q_1^{antisymm}$ is smaller than q_1^{snap}, nevertheless the comparison of Eqs (8-30) and (8-33) shows that to reach $q_1^{antisymm}$ a larger deformation is needed than to reach q_1^{snap}. Consequently, the arch first snaps through, and the antisymmetric buckling bifurcates from the descending branch of the snapping deformation. This phenomenon has, however, only theoretical importance.

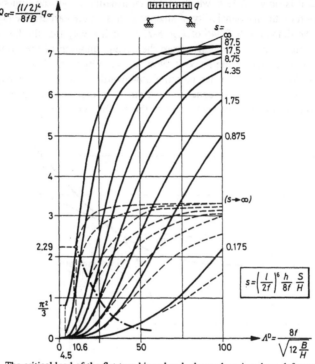

Fig. 8-14 The critical load of the flat two-hinged arch also undergoing shear deformation

Eq. (8-34) shows that at values of k smaller than 2 the load $q_1^{antisymm}$ becomes complex, so that in the domain $k < 2$ the arch does not buckle antisymmetrically at all. Here only snapping occurs, down to $k = 1$. At this value also Eq. (8-28) yields a complex value for q_1^{snap}, so that also the snapping phenomenon ceases to exist in the sense that the $q_1 (w_1)$ curve has still a horizontal tangent at $k = 1$, but also an inflexion point, i.e. it has no peak, and at $k < 1$ it has no more a horizontal tangent, but it ascends continuously: the arch deflects as a bent beam, without loss of stability.

Flat sine shaped arches under other types of loading are dealt with in [TIMOSHENKO and GERE, 1961]. They found that if the deflection of the arch considered as a simply supported beam with the span l is close to a sine curve, then the snapping load can be computed in every case from the equation

$$q_{snap} = \frac{f}{w_m} \left[1 + \frac{2}{3\sqrt{3}k} \left(k^2 - 1 \right)^{3/2} \right], \tag{8-36}$$

where w_m is the deflection of the middle cross section of the simply supported beam of the span l, caused by the load of unit intensity. So e.g. in the case of a uniform load:

$$w_m = \frac{5l^4}{384EI};$$

and if a concentrated load P acts at the middle of the span:

$$w_m = \frac{l^3}{48EI},$$

but in this latter case we have to write on the left-hand side of Eq. (8-36) P_{snap}/l instead of q_{snap}.

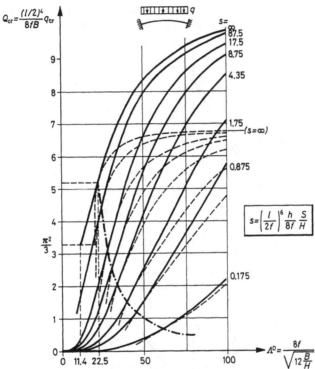

Fig. 8-15 The critical load of the flat clamped arch also undergoing shear deformation

Arches of other shapes

The snapping of flat parabolic two-hinged arches loaded by uniformly distributed forces has been investigated by FEDERHOFER [1934]. The loss of stability of flat circular arches with various support conditions and loadings was treated by SCHREYER and MASUR [1966], DICKIE and BROUGHTON [1971], MASUR and LO [1972], DYM [1973], and ORAN and BAYAZID [1978]. Their investigations yielded – apart from some minor numerical differences – essentially the results shown in *Fig. 8-12*. The snapping of parabolic arches with variable stiffness has been dealt with by SIMITSES and RAPP [1977]. AMAZIGO [1978] determined the optimum variation of the cross section of circular two-hinged arches under uniform pressure which yields – at a given volume of material – the highest snapping load. The saving related to the uniform cross section is about 10%.

The post-critical load-bearing capacity of flat circular two-hinged arches has been investigated by MASUR and LO [1972]. They found that the behaviour of the arch after antisymmetric buckling has the 'symmetric unstable' character, in accordance with what has been said earlier on two-hinged arches.

MASUR and LO [1972] – based on the investigations of KOITER – present also 'asymptotic' formulas to take into account inaccuracies in the load and imperfections of shape, which become exact if the imperfections tend to zero (see the curves in *Figs. 8-16* and *8-17*, and the explanation at the end of this section). SCHREYER [1972] performed similar investigations on clamped arches under a concentrated load.

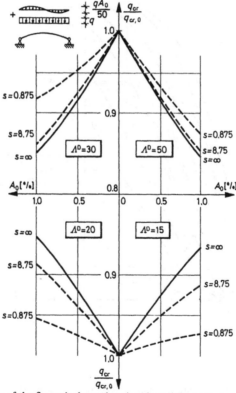

Fig. 8-16 Sensitivity of the flat arch also undergoing shear deformation to non-symmetric distribution of the load

Taking into account the shear deformation of the arch.

We have tacitly assumed so far that the shear stiffness of the arch is infinitely great, which is fulfilled with a good approximation in the case of solid cross sections. Arches with complex (sandwich, braced) cross sections, however, undergo, as a rule, a considerable shear deformation, so that it is reasonable to take this effect into consideration. In the following we shall present some important results of NAGY [1978] on this problem.

The flat circular arch is geometrically characterized by the quantities f and l, to be seen also in *Fig. 8-5*. (The radius of curvature R can be expressed by l and f with the relation $R \approx l^2 / (8f)$.) The height of the cross section is denoted by h. The arch has the bending, tension and shear stiffnesses B, H and S, respectively.

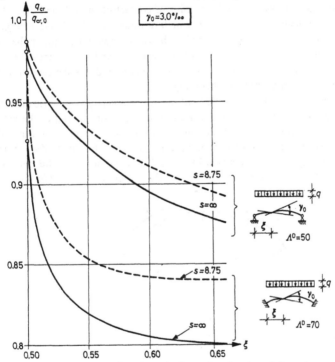

Fig. 8-17 Imperfection sensitivity of the flat arch also undergoing shear deformation

The load $q_{cr} = q_{snap}$ causing snapping is plotted by a full line as a function of the rise f for various shear rigidities in *Figs 8-14* and *8-15*. The critical load $q_{cr} = q_{antisymm}$ causing antisymmetric buckling is represented by a dashed line. On the vertical and the horizontal axes the dimensionless quantities

$$\Lambda^D = \frac{8f}{\sqrt{12\dfrac{B}{H}}} \tag{8-37}$$

and

$$Q_{cr} = \frac{\left(\dfrac{l}{2}\right)^4}{8f\,B}\,q_{cr} \tag{8-38}$$

are plotted. The shear stiffness is represented by the quantity

$$s = \left(\frac{l}{2f}\right)^6 \frac{h}{8f} \frac{S}{H}. \tag{8-39}$$

The diagrams are similar to those of *Fig. 8-13*: for arches with solid cross section (with no shear deformation)

$$\Lambda^D = \frac{8f}{\sqrt{12}i} = 4.62k,$$

so that in *Fig. 8-14* the abscissa 10.6 of the point of contact of the full and dashed lines pertaining to $s = \infty$ corresponds to $k = 2.29$, which deviates from $k = 2.345$ of *Fig. 8-13* only because the latter referred to a sine shaped arch. The dashed-dotted line in the figures connects the points where the critical loads causing snapping and antisymmetric buckling coincide. The curves clearly show how the shear deformation reduces the critical loads of both the snapping and the antisymmetric buckling.

NAGY [1978] also investigated to what extent the not completely symmetric distribution of the load and the imperfections of the arch axis reduce the critical load of the antisymmetric buckling. His results are plotted in *Figs 8-16* and *8-17*, using the shear stiffness (8-39) as a parameter. The critical loads pertaining to the symmetric load ($A_0 = 0$) and to the perfect arch shape ($\gamma_0 = 0$) are denoted by $q_{cr,0}$. The diagrams are valid only for the critical loads lying in the domain to the right of the dashed-dotted line of *Fig. 8-14*.

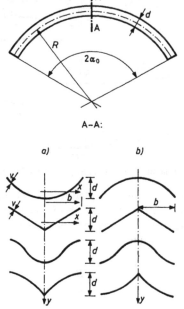

Fig. 8-18 Shell arches. *a*) The four cross sections, *b*) 'inverted' cross sections

8.1.4 Buckling of arches with thin-walled, open cross sections

Arches with cross sections according to *Fig. 8-18*, the so-called *shell arches*, have to be treated separately, because their static behaviour fundamentally differs from that of common arches (having solid cross sections).

The cross sections of shell arches – due to their flatness and thin wall – deform during bending of the arch. This can be seen from *Fig. 8-19*. The tensile and compressive internal forces parallel to the arch axis give resultants perpendicular to this axis, which bend the cross section in the transverse direction. Consequently, the thin cross section undergoes considerable deformation, it 'evades' this bending, and the stiffness of the arch against bending becomes much smaller than that of common arches whose cross sections do not deform. (In this section the bending moment will be considered positive as shown in *Fig. 8-19*.)

This phenomenon has been treated for closed tubes with curved axis by KÁRMÁN [1911] and BRAZIER [1927], and for arches with open circular cross section by WEINEL [1937]. Assuming that the cross section of the arch is flat and has a constant wall thickness, and that *Poisson*'s ratio is zero, a general theory has been developed by KOLLÁR, [1973], see also [KOLLÁR 1961a and 1961b], the main results of which will be presented in the following.

It should be noted that the 'inverted' cross sections shown in *Fig.8-18b* behave statically in the same way as those of *Fig. 8-18a*.

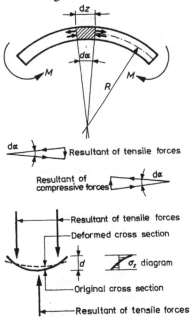

Fig. 8-19 Deformation (flattening) of the cross section of the shell arch

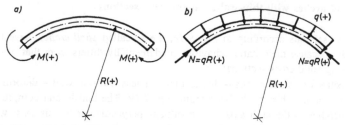

Fig. 8-20 The two basic loading cases. *a*) Pure bending, *b*) central compression

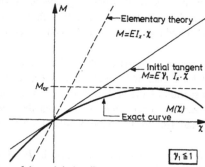

Fig. 8-21 The behaviour of the arch in bending

The aforementioned deformation of the cross section means that during bending (*Fig. 8-20a*) the relation between the bending moment M and the change in curvature χ assumes the shape shown in *Fig. 8-21*. It can be clearly seen that the arch is also initially much softer than expected on the basis of the elementary theory, since the slope of the initial tangent of its $M(\chi)$ curve is flatter. Hence we have to use this flatter slope in the investigation of the centrally compressed arch (*Fig. 8-20b*), i.e. instead of the moment of inertia I_x calculated by the elementary theory we have to take the 'effective moment of inertia' $\gamma_1 I_x$ when computing the critical compressive force described in the foregoing sections.

The expressions of the moments of inertia I_x for the four cross sections are to be seen in *Fig. 8-40*, but we have to keep in mind that – since the co-ordinate system used in Section 8.2 is different from that used in this section – the moments of inertia I_y of *Fig. 8-40* correspond to I_x used in this section.

The numerical values of the factor γ_1 necessary to calculate the effective moment of inertia of the four cross sections of *Fig. 8-18* are given in *Fig. 8-22* as functions of the geometric parameter

$$\beta_0 b = \frac{\sqrt[4]{3}b}{\sqrt{\nu R_0}} = \frac{1.316 b}{\sqrt{\nu R_0}} \tag{8-40}$$

computed with the original radius of curvature R_0 of the arch (*Fig. 8-18*). As can be seen, the γ_1 curve of every cross section has the same character, and rapidly decreases with increasing $\beta_0 b$.

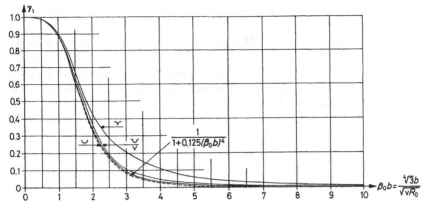

Fig. 8-22 The factor γ_1 of the effective moment of inertia

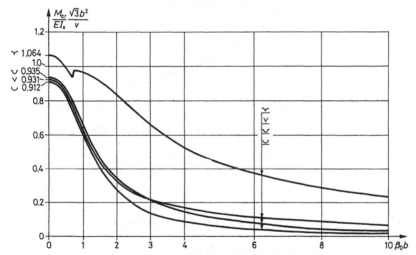

Fig. 8-23 The critical bending moment causing snapping

The γ_1 curve of the circular cross section (which can be approximated, due to its flatness, by a parabola) gives the lowest values among all cross sections, so that – with an error on the safe side – we can take these for the other cross sections too. We also give an approximate expression for γ_1, which is very close to that of the circular cross section, and at greater values of $\beta_0 b$ it deviates to the safe side:

$$\gamma_1^{\text{approx}} = \frac{1}{1 + 0.125\,(\beta_0 b)^4}. \tag{8-41}$$

Fig. 8-21 also shows that – considering the bending moment positive if it corresponds to that depicted in *Fig. 8-19* or *8-20a* – the arch becomes softer and softer with increasing positive bending moment, due to the flattening of the cross section and, at a certain value M_{cr}, loses its stability and snaps through. This does not happen under a negative bending moment, since it causes the cross section to become steeper, i.e. stiffer, see the lower section of the curve of *Fig. 8-21*.

The values of the bending moment M_{cr} causing snapping are given in *Fig. 8-23* for the four cross sections of *Fig. 8-18*, as functions of the parameter $\beta_0 b$ (8-40). Of these, the curve pertaining to the wing-shaped cross section gives much greater values than the others. It is, however, not advisable to use these high values, because they are connected with rather large deformations, see in [KOLLÁR, 1973].

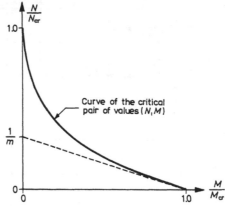

Fig. 8-24 Snapping of the shell arch due to the simultaneous action of central compression and bending

We present an approximate relation for M_{cr} which deviates slightly to the safe side from that valid for the circular cross section:

$$0 \le \beta_0 b \le 3: \quad M_{cr} = \frac{EI_x v}{\sqrt{3}b^2} \frac{0.912}{1 + 0.488\,(\beta_0 b)^2 + 0,0155\,(\beta_0 b)^4}, \qquad (8\text{-}42a)$$

$$\beta_0 b \ge 3: \quad M_{cr} = \frac{EI_x v}{\sqrt{3}b^2} \frac{1.235}{(\beta_0 b)^2}; \qquad (8\text{-}42b)$$

where b and v are defined in *Fig. 8-18*. Since the critical bending moments valid for the circular cross section are lower than those of the other cross sections, we are always on the safe side when using the approximate relations (8-42).

The compressed arch buckles antisymmetrically, so that its bending rigidity always decreases along its one half, and this decrease is greater than the increase along the other half, see *Fig. 8-21*. Consequently, the post-buckling load-bearing capacity of the centrally compressed arch is decreasing.

We have still to investigate the load-bearing capacity of the eccentrically compressed arch. To this purpose we need the interaction curve of compressive force and bending moment, which has been established by DULÁCSKA [1964], and is shown in *Fig. 8-24*. This curve can be approximated by the formula

$$\frac{N}{N_{cr}} + \sqrt[m]{\frac{M}{M_{cr}}} = 1 \qquad (8\text{-}43)$$

which always deviates to the safe side. Here N_{cr} is the critical force of the arch computed with the effective moment of inertia $\gamma_1 I_x$, and N is the actual compressive force; it is advisable to determine both at the quarter point of the arch, since the derivation of

(8-43) also took the quantities valid here as a basis. M is the initial bending moment (not increased by the bending effect of the compressive force), also valid at the quarter point of the arch. (The relation (8-43) takes into account the increase of the bending moment due to the compressive force.) The value of the root exponent m can be taken from *Fig. 8-25*, which again deviates to the safe side from the exact curves of the cross sections, except for the wing-shaped cross section, whose curve strongly deviates from the others in the range $0 \leq \beta_0 b \leq 2.3$; this section is plotted by dashed line in *Fig. 8-25*.

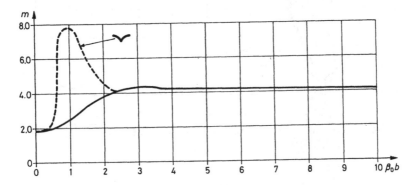

Fig. 8-25 The root exponent m appearing in Eq. (8-43)

The curve of *Fig. 8-24* starts from point $N = N_{cr}$ with a vertical tangent, so that even the smallest (accidental) eccentricity strongly reduces the value N_{cr} computed with the effective moment of inertia $\gamma_1 I_x$; and the finite bending moment of the arch loaded on one side reduces even more the compressive force which can be taken by the arch.

It follows from all these that, on the one hand, the stability of the centrally compressed arch has to be checked in such a way that we assume an appropriate accidental eccentricity or bending moment; and, on the other hand, we have to check the stability for eccentric compression too, as contrasted to common arches (with solid cross sections) with which it is sufficient to check the stability for central compression, and for eccentric compression we have to check the strength of the arch only, taking into account the increase of the bending moment due to the compressive force (cf. Eq. (1-14)).

If the radius of curvature (and the cross section) of the arch varies, then we have to compute the effective moments of inertia at several points, and to determine the critical compressive force N_{cr} of this arch of variable stiffness. The bending moment M_{cr} causing snapping has to be computed for a characteristic ('average') cross section, which may be also that of the quarter point of the arch.

The dimensioning of the arch also includes the determination of the longitudinal stresses and the transverse bending moments, so we have to determine these quantities too.

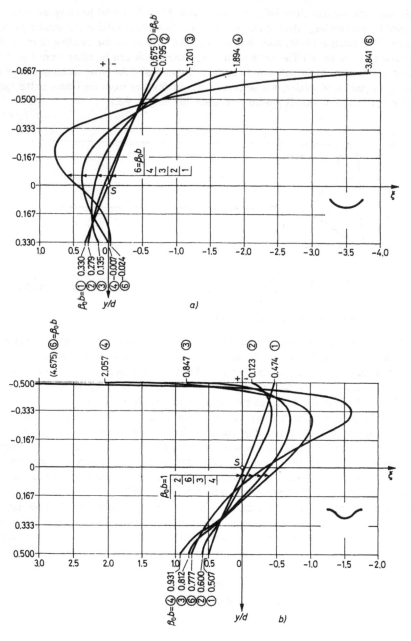

Fig. 8-26 The factor ξ of the stress distribution of bent

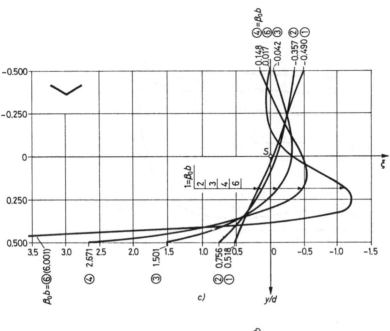

c)

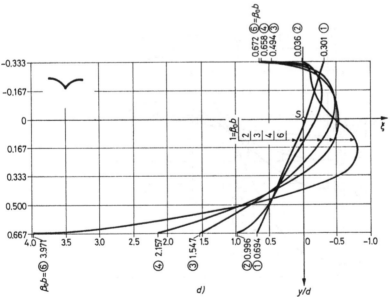

d)

shell arches (Eq. (8-44)) for the four cross sections.

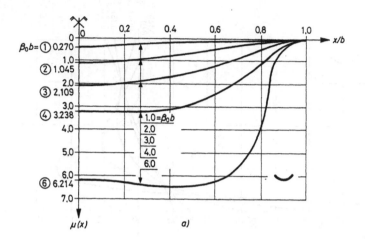

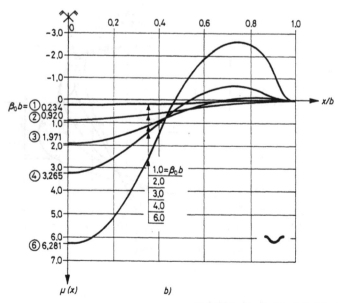

Fig. 8-27   The factor $\mu(x)$ of the transverse bending

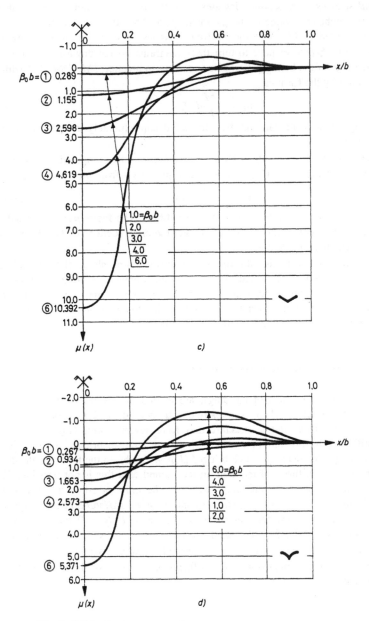

$\mu(x)$ c)

$\mu(x)$ d)

moment [Eq. (8-45)] for the four cross sections

Due to the deformation of the cross sections shown in *Fig. 8-19* the distribution of the *longitudinal stresses* from bending deviates from the linear diagram. This deviation is the greater, the higher the value of the parameter $\beta_0 b$. The stress distribution is given by the factor ξ, plotted in *Fig. 8-26* for the four cross sections of *Fig. 8-18a*. (For the cross sections of *Fig. 8-18b* we have to turn these diagrams upside down.) The stresses are given in the range of the bending moments occurring in practice ($M \leq M_{\mathrm{cr}}/3$) with a good approximation by the formula

$$\sigma_z = -\frac{M}{I_x} d \cdot \xi \tag{8-44}$$

since the bending moment diagram can be considered approximately as straight in this range, see *Fig. 8-21*.

The *transverse bending moments of the cross sections* can be computed – also in the range of bending moments $M \leq M_{\mathrm{cr}}/3$, allowable in practice – from the expression

$$m_k = -\frac{M}{b} \frac{v}{d} \mu(x), \tag{8-45}$$

considered positive if they cause tension in the lower fibres. The factor $\mu(x)$ is plotted in *Fig. 8-27* for the four cross sections of *Fig. 8-18a*. For the cross sections of *Fig. 8-18b* we have to turn the diagrams upside down.

All that has been said so far is valid for shell arches standing alone. If they are built beside each other and are interconnected, they prevent each other's deformation shown in *Fig. 8-19*, so that the structure behaves as a common arch with the moment of inertia I_x. Thus the method of computation shown in the foregoing has to be applied only in the stage of erection, as long as the arches do not stiffen each other.

8.1.5 Buckling of arches with hangers or struts

Arches of bridges, but often also of buildings, are loaded mostly indirectly: the loads act onto a horizontal structure which is supported with the aid of vertical bars (hangers or struts) by the arch. This kind of structure may have several forms (*Fig. 8-28*). The bridge deck (stiffening beam) may be under or above the arch; in bridge building, these are called through-type and deck-type arch bridges (*Fig. 8-28a,b*). In the first case the deck mostly takes the horizontal thrust of the arch; in the second case this is not possible. The own bending rigidity $E_b I_b$ of the stiffening beam may be finite or zero. In the case of *Fig. 8-28a* the beam has a finite bending stiffness and is called *stiffening beam*. A special case of this is the *tie* with zero bending stiffness. The case $E_b I_b = 0$ may also occur with the arrangement of *Fig. 8-28b*. The beam (or tie) has, of course, also in the case of $E_b I_b = 0$ a bending stiffness necessary to carry the loads between two hangers, but either is $E_b I_b$ negligibly small in comparison to the stiffness of the arch, or there are hinges at the joints of the hangers (or struts).

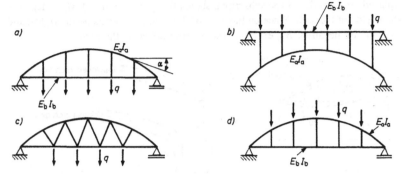

Fig. 8-28 Arches with hangers or struts. *a*) Vertical hangers, *b*) vertical struts, *c*) inclined hangers, *d*) arch with stiffening beam

We will not deal with arches whose hangers are connected to the tie by sliding joints, nor with structures having inclined hangers (*Fig. 8-28c*), because this is rather a truss.

Finally, *Fig. 8-28d* shows a roof structure whose arch has a stiffening beam (or tie), but which is loaded directly on the arch.

In the following, we will investigate, how the stiffening beam (or tie) modifies the critical load of the arch of these structures as compared to that of free arches.

Investigation of the internal forces during buckling

For a better understanding let us consider the internal forces of the *tied arch* of *Fig. 8-28a* during buckling. To simplify the problem we assume the individual hangers to be 'smeared out', i.e. we replace the hangers by a 'hanger veil'. For the time being we assume that $E_b I_b = 0$, i.e. that we have to deal with a tied arch, and that the hangers are inextensional, and furthermore that, since they undergo tension from the external load, they can take the small compressive forces arising during buckling of the whole structure without buckling themselves.

Using the equilibrium method it is sufficient to write the first-order terms of the internal forces. Consequently, we have to let the forces of finite value act on the deformed structure, but to let the infinitely small forces arising during buckling act on the undeformed structure. Let us investigate from which parts the bending moment M acting on the arch during buckling is composed.

The buckled shape of the structure is to be seen in *Fig. 8-29a*. Since the cross sections of the arch displace not only vertically (w), but also horizontally (u), but the tie displaces only vertically, the hangers get inclined.

The distributed external forces do not change their line of action nor their magnitude during buckling, so that their pressure line (the line of action of the compressive force in the arch) coincides with that valid before buckling, i.e. with the original axis $z(x)$ of the arch (*Fig. 8-29b*). Denoting differentiation with respect to x by $'$, in the section $A - A$ a horizontal force H and a vertical force $-Hz'$ act, since z' is negative at the cross section shown in *Fig. 8-29b*. These forces exert a bending moment

$$M_p = -Hw + Hz'u \tag{8-46}$$

onto the displaced cross section. (The subscript p stands for 'pressure line'.) M_p is negative on the left-hand side of the arch. Since on this side of the arch both terms on the right-hand side of Eq. (8-46) are negative, M_p exerts a destabilizing effect on the arch.

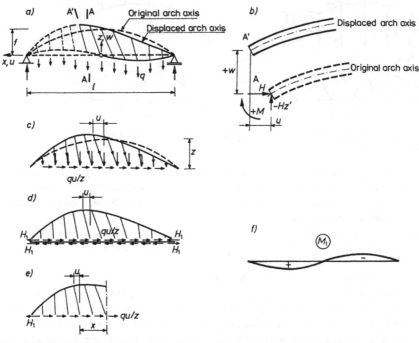

Fig. 8-29 Buckling of the tied arch. *a*) The buckling shape, *b*) the original and the displaced arch axis, *c*) the horizontal components of the forces acting in the hangers, *d*) the forces acting on the structure cut through above the tie, *e*) cutting the structure across the middle cross section, *f*) the diagram of the moment M_1

The external forces, however, do not act directly on the arch, but through the inclined hangers. We could decompose these inclined forces into vertical and horizontal components at the height of the arch axis, but it is much simpler to decompose them at the height of the tie, since then the points of action of the forces do not displace horizontally. The magnitude of the vertical components remains q, and that of the horizontal components is qu/z, where z is the height of the arch axis above the tie (*Fig. 8-29c*). These horizontal forces exert an additional moment on the arch, the magnitude of which can be obtained in the following way.

If we cut the hangers and the arch immediately above the tie, then – beside the vertical forces, now left out of consideration – on this section the forces sketched in *Fig. 8-29d* act, where H_1 is equal to the sum of the horizontal forces qu/z acting on one half of the structure. The forces qu/z acting on the arch are antisymmetric with respect to the symmetry axis of the structure, thus they cannot exert a horizontal force or a bending moment in the middle cross section of the arch. A vertical force could act here, but no

such forces act on the structure anywhere. So the middle cross section of the arch is free
from any forces or moments, thus we can cut it through (*Fig. 8-29e*). The moment M_1
acting on the arch can be determined starting from its middle cross section, and since we
thus obtain a moment acting on the right-hand side of the arch, it has to be multiplied by
(-1). At an arbitrary cross section x, where the height of the arch is z, the moment M_1
can be obtained as

$$M_1(x) = qz \int_0^x \frac{u}{z} \mathrm{d}x. \tag{8-47}$$

This moment is positive on the left-hand side of the arch, and negative on the right-hand
side, so we finally obtain the diagram of *Fig. 8-29f*. Thus the moment M_1 tries to bend the
arch in a sense opposite to the buckling deformation, i.e. it exerts a stabilizing effect on
the arch.

The bowing out of the tie gives rise, due to the force H acting in it, to additional
(distributed) forces p in the hangers. Its magnitude is $p = -Hw''$. The moment M_2 in the
arch, caused by p, can be computed, using the relation $M'' = -p$, from the expression

$$M_2 = Hw \tag{8-48}$$

(the integration constants are equal to zero). The diagram of this additional moment has
the same shape as that of M_1 (*Fig. 8-29f*), hence M_2 also stabilizes the arch.

The moment M_2 according to (8-48) and the first term on the right-hand side of (8-46)
are equal, but of opposite sign, so they cancel each other. The reason for that is that the
vertical distance between arch and tie is constant.

Summing up, the arch is loaded by the moment

$$M_p + M_1 + M_2 = Hz'u + M_1. \tag{8-49}$$

The first term on the right-hand side means a negative, the second term a positive moment
on the left-hand side of the arch, but since the absolute value of the first term is greater than
that of the second, the destabilizing effect prevails, so that the structure loses its stability
under a sufficiently high load.

Let us investigate how the situation changes in the case of the *deck-type* arrangement
shown in *Fig. 8-28b* (assuming, for the time being, $E_b I_b = 0$ also here).

The bending moment M_p caused by the deviation of the pressure line of the load from
the arch axis is again given by Eq. (8-46). The struts, however, tilt in a sense opposite to
that of the previous case, so that their bending moments M_1 overload the arch, exerting a
destabilizing effect. In addition, no horizontal force acts in the deck, so that the moment M_2
ceases to exist, which means a further destabilizing effect. Consequently, the critical load
of the deck-type arch bridge of *Fig. 8-28b* is less than that of the structure of *Fig. 8-28a*.

Let us investigate another case, when *the load acts directly on the arch* (*Fig. 8-28d*).
We again set $E_b I_b = 0$, and assume that the hangers can take the compressive forces
arising during buckling. The deviation from the case of *Fig. 8-28a* is then, on the one
hand, that the loads q act directly on the arch, so that the horizontal forces qu/z arising
from the inclination of the hangers do not come about and thus they cannot give rise to the
stabilizing moment M_1. On the other hand, the points of action of the forces q displace
horizontally together with the points of the arch, thus changing their pressure line, which
means that the load exerts distributed moments $m = qu$ on the arch (*Fig. 8-30a*). These

give rise to vertical forces

$$V = \frac{1}{l} \int\limits_{-l/2}^{+l/2} qu\,dx = \frac{2q}{l} \int\limits_{0}^{l/2} u\,dx \qquad (8\text{-}50)$$

at the supports. According to what has been said in connection with *Fig. 8-29d*, in the middle cross section of the arch only a vertical force V can arise. We thus can cut the arch into two parts according to *Fig. 8-20b*, and the bending moments M_3 arising in the cross sections of the left-hand part of the arch can be computed by the expression

$$M_3 = \int\limits_{0}^{x} qu\,dx - Vx = q \int\limits_{0}^{x} u\,dx - 2q\frac{x}{l} \int\limits_{0}^{l/2} u\,dx, \qquad (8\text{-}51)$$

see *Fig. 8-30c*. These moments M_3 replace the moments M_1 arising from the inclination of the hangers. The moments M_3 stabilize the arch, but are smaller than M_1. Hence the arch is in a worse situation: its critical load is less than that of the structure of *Fig. 8-28a*.

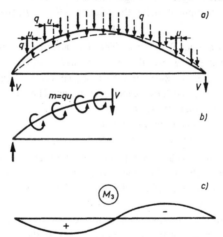

Fig. 8-30 The effect of the horizontal displacement of the load acting on the arch. *a*) The horizontal displacements, *b*) the moments caused by these displacements, *c*) the diagram of the moment M_3

The *finite bending stiffness* $E_b I_b$ *of the stiffening beam* can be taken into account in the following way. An elementary arch section ds, the horizontal projection of which is dx, has the bending stiffness referred to unit horizontal length $E_a I_a \cos\alpha$, since $dx = ds\cos\alpha$ (*Fig. 8-5*), and the bending resistance of a bar section is inversely proportional to its length. Since we assumed inextensional hangers, the vertical deflection of the arch and the beam are identical. Consequently, we have to add the bending stiffness $E_b I_b$ of the beam to the 'effective' bending stiffness $E_a I_a \cos\alpha$ of the arch, or, since all formulas have been derived using $E_a I_a$, we can add the 'reduced' bending stiffness $E_b I_b / \cos\alpha$ to $E_a I_a$. This is true independently of whether the beam is under or above the arch.

We have assumed so far that the hangers of the structure of *Fig. 8-28a* do not elongate due to the forces arising during buckling. Let us investigate how the *elongation of the hangers* influences the critical load.

The hangers always exert a stabilizing effect on the arch, see the moments M_1 and M_2 (Eqs (8-47) and (8-48)). This stabilizing effect consists essentially of the fact that the hangers 'pull back' the buckling arch. Thus if the hangers elongate, their 'pulling back' effect weakens, so that the critical load diminishes.

We also have to mention that *the arch may also buckle between the hangers*; in fact, this buckling mode may combine with the global buckling treated so far (always dealt with by assuming a 'hanger veil'), and, consequently, the critical load of the global buckling may be reduced. We shall deal with this problem at the end of this section.

Survey of the literature

It was MAYER-MITA [1913] who first dealt with the buckling of tied arches (*Fig. 8-28a*) and came to the erroneous conclusion that the tied arch as a whole cannot lose its stability, the arch can only buckle between the hangers. The reason for this error was that MAYER-MITA, although he considered the arch as interrupted by hinges at the hangers (which is an approximation to the safe side), neglected the horizontal displacements of the points of the arch ($u = 0$). This means that all terms appearing in Eq. (8-49) become zero, so that no moments act on the bowed-out arch, and the structure remains in equilibrium also in the 'buckled' state as it was in the undeformed state, which means an indifferent state of equilibrium. Since the arch is, in fact, continuous, the deformation gives rise to bending moments which try to restore the undeformed shape: the whole structure would thus be stable. The same erroneous reasoning can be found in [DISCHINGER, 1937].

SCHIBLER [1948] presents a method suitable for determining the critical load of the arch interrupted by hinges at the hangers (*Fig. 8-28b*) and stiffened by a beam with finite bending rigidity, but only for a structure with three hangers, so that this cannot be regarded as a general solution.

The problem of global buckling of the arch with stiffening beam (or tie) has been solved by PFLÜGER [1951] with the energy method, assuming a hanger veil. Although his equations are exact, his solution (which contains also closed formulas) is inaccurate, because he took only one term of the *Fourier* series of the buckling shape into account.

We find practical diagrams in [PETERSEN, 1980] for the critical load of deck-type arch bridges.

DULÁCSKA and KOVÁCS [1971] used the equilibrium method to solve the buckling problem of the arch with stiffening beam or tie (*Fig. 8-28a*). From the effects described earlier, they did not take into account the moments M_1 caused by the inclination of the hangers (which can be considered as an approximation on the safe side), so that they obtained results smaller than the exact ones. Assuming an inextensional hanger veil they obtained for the critical horizontal force

$$H_{cr} = \frac{5.15}{f^2} \frac{E_a I_a}{\sqrt{1 + \dfrac{4f^2}{l^2}}} + E_b I_b. \tag{8-52}$$

In the case of a tie we have to set $E_b I_b = 0$.

This formula has been derived for an arch with constant cross section. The factor $1 \Big/ \sqrt{1 + 4f^2 / l^2}$ beside $E_a I_a$ is the cosine of the angle of the straight line connecting the support with the crown of the arch, i.e. the 'average' $\cos \alpha$ value of the arch (*Fig. 8-28a*).

The formula has the peculiarity that in its denominator, instead of l^2, f^2 appears, so that the critical load is many times higher than that of a free arch.

KOVÁCS [1974] took also the moment M_1 into account, and his solution is thus exact. Assuming an inextensional hanger veil he obtained for an arch with the moment of inertia varying according to the law $I_a = I_{a0}/\cos\alpha$ (I_{a0} is the moment of inertia at the crown of the arch) the critical horizontal force

$$H_{cr} = \frac{14.59}{f^2}\left(E_a I_{a0} + E_b I_b\right).$$
(8-53)

If the moment of inertia of the arch varies according to the law

$$E_a I_a = E_a I_{a0} \left(\frac{1}{\cos\alpha}\right)^\mu,$$
(8-54)

then in (8-53) $E_a I_{a0}$ has to be multiplied by the factor $\bar{\alpha}$:

$$H_{cr} = \frac{14.59}{f^2}\left(\bar{\alpha} E_a I_{a0} + E_b I_b\right).$$
(8-55)

The factor $\bar{\alpha}$ is given by the curves of *Fig. 8-31* for some values of the exponent μ.

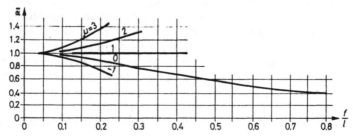

Fig. 8-31   The factor $\bar{\alpha}$ appearing in Eq. (8-55)

The correction due to the *elongation of the hangers* (or the struts) during buckling has been dealt with by DULÁCSKA and KOVÁCS [1971]. They gave simple approximate formulas for taking this effect into consideration, but they also showed that, in the case of through-type arch bridges with a tie, practically no such correction is needed.

KOVÁCS [1974, 1984] investigated the effect of the elongation of the hangers of the structures shown in *Fig. 8-28a*. He found that if the moment of inertia of the arch varies according to $I_a = I_{a0}/\cos\alpha$, then the critical horizontal force H_{cr} yielded by (8-53) has to be multiplied by a factor γ:

$$H_{cr} = \frac{14.59}{f^2}\gamma\left(E_a I_{a0} + E_b I_b\right).$$
(8-56)

The factor γ is plotted in *Fig. 8-32* as a function of the parameter ρ and the ratio f/l for two stiffness ratios: $E_b I_b/(E_a I_{a0}) = 0$ (tied arch) and $E_b I_b/(E_a I_{a0}) = \infty$ (very rigid stiffening beam). The parameter ρ is defined by the expression

$$\rho = 10^4 \frac{4e}{l^2 f}\frac{E_a I_{a0} + E_b I_b}{E_h A_h},$$
(8-57)

where

$E_h A_h$ – the extensional rigidity of one hanger,
e – the distance between the hangers (*Fig. 8-33*).

For intermediate $E_b I_b / (E_a I_{a0})$ values we can linearly interpolate between the two curves.

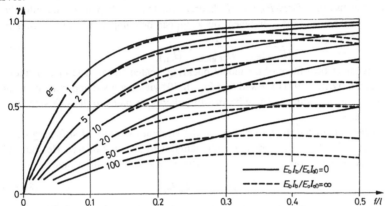

Fig. 8-32 The factor γ appearing in Eq. (8-56)

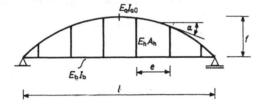

Fig. 8-33 Notations

KOVÁCS [1984] states that this correction due to the elongations of the hangers can be used also for arches with constant cross section, as an approximation, in the following way. We determine the arch with a cross section varying according to the law $E_a I_a = E_a I_{a0} / \cos\alpha$, whose moment of inertia at the horizontal distance $0.2\, l$ measured from the support has the same moment of inertia as the arch with constant cross section, and compute the reduction of the critical load due to the elongations of the hangers. This reduction ratio is then valid for the arch with constant cross section too.

It should be mentioned that in the case of very weak hangers the minimum critical load may belong not to the buckling shape with two half waves, but to that with three or even four half waves. However, KOVÁCS [1974] showed that if the material quality of the hangers is not higher than St 52, and they has been dimensioned in such a way that the the load causing their allowable force is greater then 0.4 times the critical load of the structure, then it is always the buckling shape with two half waves which yields the minimum critical load. If this inequality is reversed, then the load that can be carried by the structure is limited not by buckling, but by the load-carrying capacity of the hangers.

We still have to deal with the *buckling of the arch between the hangers*. The horizontal component of the critical force of such a section is

$$H_{cr,1} = \frac{\pi^2 E_a I_a}{(e / \cos\alpha)^2} \cos\alpha. \tag{8-58}$$

The combination of this 'local' buckling with the global buckling of the structure (computed with a hanger veil) has not yet been clarified theoretically, so that – to be on the safe side – we can use the *Föppl-Papkovich* theorem treated in Chapter 2. Accordingly, the critical horizontal force of the 'combined' buckling can be computed from the relation:

$$\frac{1}{H_{cr,comb}} = \frac{1}{H_{cr,glob}} + \frac{1}{H_{cr,1}}, \tag{8-59}$$

where $H_{cr,glob}$ is the critical force of the global buckling treated so far. The investigations of DULÁCSKA and KOVÁCS [1971] showed that Eq. (8-59) yields a good approximation for tied arches, but for arches with stiffening beams it markedly deviates to the safe side from the more exact value.

KOVÁCS [1984] calculated the critical load of the structure with tie of stiffening beam loaded on the arch (*Fig. 8-28d*), assuming that the hanger veil does not buckle. He obtained for arches with the variation of the moment of inertia $I_a = I_{a0}/\cos\alpha$ the value

$$H_{cr} = \frac{5.81}{f^2}\left(E_a I_{a0} + E_b I_b\right), \tag{8-60a}$$

i.e. 40% of the critical force (8-53) of the structure loaded underneath (*Fig. 8-28a*).

In the case of an arch with constant cross section the formula (8-60*a*) modifies to

$$H_{cr} = \frac{5.81}{f^2}\left(\alpha^* E_a I_{a0} + E_b I_b\right). \tag{8-60b}$$

The factor α^* is given in *Table 8-4*.

Table 8-4 The multiplying factor α^* of the bending rigidity $E_a I_a$ of an arch with constant cross section (Eq. (8-60*b*)), loaded on the arch

f/l	a^*
0.1	0.97
0.2	0.89
0.3	0.80
0.4	0.71
0.5	0.62

The effect of the elongation of the hanger veil during buckling can be taken into account – with an error to the safe side – by the factor γ given in *Fig. 8-32*.

The hangers of the structure loaded on the arch can be weaker than those of the structure loaded on the tie (or stiffening beam), because they do nor carry the load. Thus it may happen that a buckling shape with three or more half waves yields a lower critical load than that with two half waves (Eq. (8-60)). Investigations carried out so far show, however, that this is possible only in the range of $f/l < 0.3$.

The critical load of global buckling, computed with a hanger veil, is in most cases very high, see the upper curve in *Fig. 8-36*. The arch buckles between the hangers at a much lower load. Hence the 'combined' critical load calculated from the *Föppl-Papkovich* formula (8-59) hardly deviates from $H_{cr,1}$. We can thus say that for dimensioning the arches with stiffening beams (or with ties) buckling between the hangers is the most critical.

On the basis of what has been said so far we can easily find out how the critical loads of the structures of *Fig. 8-34* relate to each other. The upper and lower chord of every structure is connected by an inextensional veil which does not buckle; the bending stiffness of the upper chord is the same in each structure; furthermore we assume, for simplicity, that the lower chords have no bending stiffnesses.

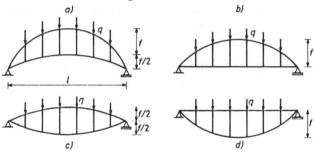

Fig. 8-34 Four arrangements of the 'tied arch'

We have investigated the structure of *Fig. 8-34b* in the foregoing and found that the upper chord, due to the horizontal displacements u of its cross sections, undergoes destabilizing external moments (Eqs (8-49)), (8-47), (8-51)), so that the structure buckles at a certain value of the load q.

The internal forces of the structure of *Fig. 8-34d* can be obtained from those of *Fig. 8-34b* if we let the load act on the lower chord and multiply every quantity by (-1). Since the structure of *Fig. 8-34b* undergoes, after all, destabilizing moments due to the loads pointing downwards and acting on the lower chord, the structure of *Fig. 8-34d* is acted upon by stabilizing moments. Consequently, there is nothing to destabilize the structure, so that it has no finite critical load. To put it in another way: during buckling deformation only the cross sections of the chord in tension displace horizontally, causing stabilizing moments, so that there is no danger of buckling.

The structure of *Fig. 8-34c* obviously represents a transition between those of *Figs 8-34b* and *8-34d*. Consequently, there is a limit position of the upper and lower chords (in the vicinity of the shape depicted in *Fig. 8-34c*) where the destabilizing effect of the displacements of the upper chord and the stabilizing effect of the displacements of the lower chord just cancel each other. This structure has again no critical load, since no external moments act on it, and the internal bending moments try to restore the original shape of the structure.

Finally, the structure of *Fig. 8-43a* has obviously a lower critical load than that of *Fig. 8-34b*. That is, the upper chord is steeper than in *Fig. 8-34b*, so that the displacements u and also the destabilizing moments are greater. On the other hand, the lower chord has a curvature contrary to that in *Fig. 8-34d*, hence the moments resulting from the displacements u have a destabilizing character also on the lower chord.

The critical load of the structure loaded on the arch, having hangers which buckle themselves

Tied arches used as roof structures are loaded, as a rule, directly. Their hangers are mostly not designed to reliably withstand compressive forces arising during buckling. These structures has thus to be designed for buckling by assuming that their hangers buckle when un-

Buckling of arches and rings

dergoing compression, i.e. they cease to exist as structural elements. Hence we can assume the static model of the buckled structure according to *Fig. 8-35*. The additional difficulty to the problems treated so far is that we do not know the boundary between the regions of the hangers in tension and in compression, since this depends on the buckling shape, which again depends on the position of this boundary.

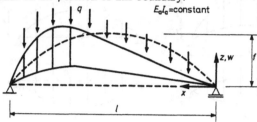

Fig. 8-35 Buckling of the tied arch structure loaded on the arch

This problem has been treated by EIBL [1963]. In the following we present his method.

Due to the complexity of the problem EIBL introduced several approximations. First he neglected the horizontal displacements of the points of the arch. We have seen in Section 8.1.2 that this results in a critical load significantly higher than the exact value. To compensate this error EIBL introduced another approximation which we show on the free arch having a constant cross section. Its differential equation is given, if we confine ourselves to vertical displacements only, by Eq. (8-23). In order to obtain constant coefficients let us replace the variable $\cos \alpha$ by its average value valid at $x = l/4$:

$$\cos \alpha_{\text{aver}} = \frac{1}{\sqrt{1 + \dfrac{4f^2}{l^2}}} \tag{8-61}$$

Thus Eq. (8-23) assumes the form:

$$H\sqrt{1 + \frac{4f^2}{l^2}}\,w + E_a I_a w'' = 0, \tag{8-62}$$

which has the solution in the case of a two-hinged arch:

$$H_{\text{cr}} = \frac{4\pi^2 E_a I_a}{l^2 \sqrt{1 + \dfrac{4f^2}{l^2}}}. \tag{8-63}$$

We plotted this H_{cr} curve in *Fig. 8-36* with dashed line. As can be seen, it lies significantly higher than the exact H_{cr} curve (full line), taken from *Fig. 8-6*. It is thus desirable to reduce the values of H_{cr} obtained from (8-63). A possible way for this reduction is – which cannot be proved correctly – to substitute the arch length

$$s \approx l \left(1 + \frac{8f^2}{3l^2}\right), \tag{8-64}$$

approximately valid for flat arches, for l appearing in the denominator of (8-63). We thus arrive at the expression:

$$H_{cr} = \frac{4\pi^2 E_a I_a}{l^2 \left(1 + \dfrac{8f^2}{3l^2}\right)^2 \sqrt{1 + \dfrac{4f^2}{l^2}}}. \tag{8-65}$$

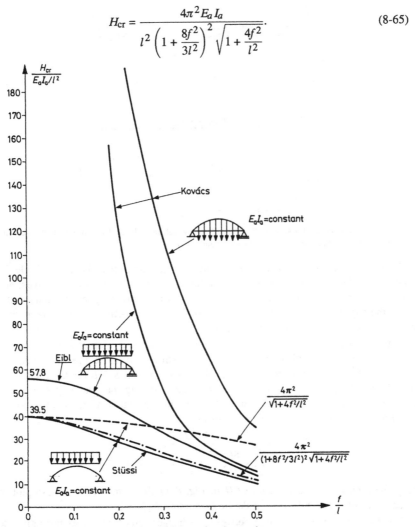

Fig. 8-36 Comparison of the critical horizontal forces of various arches

This curve, plotted with dash-dotted line in *Fig. 8-36*, is pretty close to the exact curve, above all in the domain of smaller f/l values. Hence this correction rendered the approximate method, based on neglecting the horizontal displacements, sufficiently accurate for practical purposes.

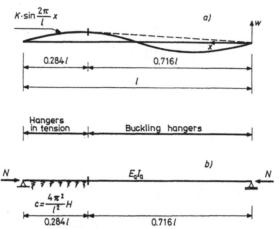

Fig. 8-37 Approximate assumptions in the stability investigation of the tied structure loaded on the arch. *a*) The assumed buckling shape, *b*) the elastic foundation

EIBL 'smeared out' the hangers, i.e. he used an (inextensional) hanger veil. He further assumed that the buckling shape of the arch has a sine form with two half waves; and the tie has the same form on its section bowing upwards, continuing in a tangential straight line to the support (dashed line in *Fig. 8-37a*). Thus the junction point between sine and straight line becomes uniquely determined.

EIBL replaced the stabilizing effect of the tie on the arch along the left-hand section $0.284l$ by distributed vertical springs, for whose stiffness c he obtained a constant value $c = 4\pi^2 H/l^2$. He thus arrived at the static model of *Fig. 8-37b* for the arch. Writing the differential equations for the two sections and fulfilling the boundary and continuity conditions, he obtained an expression for H_{cr}, in which he again substituted the expression (8-64) for l. He thus arrived at the final result:

$$H_{cr} = \frac{4\pi^2 E_a I_a}{l^2 \left(1 + \frac{8f^2}{3l^2}\right)^2 \left(\sqrt{1 + \frac{4f^2}{l^2}} - 0.317\right)}.$$

(8-66)

We plotted this critical force against f/l in *Fig. 8-36*. For comparison we also plotted the critical force of the arch loaded on the tie (assuming $E_b I_b = 0$).

The local buckling of the arch between the hangers practically does not influence the critical load of the global buckling, since we considered the hangers along the right-hand side of the arch as nonexistent. Model tests performed by EIBL were in rather good agreement with Eq. (8-66), so that, in spite of the approximate assumptions introduced, this formula can be considered sufficiently accurate for practical purposes, and there is no need to consider the buckling of the arch between the hangers.

EIBL also investigated the influence of a stiffening beam having a finite bending rigidity. Assuming that $E_b I_b << E_a I_a$, he obtained the critical force

$$H_{cr} = \frac{\pi^2 (4E_a I_a + 1.27 E_b I_b)}{l^2 \left(1 + \frac{8f^2}{3l^2}\right)^2 \left(\sqrt{1 + \frac{4f^2}{l^2}} - 0.317\right)}. \tag{8-67}$$

EIBL also found that the elongation of the hangers reduce the critical load only by about 1-2 % in practical cases, so that this effect can be neglected. He also determined the optimum rise of the arch and obtained that the critical load is maximum in the vicinity of $f/l = 0.4$ (as contrasted to $f/l \approx 0.3$ valid for free arches).

8.2 LATERAL BUCKLING OF RINGS AND ARCHES

The lateral buckling of bars with curved plane axis represents a rather complicated phenomenon, because the buckling deformation is spatial and consists of two parts: lateral curvature and twist.

We shall treat only arches having cross sections with at least one axis of symmetry lying in the plane of the arch. The consequence of this assumption is that the lateral deformations of the arch are independent of those occurring in its own plane, so that we can write the equations of lateral buckling taking only lateral displacement and rotation of the cross sections into consideration.

We assume that the cross sections do not distort in their own planes, and – if not said the contrary – the cross section is constant all along the length of the arch. In accordance with what has been said at the beginning of Chapter 8, we shall use the assumption that all dimensions of the cross section are much smaller than the radius of curvature of the arch.

Since arches with circular axis can again be treated mathematically most easily, we begin the treatment of centrally compressed arches with them. Arches with other axes will be dealt with next. Through-type and deck-type arches constitute a separate topic, to which also twin arches with transverse bracing belong. Post-critical load-bearing capacity of laterally buckling arches will be shortly described. Finally, lateral buckling of bent arches will be treated.

8.2.1 Lateral buckling of centrally compressed arches with circular axis

For the sake of a general treatment we investigate arches with thin-walled cross section (*Fig. 8-38*). These differ from those with solid cross section in several respects. First, the *shear centre* T does not coincide with the centroid C. As a consequence, the deformations (curvature and twist) of the arch have to be expressed by the displacement v_T of the shear centre [TIMOSHENKO and GERE, 1961]. Second, we have to take the *warping rigidity* EI_ω also into account, which is generally zero for solid cross sections. Third, we have to consider that the *longitudinal compressive stresses have to follow the twist of the cross sections*, since they cannot retain their original line of action, being not able to 'step out' of the thin wall. Consequently, their lines of action become curved, and they produce resultants which act as 'external loads' on the arch. It is to be remarked that this phenomenon appears with solid cross sections also, but is in most cases neglected, which is equivalent to leaving the torsional buckling out of consideration, which is mostly allowable wih solid cross sections.

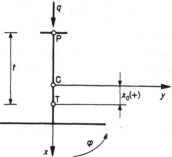

Fig. 8-38 Thin-walled cross section

To be general we also have to consider that the *load may act outside the centroid*, at a point P (*Fig. 8-38*), which does not coincide with the shear centre either.

It shold be emphasized that in Section 8.2 we use the co-ordinate system shown in *Fig. 8-38* (the axis z is parallel to the axis of the arch), so that x and y are 'interchanged' with respect to the other sections of this book.

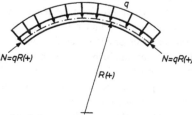

Fig. 8-39 Centrally compressed arch

Taking all these into consideration we can write the differential equations of a centrally compressed free arch (*Fig. 8-39*) with thin-walled cross section as follows [KOLLÁR, 1962; 1973]:

$$GI_t \left(\varphi'' + \frac{v_T''}{R} \right) - EI_\omega \left(\varphi'''' + \frac{v_T''''}{R} \right) - \frac{EI_x}{R} \left(\frac{\varphi}{R} - v_T'' \right) + \tag{8-68a}$$

$$+ N \left[x_0 v_T'' - x_0^2 \varphi'' + \left(i_x^2 + i_y^2 \right) \varphi'' + \left(\frac{t - x_0}{R} - \frac{i_x^2}{R^2} \right) \varphi \right] = 0,$$

$$EI_x \left(\frac{\varphi''}{R} - v_T'''' \right) + \frac{GI_t}{R} \left(\varphi'' + \frac{v_T''}{R} \right) - \tag{8-68b}$$

$$\frac{EI_\omega}{R} \left(\varphi'''' + \frac{v_T''''}{R} \right) + N \left(x_0 \varphi'' - v_T'' \right) = 0.$$

The notations are:

GI_t torsional rigidity

EI_x lateral bending rigidity

v_T lateral displacement of the shear centre

φ specific angle of twist

i_x, i_y radiuses of inertia of the cross section

x_0 and t are defined in *Fig. 8-38*

$'$ means differentation with respect to the arc length s.

The boundary conditions define also the warping, so that we have to write four conditions for both ends of the arch, corresponding to the fact that the system (8-68) contains two differential equations of the fourth order. These boundary conditions can be the following [TIMOSHENKO and GERE, 1961], [KOLLÁR, 1973]:

– the rotation of the cross section around the axis z is zero:

$$\varphi = 0, \tag{8-69}$$

– the lateral displacement is zero:

$$v_T = 0, \tag{8-70}$$

– in the case of a support hinged in lateral direction:

$$\kappa_{Tx} = \frac{\varphi}{R} - v_T'' = 0, \tag{8-71a}$$

(the curvature in the lateral direction, κ_{Tx}, is zero),

or in the case of clamping in the lateral direction:

$$v_T' = 0, \tag{8-71b}$$

– the cross section is free to warp:

$$\vartheta_T' = \varphi'' + \frac{v_T''}{R} = 0, \tag{8-72a}$$

or the warping is prevented:

$$\vartheta_T = \varphi' + \frac{v_T'}{R} = 0, \tag{8-72b}$$

The equation system (8-68) can be solved analytically if the arch with the central angle $2\alpha_0$ rests on fork-like supports (boundary conditions (8-69), (8-70), (8-71a), (8-72a)). Assuming for the deformation functions the expressions

$$\varphi = \varphi_n \sin \lambda_n s, \tag{8-73a}$$

$$v_s = v_{s_n} \sin \lambda_n s, \tag{8-73b}$$

taking only the first term with

$$\lambda_1 = \frac{\pi}{2R\alpha_0}, \tag{8-74}$$

and introducing them into (8-68) we obtain the following quadratic equation for the critical compressive force $N_{cr} = q_{cr}R$ [KOLLÁR, 1973]:

$$a_2 N_{cr}^2 + a_1 N_{cr} + a_0 = 0 \tag{8-75}$$

with the coefficients:

$$a_2 = \left(i_x^2 + i_y^2 \right) \lambda_1^2 + \frac{t - x_0}{R} - \frac{i_x^2}{R^2}, \tag{8-76a}$$

$$a_1 = -EI_x \left[\frac{1}{R^2} + \lambda_1^2 \left(\frac{x_0 + t}{R} - \frac{i_x^2}{R^2} \right) + \left(i_x^2 + i_y^2 + x_0^2 \right) \lambda_1^4 \right] - \quad (8\text{-}76b)$$

$$- \left(GI_t + EI_\omega \lambda_1^2 \right) \left[\lambda_1^2 \left(1 + \frac{i_x^2}{R^2} \right) + \frac{1}{R^2} \left(\frac{1 - x_0}{R} - \frac{i_x^2}{R^2} \right) \right],$$

$$a_0 = EI_x \left(GI_t + EI_\omega \lambda_1^2 \right) \left(\lambda_1^2 - \frac{1}{R^2} \right)^2 . \quad (8\text{-}76c)$$

If the arch is supported in another way, we have to resort to some approximate or numerical method. However, in the case of shell arches, the cross sections of which are shown in *Fig. 8-40*, we can introduce a further simplification: we can assume their lateral bending stiffness to be infinitely great (in comparison to their torsional stiffness), so that their curvature κ_{Tx} is equal to zero:

$$\kappa_{Tx} = \frac{\varphi}{R} - v_T'' = 0, \quad (8\text{-}77)$$

which yields a unique relation between φ and v_T.

If the end cross sections of the arch are clamped, we can assume a deformation function containing two unknown parameters, and applying the energy method we obtain a quadratic equation for the critical force [KOLLÁR and IVÁNYI, 1966], which yields a result very close to the exact one. For practical purposes it is generally sufficient to take a deformation function with only one unknown parameter, which leads to the simple formula:

$$N_{cr} = \frac{\Psi_1}{\Psi_4}, \quad (8\text{-}78)$$

where the coefficients are:

$$\Psi_1 = GI_t \left[\lambda_2^2 + \lambda_4^2 - \frac{4}{R^2} + \frac{1}{R^4} \left(\frac{1}{\lambda_2^2} + \frac{1}{\lambda_4^2} \right) \right] + \quad (8\text{-}79)$$

$$+ EI_\omega \left[\lambda_2^4 + \lambda_4^4 - \frac{2}{R^2} \left(\lambda_2^2 + \lambda_4^2 \right) + \frac{2}{R^4} \right],$$

$$\Psi_4 = 2 \left(\frac{t - x_0}{R} - \frac{i_x^2}{R^2} \right) + \left(i_x^2 + i_y^2 + x_0^2 \right) \left(\lambda_2^2 + \lambda_4^2 \right) + \frac{1}{R^2} \left(\frac{1}{\lambda_2^2} + \frac{1}{\lambda_4^2} \right) + 4\frac{x_0}{R}, \quad (8\text{-}80)$$

and

$$\lambda_k = \frac{k\pi}{2\alpha_0 R}, \quad k = 1, 2, 3, 4, \ldots . \quad (8\text{-}81)$$

The error committed by Eq. (8-78) is only some percents (to the unsafe side), see in [KOLLÁR, 1973].

In *Fig. 8-40a* we indicated the moments of inertia of the cross sections, which are valid also for those of *Fig. 8-40b*. (In this latter case only the sign of x_0 becomes negative).

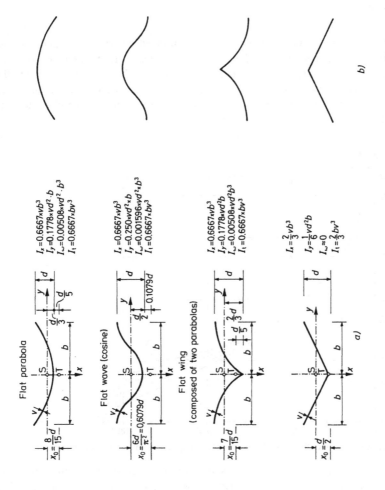

Flat parabola

$x_0 = \dfrac{8}{15}d$

$I_x = 0.6667 \varkappa v b^3$
$I_y = 0.1778 \varkappa v d^2 \cdot b$
$I_\omega = 0.00508 \varkappa v d^2 \cdot b^3$
$I_t = 0.6667 \varkappa b v^3$

Flat wave (cosine)

$x_0 = \dfrac{6d}{\pi^2} = 0.6079d$

$I_x = 0.6667 \varkappa v b^3$
$I_y = 0.250 \varkappa v d^2 \cdot b$
$I_\omega = 0.001596 \varkappa v d^2 \cdot b^3$
$I_t = 0.6667 \varkappa b v^3$

Flat wing
(composed of two parabolas)

$x_0 = \dfrac{7}{15}d$

$I_x = 0.6667 \varkappa v b^3$
$I_y = 0.1778 \varkappa v d^2 b$
$I_\omega = 0.00508 \varkappa v d^2 b^3$
$I_t = 0.6667 \varkappa b v^3$

$x_0 = \dfrac{d}{2}$

$I_x = \dfrac{2}{3} v b^3$
$I_y = \dfrac{1}{6} v d^2 b$
$I_\omega = 0$
$I_t = \dfrac{2}{3} b v^3$

a)

b)

Fig. 8-40 Cross sections of shell arches. *a*) Cross sectional quantities, *b*) cross sections turned upside down

Arches supported elastically all along their lengths

With certain arrangements we can assume that the roofing elastically supports the arches in lateral direction. This elastic support may hinder the lateral displacements v of the cross sections, or their rotations v' in the yz plane [KOLLÁR, 1982]. In the first case we assume distributed springs in lateral direction, which have a spring constant c, exerting a force in the y direction

$$q_y^c = -cv. \tag{8-82}$$

In the second case we may assume that the distributed moments m_x, hindering the rotation, consist of horizontal pairs of forces acting perpendicularly to the arch axis. The difference of these pairs of forces appear as distributed forces q_y^g, which are similar to the so-called *Kirchhoff* forces, acting on the edges of bent plates, arising from the difference of torsional moments. So the elastic restraint hindering the rotation can be characterized, instead of the relation $m_x = gv'$, by the relation

$$q_y^g = gv'' \tag{8-83}$$

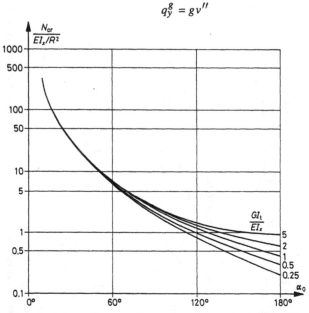

Fig. 8-41 The critical force of the centrally compressed clamped arch with solid cross section

The problem has been treated in detail and solved in the cases of fork-like support in [KOLLÁR and GYURKÓ, 1982; KOLLÁR and BÓDI, 1982] and for clamped arch ends in [BÓDI, 1985]. HEGEDŰS [1984] set up a simple formula which yields a fairly good approximation:

$$N_{cr}^{elast} \approx N_{cr}^{free} + c\frac{(2\alpha_0 R)^2}{\pi^2 k^2} + g \tag{8-84}$$

Here N_{cr}^{free} is the critical force of the arch without elastic restraints, and k is the number of half waves of the buckling shape, see also Eq. (8-80). We have to find the value of k yielding the smallest critical force. It should be remarked that if the arch is elastically supported by a membrane, then the spring constant c has to be taken equal to zero if the arch buckles in two or more half waves, because in these cases the membrane can follow the buckling deformation with inextensional deformation, see [KOLLÁR, 1982].

Arches with solid cross sections
We can obtain the critical forces of arches having solid cross sections from the above results, valid for thin-walled cross sections, by setting $x_0 = 0$, $I_\omega = 0$, and if the load acts at the centroid of the cross sections, by taking also $t = 0$ (*Fig. 8-38*). These results are to be found e.g. in [TIMOSHENKO and GERE, 1961], so we do not repeat them here. We only give the critical compressive force of the clamped arch to be found in [PETERSEN, 1980] and reproduced in *Fig. 8-41*.

Various kinds of loading
So far we assumed that the loads maintain their original direction during buckling. As mentioned in Section 8.1.1, there are also other kinds of loads: central and hydrostatic ones. We show them in *Fig. 8-42*. As can be seen, both the hydrostatic and the central loads hinder the lateral buckling deformation, so that they are more 'benevolent' than the constant directional load, as contrasted to the case of buckling in the plane of the arch. This helping effect of the central load becomes important if the arch is loaded by hangers, see Section 8.2.3.

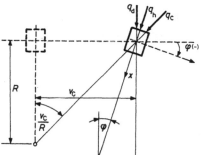

Fig. 8-42 The three kinds of radial load

8.2.2 Buckling of centrally compressed arches with axes other than circular

The critical force of an arch with a solid cross section and a parabolic axis under a constant directional vertical load uniformly distributed in ground plan has been solved by STÜSSI [1943-44]. He presented numerical results for the clamped arch with two different ratios of GI_t/EI_x in the cases of constant and variable cross sections, shown in *Fig. 8-43*.

TOKARZ and SANDHU [1972] computed the critical horizontal force of the parabolic arch with constant cross section under uniformly distributed constant directional load, whose end cross sections are prevented from rotating around the arch axis, and are either free to rotate laterally or are clamped. Of their results we show four characteristic curves in *Fig. 8-44*.

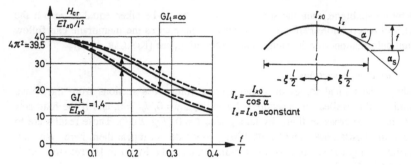

Fig. 8-43 The critical horizontal force of the clamped arch with parabolic axis loaded by forces uniformly distributed in ground plan

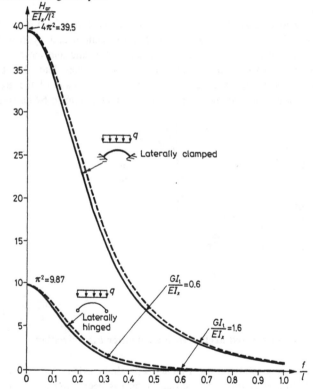

Fig. 8-44 Critical forces of parabolic arches with constant cross section

Since the problem of lateral buckling of circular arches has been solved for many more cases than that of parabolic arches, it is worth while to investigate the question: is it allowed to replace the parabolic arch by a circular one having the same dimensions f and l and loaded by radial loads of the same magnitude as the vertical load q of the parabolic arch? In *Fig. 8-45* we compared the 'average' value of the critical compressive force of the parabolic arch valid at $l/4$ and the critical force of the circular arch, assuming complete clamping at both ends, for the stiffness ratio $GI_t/(EI_x) = 1.6$. As can be seen, up to about $f/l \leq 0.2$ we obtain smaller critical forces with the circular axis, but when the arch is steeper, the circular axis yields results which lie on the unsafe side.

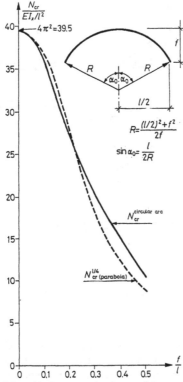

Fig. 8-45 Comparison of the critical compressive forces of the parabolic and circular arches with the same ratio f/l

8.2.3 Lateral buckling of centrally compressed arches loaded by hangers or struts

The arch types described in Section 8.1.5 behave differently from the free arches also with respect to lateral buckling. If the arch is loaded by hangers (through-type structures), then – considering the deck as infinitely rigid in the horizontal plane – the horizontal components H_h of the tensile forces arising in the slanting hangers exert a stabilizing effect on the arch (*Fig. 8-46*), similarly to the effect of the 'central' load (see *Fig. 8-42*). In the case of arches loaded by struts (deck-type structures) the horizontal components H_s of the strut forces destabilize the arch (*Fig. 8-47*). In this latter case another effect may arise: if the crown A of the arch is connected to the deck in horizontal direction, then the deck – whose two ends are assumed to be rigidly supported against horizontal displacements – laterally supports the crown of the arch, i.e. stabilizes the arch. This supporting effect can not always be considered as infinitely rigid, above all if the deck is longer than the span of the arch, cf. the data presented in [HAVIÁR, GÁLLIK and MAGYAR, 1954]. In theoretical investigations, however, this support is often assumed as being infinitely rigid, so that the critical force may lie on the unsafe side.

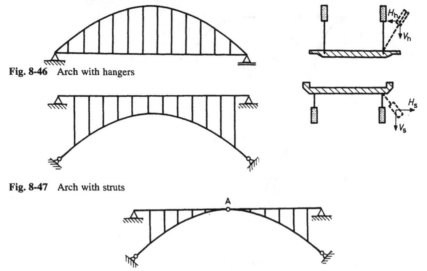

Fig. 8-46 Arch with hangers

Fig. 8-47 Arch with struts

Fig. 8-48 Arch with struts; the deck is connected to the crown of the arch

Arches with hangers

Lateral buckling of arches with hangers was first investigated by GODDEN [1954]. He treated parabolic arches under a load uniformly distributed in the horizontal plane, having a doubly symmetric, constant cross section, neglecting the warping rigidity ($EI_\omega = 0$). He 'smeared out' the hangers, i.e. he assumed a hanger veil, considered as inextensional and connected to the centroid of the arch. The displacement and lateral rotation of the end cross sections of the arch were prevented, but they were free to rotate around the arch axis. He found that, the end cross sections being laterally clamped, preventing their rotation around the arch axis practically does not increase the critical force.

GODDEN [1954] computed the critical horizontal force H in the range $f/l = 0 \ldots 0.24$, varying the ratio $GI_t/(EI_x)$ between 0.1 and ∞. His results are shown in *Fig. 8-49*. He also performed about 100 model tests which were in good agreement with the theoretical results.

We find the description of a further model test in [GODDEN and THOMPSON, 1959] which yielded a similar agreement.

STEIN [1961] solved the problem by another method and gave results for several geometic ratios, being in good agreement with those of [GODDEN, 1954].

SHUKLA and OJALVO [1971] determined the critical load of the clamped arch for a broader range of f/l, using *Godden*'s principles and assumptions. Their results, converted into H_{cr}, are presented in *Table 8-5*. The two numerical values in parentheses belong to the (antisymmetrc) buckling shape with two half waves, yielding smaller critical loads than the shape with one half wave.

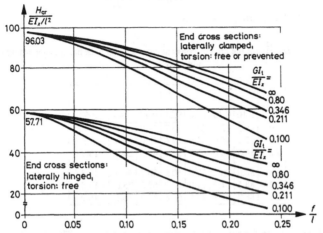

Fig. 8-49 The critical horizontal force of the arch with hangers

KOVÁCS [1974] used the theorems of *Dunkerley* and *Southwell* to determine the critical force H and showed that by using these theorems we commit a very small error on the safe side. His method makes it possible to take the warping rigidity EI_ω also into account. He considered only buckling shapes with one half wave which yield in the range of f/l investigated the minimum critical load, according to [SHUKLA and OJALVO, 1971]. The end cross sections of the arch are supposed to be clamped for every displacement, except for the warping which is free. Otherwise he uses the assumptions of GODDEN [1954].

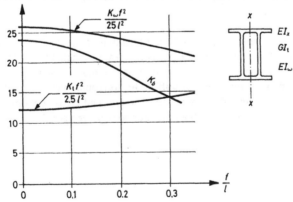

Fig. 8-50 The factors K_b, K_t, K_ω appearing in Eq. (8-85)

Kovács [1974] derived the critical horizontal force

$$\frac{1}{H_{cr}} = \frac{(l/2)^2}{EI_x K_b} + \frac{(l/2)^2}{GI_t K_t + \dfrac{4EI_\omega K_\omega}{l^2}}. \tag{8-85}$$

The factors K_b, K_t and K_ω are given in *Fig. 8-50*.

Table 8-5 The values of $H_{cr}\dfrac{l^2}{EI_x}$ for the arch of constant cross section with hangers according to Shukla and Ojalvo

$\frac{GI_t}{EI_x}$	f/l						
	0.10	0.15	0.20	0.25	0.30	0.40	0.60
0.01	49.2	32.1	21.8	15.7	11.9	7.5	(3.31)
0.02	62.1	45.0	32.9	24.6	19.0	12.3	(5.75)
0.2	84.5	74.0	62.7	52.1	42.7	28.6	13.7
0.8	88.0	80.2	70.6	60.5	50.7	34.3	15.8
1.2	88.5	81.2	71.9	62.0	52.2	35.6	16.5
1.6	88.7	81.7	72.6	62,8	53.2	36.5	16.9

All that has been said so far refers to the case of a hanger veil. The arch, however, may also buckle between the hangers. We can consider these sections of the arch as straight and calculate the critical force of this buckling mode assuming laterally hinged supports at the hangers. (If the arch has a thin-walled open cross section with only one axis of symmetry, then the buckling in the lateral direction combines with the torsional buckling, so that the critical force of this combined buckling has to be determined.) The interaction of this 'local' buckling with the 'global' buckling (treated so far, assuming a hanger veil) can be taken into account with the relation (8-59) of *Föppl-Papkovich*.

It should be mentioned that Kee [1961] extended the investigation to plastic buckling.

Arches with struts

Shukla and Ojalvo [1971] also treated the lateral buckling of the deck-type arch, shown in *Fig. 8-48*, in detail. They considered the struts as inextensional, hinged at both ends, and 'smeared out'; they supposed that the crown A of the arch is connected to the deck assumed as laterally supported and infinitely stiff in the horizontal plane, so that the lateral displacement of the cross section A is prevented. The two end cross sections of the arch are laterally clamped and their rotation about the arch axis prevented; the cross section of the arch is constant.

Table 8-6 The values of $H_{cr}\frac{l^2}{EI_x}$ of the arch of constant cross section with struts, according to SHUKLA and OJALVO

$\frac{GI_t}{EI_x}$	f/l						
	0.10	0.15	0.20	0.25	0.30	0.40	0.60
0.01	14.7	10.2	7.4	5.3	4.1	2.5	0.8
0.02	17.7	13.2	9.9	7.5	5.8	3.6	1.4
0.2	22.2	19.8	17.0	14.2	11.8	8.1	4.1
0.8	23.0	21.0	18.6	16.2	13.9	10.1	5.5
1.2	23.0	21.2	18.9	16.5	14.2	10.5	5.9
1.6	23.0	21.2	19.0	16.6	14.4	10.7	6.2

The result of these assumptions is that the arch buckles antisymmetrically. The numerical values of the critical horizontal force H_{cr} are given in *Table 8-6*. The interaction of local buckling (between the struts) and global buckling (*Table 8-6*) can be considered again by Eq. (8-59).

The values of the critical force given by Table 8-6 can also be used in the case of the arrangement of *Fig. 8-47*, provided that we connect the crown of the arch to the deck by a (rigid) bracing, i.e. we support it laterally. The critical force yielded by *Table 8-6* lies then on the safe side, since the struts are longer than those of *Fig. 8-48*, and thus the destabilizing horizontal components of the compressive forces are smaller.

If the crown of the arch is not rigidly connected to the deck (*Fig. 8-47*), the critical force significantly decreases. This effect can be taken into account only numerically.

Twin arches connected by cross beams or bracings

The arches of through-type bridges are mostly connected by cross beams (*Fig. 8-51*) or bracings (*Fig. 8-52*). Both connections result in transforming the two arches into a single built-up curved column which has a comparatively large shear-type deformation. If the arches are flat, we can disregard the curvature of this structure and treat it as a straight column whose length is equal to the arc length of the arch. The investigations of ÖSTLUND [1954] on twin arches with cross beams showed that this approximation results in an error of less than 10% (to the unsafe side) if

- the arches have a considerable torsional rigidity (having e.g. solid or closed hollow cross section); or
- if this is not fulfilled (as in the cases of I or U shaped cross sections), then the requirement

$$6\frac{(EI)_{c,u}}{(EI_0)_{a,w}}\frac{a}{b} \geq 1 \tag{8-86}$$

is fullfilled. Here $(EI)_{c,u}$ is the bending stiffness of one cross beam around the arch axis u, $(EI_0)_{a,w}$ is the lateral bending stiffness of the arch around the axis w, a and b are the horizontal projections of the distance of the cross beams and the distance of the two arches, respectively, see *Fig. 8-51*. These two requirements are alternative, since the bending stiffness of the cross beams in the radial plane $(EI)_{c,u}$ 'replaces' the torsional stiffness of the arches.

To the error limit of 10 % it is further necessary that
- in the case of $f/l = 1/2$ and three cross beams: $\kappa \geq 6$,
- in the case of $f/l = 1/2$ and four cross beams: $\kappa \geq 2.5$,
- in the case of $f/l = 1/6$ and three cross beams: $\kappa \geq 2.5$, where

$$\kappa = 6\frac{(EI)_{c,w}}{(EI_0)_{a,w}}\frac{a}{b}, \tag{8-87}$$

and $(EI)_{c,w}$ is the bending stiffness of one cross beam around the axis w (*Fig. 8-51*).

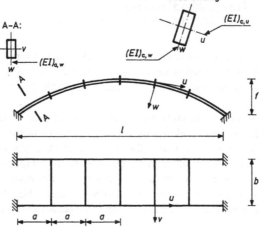

Fig. 8-51 Twin arches stiffened by cross beams

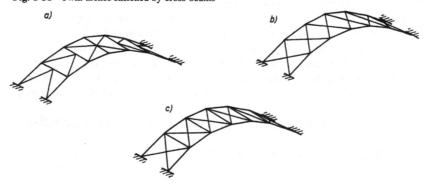

Fig. 8-52 Twin arches stiffened by bracings.

In calculating the above quantities ÖSTLUND assumed that the arches are not stiffened by e.g. an upper deck, and that we can neglect the deformation of the arches in their own planes and, consequently, also the displacements of the nodes parallel and perpendicular to the arch axis. That is, this deformation reduces the critical load, but to an appreciable extent only if f/l is large (about 0.4 to 0.5), the torsional stiffness of the arches is negligibly small (I or U sections), and the bending stiffness of the arches in the vertical plane is at the most 3 to 4 times that in the horizontal plane. Even if all these three conditions are fulfilled, the reduction in the critical load is greater in the case of antisymmetric buckling than with symmetric buckling.

Hence the stability of twin arches stiffened by cross beams, fulfilling the criteria of ÖSTLUND, can be analysed by the methods valid for plane frames (Chapters 5 and 6).

What has been said so far is essentially also valid for twin arches stiffened by bracings. However, we did not find any investigations which established the rules allowing to replace the structure by a plane braced twin column. We thus recommend for this purpose the criteria of ÖSTLUND as a guideline.

SAKIM OTO AND KOMATSU [1982, 1983] analysed the buckling of twin arches braced as shown in *Fig. 8-52a,b* in detail, taking also plasticity into account, and presented diagrams and tables for practical applications. SAKIMOTO and YAMAO [1983] treated deck-type twin arches braced as shown in *Fig. 8-52c* in a similar way.

8.2.4 Post-critical behaviour of laterally buckling arches

This problem has not been investigated to such an extent as that of buckling in the plane of the arch. Thus we can stress only some points.

Lateral buckling consists of bending and torsional deformations. Of these, bending deformation yields an – almost exactly – constant post-critical load-bearing capacity, as we know from the theory of buckling of straight bars [TIMOSHENKO and GERE, 1961]. Pure torsional buckling yields in the case of straight bars with solid cross section also a constant post-critical load-bearing capacity, but an increasing one with thin-walled open cross section, as outlined in Section 1.5. The role of pure torsional buckling is the same on curved bars (arches). This increasing post-critical load-bearing capacity can be established also from the equations (8-75) and (8-76) of arches with thin-walled open cross sections, if we divide these latter expressions by EI_x and we set $EI_x \to \infty$, as valid for shell arches. We thus obtain the equation

$$N_{cr}^{tors} = \frac{\left(GI_t + EI_\omega \lambda_1^2 \right) \left(\lambda_1^2 - 1/R^2 \right)^2}{\dfrac{1}{R^2} + \lambda_1^2 \left(\dfrac{x_0 + t}{R} - \dfrac{i_x^2}{R^2} \right) + \left(i_x^2 + i_y^2 + x_0^2 \right) \lambda_1^4}. \tag{8-88}$$

The expression $\left(i_x^2 + i_y^2 + x_0^2 \right)$ appearing in the denominator is proportional to the square of the width of the cross section. Hence if we assume that the outer parts of the cross secion cease to carry the load, and the stresses concentrate to the inner part, the terms in the denominator decrease, since we have to compute them from the inner part of the cross section which carries the load. On the other hand we can assume that the rigidity characteristics appearing in the numerator do not decrease due to the rearrangement of the stresses. So the load-bearing capacity will obviously increase.

Model tests performed on shell arches [KOLLÁR, 1973] verified the increasing post-buckling load-bearing capacity for pure torsional buckling.

In summary, in the general case when lateral buckling consists of both lateral bending and torsion, we can expect an (approximately) constant post-critical load-bearing capacity for solid cross sections, and an increasing one for thin-walled open cross sections.

8.2.5 Lateral buckling of arches bent in the plane of the arch

As contrasted to the phenomena occurring in the plane of the arch where bending generally does not cause loss of stability (except for shell arches with cross sections according to *Fig. 8-40*), arches of any cross section may buckle laterally if the bending moment reaches a critical value.

The differential equations of buckling can be derived in the same way as those describing lateral buckling due to central compression [KOLLÁR, 1964a, 1973]. Considering circular arches with constant thin-walled open cross section (*Fig. 8-38*) subjected to pure bending (*Fig. 8-20a*), and assuming that the normal stresses arising from bending are linearly distributed along the height of the cross section, the differential equations desribing lateral buckling can be written as follows:

$$GI_t\left(\varphi'' + \frac{v_T''}{R}\right) - EI_\omega\left(\varphi'''' + \frac{v_T''''}{R}\right) - EI_x\left(\frac{\varphi}{R} - v_T''\right) + \tag{8-89a}$$

$$+M\left[-\beta_1\varphi'' + \left(1 - \frac{J_2}{R}\right)\frac{\varphi}{R} - v_T''\right] = 0,$$

$$EI_x\left(\frac{\varphi''}{R} - v_T''''\right) + \frac{GI_t}{R}\left(\varphi'' + \frac{v_T''}{R}\right) - \frac{EI_\omega}{R}\left(\varphi'''' + \frac{v_T''''}{R}\right) - \tag{8-89b}$$

$$-M\left(\varphi'' + \frac{v_T''}{R}\right) = 0,$$

with the notations:

$$J_1 = \frac{\displaystyle\int_{(A)} x^3 dA}{I_y}, \tag{8-90a}$$

$$J_2 = \frac{\displaystyle\int_{(A)} xy^2 dA}{I_y}, \tag{8-90b}$$

$$\beta_1 = J_1 + J_2 - 2x_0. \tag{8-90c}$$

If the arch rests on fork-like supports, the solution can be obtained in closed form:

$$a_2 M_{cr}^2 + a_1 M_{cr} + a_0 = 0 \tag{8-91}$$

with the coefficients:

$$a_2 = -\frac{1}{R^2}\left[\left(\frac{\pi}{2\alpha_0}\right)^2 - 1\right]\left(1 - \frac{J_2}{R}\right), \tag{8-92a}$$

$$a_1 = \frac{EI_x}{R^3}\left[-1 - \frac{1}{R}\left(\frac{\pi}{2\alpha_0}\right)^4 \beta_1 + \left(\frac{\pi}{2\alpha_0}\right)^2\left(1 + \frac{J_2}{R}\right)\right] + \tag{8-92b}$$

$$+\frac{1}{R^3}\left[GI_t + \frac{EI_\omega}{R^2}\left(\frac{\pi}{2\alpha_0}\right)^2\right]\left[\left(\frac{\pi}{2\alpha_0}\right)^2 - 1\right]\left(1 - \frac{J_2}{R}\right),$$

$$a_0 = \frac{EI_x}{R^4}\left[GI_t + \frac{EI_\omega}{R^2}\left(\frac{\pi}{2\alpha_0}\right)^2\right]\left[\left(\frac{\pi}{2\alpha_0}\right)^2 - 1\right]^2. \qquad (8\text{-}92c)$$

In the case of shell arches with cross sections according to *Fig. 8-40*, due to the deformation of the cross sections, the distribution of the normal stresses ceases to be linear (see *Fig. 8-26*). Consequently, the quantities (8-90) must be modified in the following way [KOLLÁR, 1964*b*, 1973]:

$$J_1^{\text{shell arch}} = dj_x, \qquad (8\text{-}93a)$$

$$J_2^{\text{shell arch}} = \frac{b^2}{d}j_y, \qquad (8\text{-}93b)$$

$$\beta_1^{\text{shell arch}} = J_1^{\text{shell arch}} + J_2^{\text{shell arch}} - 2x_0 = dj_x + \frac{b^2}{d}j_y - 2x_0. \qquad (8\text{-}93c)$$

The quantities b and d are defined in *Fig. 8-40*, and the factors j_x and j_y are given in *Fig. 8-53* for the cross sections of *Fig. 8-40a*. In the case of the cross sections of *Fig. 8-40b* the quantities x_0, J_1, J_2 change sign.

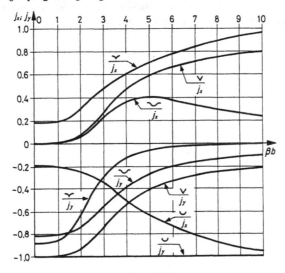

Fig. 8-53 The factors j_x and j_y appearing in Eqs. (8-93)

Of the positive and negative critical bending moments that one is smaller in absolute value, as a rule, which causes compression in the free edge of the cross section.

If the arch is supported in other ways, the determination of the critical bending moment becomes comparatively easy if we have to deal with shell arches, because due to their great lateral stiffness and the resulting relation (8-77) we have to assume only one deformation function to be used in the energy method [KOLLÁR and IVÁNYI, 1966]. If the end cross sections are clamped, and we take a deformation function containing only one unknown parameter (which yields results sufficiently accurate for most practical purposes, see [KOLLÁR, 1973]), the critical moment becomes:

$$\frac{M_{cr}}{R} = \frac{\Phi_1}{\Phi_4}. \tag{8-94}$$

with the coefficients:

$$\Phi_1 = GI_t \left[\lambda_2^2 + \lambda_4^2 - \frac{4}{R^2} + \frac{1}{R^4} \left(\frac{1}{\lambda_2^2} + \frac{1}{\lambda_4^2} \right) \right] + \tag{8-95}$$

$$+ EI_\omega \left[\lambda_2^4 + \lambda_4^4 - \frac{2}{R^2} \left(\lambda_2^2 + \lambda_4^2 \right) + \frac{2}{R^4} \right],$$

$$\Phi_4 = -2 \left(1 + \frac{b^2}{Rd} j_y \right) - R \left(2x_0 - dj_x - \frac{b^2}{d} j_y \right) \left(\lambda_2^2 + \lambda_4^2 \right) + \frac{1}{R^2} \left(\frac{1}{\lambda_2^2} + \frac{1}{\lambda_4^2} \right), \tag{8-96}$$

and λ_k is defined by Eq. (8-81).

The lateral buckling of arches with thin-walled cross section loaded by a compressive force and a simultaneously acting bending moment is treated in [KOLLÁR and GÁRDONYI, 1967] and [KOLLÁR, 1973]. This simultaneous action can be approximately taken into account by the theorem of Dunkerley (see Chapter 2), with an error to the safe side.

For the post-critical behaviour of laterally buckling bent arches the same statements are valid as for centrally compressed ones (Section 8.2.4).

9

Special stability problems of beams and trusses

Lajos KOLLÁR

In this chapter we intend to deal with three topics. First, we treat some problems of the lateral stability of bent beams. Second, we investigate in which cases the nodes of a truss may become laterally unstable. Third, we show that a certain kind of beams may lose their stability in the plane of bending by snapping through.

9.1 PROBLEMS OF LATERAL STABILITY OF BEAMS

Lateral stability of beams is extensively treated in several books [BLEICH, 1952; STÜSSI, 1954; TIMOSHENKO and GERE, 1961; PFLÜGER, 1964; BÜRGERMEISTER, STEUP and KRETSCHMAR, 1966; TRAHAIR, 1977], with solutions for many cases, but they do not contain all cases occurring in practice. These may differ in the following points:
- the *support conditions* of the beam: simply supported, simply supported with cantilever(s), continuous, cantilever, beam of a frame; the support itself can be: 'fork-like' (see later), elastic or rigid clamping, suspension by vertical or inclined cables;
- the *cross section* of the beam: the shear centre coincides with the centroid or not? does it have torsional or warping rigidity? can its lateral bending stiffness be considered as infinitely great? does it maintain its shape in the plane of the cross section? whether the cross section varies along the beam length or not;
- the *distribution of the load* (and its bending moment) along the beam axis: is it uniformly (or otherwise) distributed or concentrated? is the bending moment diagram constant (due to two end moments) or variable? whether a compressive force also acts on the beam, etc.
- the *position of the load in height*: whether it acts above, at, or under the shear centre.

As can be seen, the number of the possible cases is very high. Although it is possible to develop the differential equation system for the general case and to solve it (mostly numerically), nevertheless it seems to be expedient to develop a straightforward method which is simple enough to be used in practice, but which yields a solution for any case, even if the results are approximate.

In the following, we derive the solution of a special problem from the general differential equations of lateral buckling. After that we present an approximate method suitable for obtaining the critical load for any of the aforementioned cases.

We assume that the cross section of the beam has one plane of symmetry, and the load acts in this plane. The beam may be subjected to a compressive force too. The cross section be thin-walled and open, i.e. the lines of action of the longitudinal stresses get inclined or curved together with the longitudinal fibres of the beam.

The curvature of the originally straight beam axis due to the load, occurring in the loading plane before lateral buckling, will always be neglected. This is acceptable only if the bending rigidity in the loading plane is much greater than the lateral rigidity ($I_x \gg I_y$, see also *Fig. 9-1*). As LOVASS-NAGY [1953] and later, independently, TRAHAIR and WOOLCOCK [1973] have shown, the critical load of lateral buckling, acting at the shear centre, becomes infinitely great if $I_x = I_y$; thus beams with such cross sections do not buckle laterally.

If not said the contrary, we assume that the cross section is constant along the beam length, and that the beam is geometrically perfect.

9.1.1 The governing differential equations of lateral buckling and some conclusions

Let us consider the cross section shown in *Fig. 9-1*. C, T and P are the centroid, the shear centre and the point of action of the vertical force P, respectively. The distances t and y_0 are positive as indicated in the figure.

The co-ordinate axis z lies in the axis of the beam.

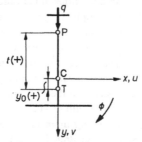

Fig. 9-1 Thin-walled, simply symmetric cross section

During lateral buckling the beam bows out in lateral direction and gets twisted. Since T is the point around which the cross sections rotate without causing bending in the beam, we have to take the fibre passing through the shear centres as the 'bar axis' to describe bending.

The co-ordinate system is attached to the cross section, hence it displaces and rotates together with the cross section.

The differential equations of lateral buckling when the bending moment is constant and a compressive force also acts on the beam are identical with those valid for torsional buckling of eccentrically compressed bars, and can be found e.g. in [TIMOSHENKO and

GERE, 1961]. It is easily possible to generalize them for variable bending moment. We do not want to give the derivation in detail; this can be found e.g. in [KOLLÁR, 1991]. The differential equation system has the following shape:

$$EI_y u_T'''' + (M_x \Phi)'' + N \left(u_T'' + y_0 \Phi'' \right) = 0, \tag{9-1a}$$

$$EI_\omega \Phi'''' - GI_t \Phi'' + M_x u_T'' + t M_x'' \Phi - \beta_1 \left(M_x \Phi' \right)' + N \left(y_0 u_T'' + i_{pT}^2 \Phi'' \right) = 0, \tag{9-1b}$$

with the notations (see also *Fig. 9-1*):

$' = d/dz,$

EI_y bending rigidity in the lateral direction,

EI_ω warping rigidity,

GI_t torsional rigidity,

u_T displacement of the shear centre in the x direction,

Φ twist,

M_x bending moment around the axis x,

N compressive force,

$$i_{pT}^2 = i_x^2 + i_y^2 + y_0^2 \tag{9-2}$$

the radius of gyration of the cross section referred to the shear centre,

$$\beta_1 = J_1 + J_2 - 2y_0 \tag{9-3}$$

is a geometric characteristic of the cross section, where the integrals

$$J_1 = \frac{\int\limits_{(A)} y^3 dA}{I_x}, \tag{9-4a}$$

and

$$J_2 = \frac{\int\limits_{(A)} x^2 y dA}{I_x} \tag{9-4b}$$

appear.

The values of the integrals J_1 and J_2 (and of other geometric quantities) are given in *Fig. 9-2* for some cross sections. We draw the attention to the fact that if the cross sections of *Fig. 9-2* are turned upside down (as in *Fig. 8.1-18b* or *8.2-3b*), then the signs of J_1, J_2 and y_0 change, and, consequently, also that of β_1.

The differential equation system (9-1a,b) can be easily solved if the beam is acted upon by a bending moment constant along its length (*Fig. 9-3*), the compressive force is zero, and the beam rests on 'fork-like' supports, i.e. the rotation of the end cross sections about the z axis is prevented, but they are free to warp and free to rotate about the axis x. In this case the differential equations have constant coefficients, and we can assume the exact solution in the form of sine functions:

$$u_T = U_1 \sin \frac{\pi}{l} z, \tag{9-5a}$$

$$\Phi = \Phi_1 \sin \frac{\pi}{l} z. \tag{9-5b}$$

$$\beta_1 = J_1 + J_2 - 2y_0; \qquad J_1 = \frac{\int\limits_{(A)} y^3 dA}{I_x}; \qquad J_2 = \frac{\int\limits_{(A)} x^2 y\, dA}{I_x};$$

I CROSS SECTION WITH UNEQUAL FLANGES

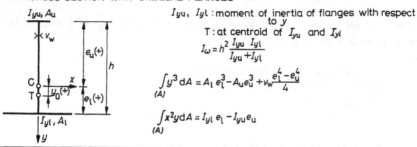

I_{yu}, I_{yl} : moment of inertia of flanges with respect to y

T : at centroid of I_{yu} and I_{yl}

$$I_\omega = h^2 \frac{I_{yu}\, I_{yl}}{I_{yu} + I_{yl}}$$

$$\int\limits_{(A)} y^3 dA = A_l\, e_l^3 - A_u e_u^3 + v_w \frac{e_l^4 - e_u^4}{4}$$

$$\int\limits_{(A)} x^2 y\, dA = I_{yl}\, e_l - I_{yu} e_u$$

SHELL BEAMS (flat cross sections)	$\dfrac{I_x}{vd^2b}$	$\dfrac{I_y}{vb^3}$	$\dfrac{I_\omega}{vd^2b^3}$	$\dfrac{I_t}{v^3b}$	$\dfrac{J_1}{d}$	$\dfrac{J_2}{b^2/d}$
$y_0 = \dfrac{8}{15}d$	0.1778	$\dfrac{2}{3}$	0.005 08	$\dfrac{2}{3}$	−0.1905	−1.0
$y_0 = \dfrac{6}{\pi^2}d$	0.250	$\dfrac{2}{3}$	0.001 596	$\dfrac{2}{3}$	0	−0.8106
$y_0 = \dfrac{7}{15}d$	0.1778	$\dfrac{2}{3}$	0.005 08	$\dfrac{2}{3}$	+0.1905	−0.875
$y_0 = d/2$	$\dfrac{1}{6}$	$\dfrac{2}{3}$	~0	$\dfrac{2}{3}$	0	−1.0

Fig. 9-2 The geometric quantities of some cross sections

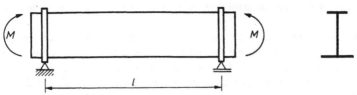

Fig. 9-3 The beam on fork-like supports, loaded by a constant bending moment

Introducing these into Eqs (9-1a,b), we obtain a homogeneous linear equation system for U_1 and Φ_1. Setting its determinant equal to zero yields the following equation of the second degree for $M_x = M_{cr}$:

$$\frac{1}{EI_y}M_{cr}^2 - \frac{\pi^2}{l^2}\beta_1 M_{cr} - \frac{\pi^2}{l^2}\left(GI_t + \frac{\pi^2}{l^2}EI_\omega\right) = 0. \tag{9-6}$$

This equation has one positive and one negative root, corresponding to a positive and a negative critical bending moment. If $\beta_1 = 0$, fulfilled for doubly-symmetric cross sections, then the absolute values of the negative and positive critical moments are equal to each other.

In the following we shall frequently encounter the so-called *shell-beams*, the cross sections of which are shown in *Fig. 8-18* (or *8-40*). These shell-beams *have very large horizontal bending stiffnesses* EI_y related to their torsional rigidity GI_t. Let us put again $N = 0$. If the condition is also fulfilled that bending causes compression in the free edges of the beams, as in the cases of the cross sections of *Fig. 8-18a*, then, according to [KOLLÁR, 1973], we can assume their bending stiffnesses EI_y as infinitely great. Thus the beam does not undergo horizontal bending, i.e. $u_T'' \equiv 0$, and the higher derivatives of u_T also disappear. Consequently, Eq. (9-1a) becomes meaningless, since after dividing it by EI_y it assumes the shape $0 = 0$. In Eq. (9-1b) we have to omit the terms containing the derivatives of u_T, and the governing equation of lateral buckling becomes:

$$EI_\omega \Phi'''' - GI_t\Phi'' + t M_x''\Phi - \beta_1\left(M_x\Phi'\right)' = 0. \tag{9-7}$$

Thus the beam undergoes only torsional deformation around its shear centre axis.

In the loading case and with the support conditions of *Fig. 9-3*, Eq. (9-7) is satisfied by the twist function (9-5b), and the solution is:

$$M_{cr} = -\frac{GI_t + \dfrac{\pi^2}{l^2}EI_\omega}{\beta_1}. \tag{9-8}$$

Hence we have only one critical bending moment, causing compression in the free edges.

If, however, the free edges of the shell beams undergo tension from bending, as in the cases of the cross sections of *Fig. 8-18b*, then, according to [KOLLÁR, 1973], we cannot assume their bending rigidity as infinitely great, and we have to use the two equations (9-1a,b).

9.1.2 The energy method for determining the critical loads of suspended beams

The energy method is very suitable for calculating the critical loads of structures [TIMO-SHENKO and GERE, 1961]. The buckling shape has to be assumed, mostly in the form of a series of functions, the strain energy (U) stored in the structure and the work (W) done by the external forces have to be calculated, the two quantities set equal, and the critical load expressed from this equality minimized according to the coefficients of the series. We thus obtain a homogeneous linear equation system for these coefficients, the determinant of which has to be equal to zero. This yields, as a rule, an algebraic equation for the critical load. The energy method approaches the critical load 'from above', thus the result is on the unsafe side, but the error can be arbitrarily reduced by taking more terms from the series.

To show the application of the energy method in the general case, we show the determination of the critical load of suspended beams.

The strain energy due to the deformation of lateral buckling, consisting of a twist $\Phi(z)$ and a lateral bending displacement $u_T(z)$, is, according to [CHWALLA, 1944; BLEICH, 1952]:

$$U = \frac{GI_t}{2} \int_0^l \left(\Phi'\right)^2 dz + \frac{EI_\omega}{2} \int_0^l \left(\Phi''\right)^2 dz + \frac{EI_y}{2} \int_0^l \left(u_T''\right)^2 dz. \tag{9-9}$$

We will deal with a beam suspended at both ends on inclined cables, loaded by uniformly distributed vertical forces acting above the shear centre (*Fig. 9-4*). The work done can be composed of five parts:

a) In the first step we prevent the bar ends from rotating about the z axis, i.e. we apply a fork-like support, relocate the load to the shear centre, and determine the work done in this case.

b) In the second step we take into consideration that the load acts above the shear centre, and since this point P displaces vertically due to the rotation of the cross section around the shear centre, the load performs work.

c) We remove the fork-like support and allow the beam to rotate as a rigid body around the points of suspension; the load again performs some work.

d) The horizontal components of the inclined cable forces act as compressive forces on the beam; we assume that these forces act at the centroids of the end cross sections which get closer to each other, due to the deformation described under a), so that the forces perform work.

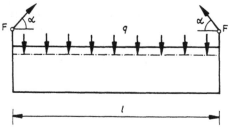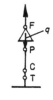

Fig. 9-4 The beam suspended at both ends on inclined cables

e) The points of suspension are above the centroid of the cross section, hence the compressive forces mentioned under *d*) act eccentrically, causing a bending moment constant along the bar length which perform work during the deformation described under *a*).

The work parts corresponding to the cases *a*) to *e*) can be written as follows.

Case a) (Fig. 9-5). The expression of the work done can be taken from [CHWALLA, 1944]:

$$W^{\text{fork}} = \frac{1}{2} \int_0^l \left[-\beta_1 M_x \left(\Phi' \right)^2 - 2 M_x \Phi u_T'' \right] dz. \tag{9-10}$$

In our case

$$M_x = \frac{qz}{2}(l - z). \tag{9-11}$$

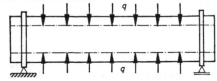
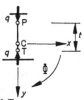

Fig. 9-5 The beam on fork-like supports, loaded at the shear centre

Case b). We can transfer the load from the shear centre to point P by applying forces q of opposite sign at both points (*Fig. 9-6*). The load q performs work due to the rotation of the cross sections only. Since the distance of its point of application from the shear centre is t, its vertical displacement becomes

$$t(1 - \cos \Phi) \approx t \frac{\Phi^2}{2} \tag{9-12}$$

and the work becomes:

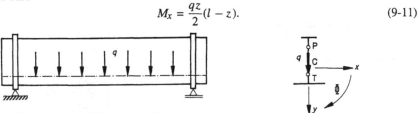

Fig. 9-6 The beam on fork-like supports, loaded by opposite forces at P and T

$$W^{\text{ecc}} = \frac{t}{2} q \int_0^l \Phi^2 dz. \tag{9-13}$$

Case c). The beam can freely rotate as a rigid body about the points of suspension F which is at a distance f above the shear centre (*Fig. 9-7*). Since the beam deformed laterally in step *a*) (*Fig. 9-7c*), the resultant of the distributed load, not acting any more in the vertical plane passing through points F, causes the beam to rotate as a rigid body around points F as long as the resultant does not get into the aforementioned vertical plane. The load thus displaces vertically and performs some work which can be computed as follows.

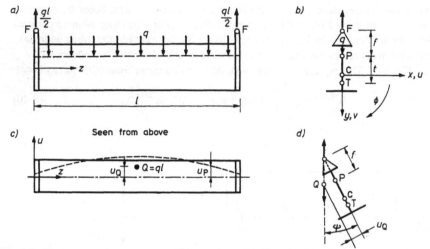

Fig. 9-7 The beam rotating as a rigid body around the suspension points F

The line passing through the points of application of the load P bows out in the horizontal plane due to the deformation components (9-5a,b), and assumes the shape

$$u_P = u_T + t\Phi.$$ (9-14)

The resultant of the load then acts at the centroid Q of this curve, whose distance from the original position of the fibre is

$$u_Q = \frac{\int_0^l u_p dz}{l},$$ (9-15)

see *Fig. 9-7c*.

To get the point Q again into the vertical plane passing through points F the beam must rotate as a rigid body by an angle Ψ around the straight line passing through points F. The magnitude of this angle Ψ is, from *Fig. 9-7d*:

$$\Psi \approx \tan\Psi = \frac{u_Q}{f}.$$ (9-16)

The work done by the load during this rotation is:

$$W = ql\left(\frac{f}{\cos\Psi} - f\right) \approx ql\frac{f\Psi^2}{2}.$$ (9-17)

Introducing here (9-16), (9-15) and (9-14) we obtain:

$$W^{\text{susp}} = \frac{q}{2fl}\left[\int_0^l (u_T + t\Phi)\,dz\right]^2$$ (9-18)

Case d). The compressive force acting on the beam is (*Fig. 9-8*):

$$N = \frac{ql}{2}\cot\alpha,$$ (9-19)

which is assumed to act at the centroid of the cross section. To compute the work done we need the apparent shortenings of the longitudinal fibres of the beam due to the deformations (9-5a,b). This is to be found in [KOLLÁR, 1973], where also the work done by a centrally applied compressive force is given:

$$W^{(N)} = \frac{N}{2} \int_0^l \left[(u_T')^2 - 2y_0 \Phi' u_T' + i_{pT}^2 (\Phi')^2 \right] dz \tag{9-20}$$

with i_{pT}^2 according to (9-2).

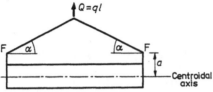

Fig. 9-8 The effect of the suspension cables

Case e). The eccentricity a of the compressive force N is (cf. *Figs. 9-7b* and *9-8*):

$$a = f + t - y_0, \tag{9-21}$$

so that the beam is acted upon at both ends by the bending moments:

$$M_N = aN. \tag{9-22}$$

The work done by bending moments is given by Eq. (9-10). Introducing here (9-22) and also using (9-19), the work becomes:

$$W^{(M_N)} = a\frac{ql}{4} \cot\alpha \int_0^l \left[-\beta_1 (\Phi')^2 - 2\Phi u_T'' \right] dz. \tag{9-23}$$

The energy method requires the more complicated computations, the more terms are assumed for the deformation functions u_T and Φ. The investigations of RAFLA [1968] showed that it is sufficient, as a rule, to assume, as in the expressions (9-5), only one term of the sine series of each deformation function (except for the shell beams if their free edges are compressed (*Fig. 8-18a*)), since the error committed is only a few percent. The above expressions of the work done hence assume the following forms:

$$U = \frac{\pi^2}{4l} \left(GI_t + \frac{\pi^2}{l^2} EI_\omega \right) \Phi_1^2 + \frac{\pi^4}{4l^3} EI_y U_1^2, \tag{9-24}$$

$$W^{\text{fork}} = ql \left(-0.1431\beta_1 \Phi_1^2 + 0.5362\Phi_1 U_1 \right), \tag{9-25a}$$

$$W^{\text{ecc}} = \frac{qlt}{4} \Phi_1^2, \tag{9-25b}$$

$$W^{\text{susp}} = \frac{2ql}{\pi^2 f} (U_1 + t\Phi_1)^2, \tag{9-25c}$$

$$W^{(N)} = \frac{q\pi^2}{8} \cot\alpha \left[U_1^2 - 2y_0 U_1\Phi_1 + i_{pT}^2 \Phi_1^2 \right], \tag{9-25d}$$

$$W^{(M_N)} = a\frac{q\pi^2}{8} \cot\alpha \left(-\beta_1\Phi_1^2 + 2\Phi_1 U_1 \right). \tag{9-25e}$$

We can now write the equality

$$\sum W_i = W = U \tag{9-26}$$

in the following form, expressing the quantity q we are looking for:

$$q = \frac{B_1\Phi_1^2 + B_2 U_1^2}{C_1\Phi_1^2 + C_2 U_1^2 + C_3\Phi_1 U_1} \tag{9-27}$$

with

$$B_1 = \frac{\pi^2}{4l}\left(GI_t + \frac{\pi^2}{l^2}EI_\omega\right), \tag{9-28a}$$

$$B_2 = \frac{\pi^4}{4l^3}EI_y, \tag{9-28b}$$

$$C_1 = -0.1431\beta_1 l + \frac{lt}{4} + \frac{2l}{\pi^2 f}t^2 + \frac{\pi^2}{8}i_{pT}^2\cot\alpha - \frac{\pi^2}{8}\beta_1 a\cot\alpha, \tag{9-28c}$$

$$C_2 = \frac{2l}{\pi^2 f} + \frac{\pi^2}{8}\cot\alpha, \tag{9-28d}$$

$$C_3 = 0.5362l + \frac{4l}{\pi^2 f}t - \frac{\pi^2}{4}y_0\cot\alpha + \frac{\pi^2}{4}a\cot\alpha. \tag{9-28e}$$

Minimizing q with respect to U_1 and Φ_1 we obtain the following linear equation system for U_1 and Φ_1:

$$2B_2 U_1 - q\left(2C_2 U_1 + C_3\Phi_1\right) = 0, \tag{9-29}$$
$$2B_1\Phi_1 - q\left(2C_1\Phi_1 + C_3 U_1\right) = 0,$$

of which the determinant has to be set equal to zero in order to obtain a nontrivial solution. This yields a quadratic equation for q:

$$q^2\left(C_3 - 4C_1 C_2\right) + q\left(4B_2 C_1 + 4B_1 C_2\right) - 4B_1 B_2 = 0. \tag{9-30}$$

In the case of shell-beams with compressed free edges (*Fig. 8-18a*) the displacement u_T can be assumed as zero, since their lateral bending stiffness can be regarded as infinitely great, see at the end of Section 9.1.2. In this case, however, we have to assume more terms in the function Φ [KOLLÁR and GÁRDONYI, 1967; KOLLÁR, 1973], see also the following.

As can be seen, the energy method yields the straightforward solution of the problem, but it does not show visually how the modification of the various parameters affects the critical load, although this would be very useful in design. However, we can draw from these results some interesting conclusions which allows us to develop a more visual method.

The relations (9-25) show that the work done can be written, also if we take more terms in the series of the deformation functions u_T and Φ into account, in the following form:

$$W = q\sum_{i=1}^{5} W_i^{(1)}, \tag{9-31}$$

where $W_i^{(1)}$ means the ith part of the work done by the load of unit intensity. $W_i^{(1)}$ is determined by the data of the beam and the assumed deformation functions. If we assumed functions representing the exact solution of the problem for u_T and Φ, then the critical load could be written as follows:

$$q_{cr} = \frac{U}{\sum\limits_{i=1}^{5} W_i^{(1)}},$$ (9-31a)

or

$$\frac{1}{q_{cr}} = \frac{\sum\limits_{i=1}^{5} W_i^{(1)}}{U} = \sum_{i=1}^{5} \frac{1}{q_{cr,i}}.$$ (9-31b)

We denoted by $q_{cr,i}$ the 'critical partial loads'

$$q_{cr,i} = \frac{U}{W_i^{(1)}}$$ (9-31c)

obtained as the solutions of the partial cases a) to e), which have to be determined by using the deformation functions u_T and Φ, representing the exact solution of the whole problem.

The equation (9-31b) represents a mixed *Dunkerley*- and *Föppl-Papkovich*-type relation, see Chapter 2 on the summation theorems. Hence if we do not determine the $W_i^{(1)}$ of each partial case with the exact deformation functions of the whole problem, but we minimize $q_{cr,i}$ in each case (i.e. we determine every $q_{cr,i}$ by using different deformation functions), then every $q_{cr,i}$ will be smaller than that determined by the exact deformation functions of the whole problem, and, consequently, the critical load q_{cr}^*, computed from these will also be smaller than the exact q_{cr}. Thus, by so doing, we are on the safe side (except for the case when the load acts under the shear centre, see the remark after Eq. (9-42). Hence this method has two advantages. First, to some extent it compensates the error to the unsafe side of the energy method and, second, it shows very clearly how the various parameters influence the critical load.

9.1.3 Determination of the critical load by the summation theorem

The critical loads of the five partial cases can be determined as follows [LÓRINCZ, 1977]. *Case a)*. To determine the critical load of the beam loaded at the shear centre and resting on fork-like supports (*Fig. 9-5*) we assume the deformation functions (9-5), and obtain the expressions (9-24) and (9-25a) for U and W. Equating them and proceeding according to the rules of the energy method we arrive at the following equation of the second degree for the critical load:

$$0.01181 \frac{l^4}{EI_y} \left(q_{cr}^{fork}\right)^2 - 0.5725\beta_1 q_{cr}^{fork} - \frac{\pi^2}{l^2}\left(GI_t + \frac{\pi^2}{l^2}EI_\omega\right) = 0.$$ (9-32)

Positive sign means downward, negative sign upward pointing load.

MEISSNER [1955] improved this result by assuming only the function (9-5b) for Φ, determined u_T from (9-1a) and introduced it into (9-1b). Thus he obtained the following equation for the improved value of q_{cr}:

$$0.01218 \frac{l^4}{EI_y} \left(q_{cr}^{fork}\right)^2 - 0.5725\beta_1 q_{cr}^{fork} - \frac{\pi^2}{l^2}\left(GI_t + \frac{\pi^2}{l^2}EI_\omega\right) = 0.$$ (9-33)

The assumption of a single-term function for Φ furnishes a result of sufficient accuracy as long as the deformation corresponding to the bending-type stiffnesses (EI_y, EI_ω) predominate with respect to the deformations corresponding to the torsional stiffness (GI_t). The structure of Eqs (9-1a,b) shows that this condition is fulfilled if EI_y is not too great and/or $EI_\omega \pi^2/l^2$ is not too small in comparison to GI_t.

Shell-beams with cross sections of *Fig. 8-18a* do not undergo lateral bending deformation (i.e. $u_T = 0$) since their lateral bending stiffness EI_y can be taken as infinitely great. In this case we have to assume for Φ a function containing several terms [KOLLÁR and GÁRDONYI, 1967; KOLLÁR, 1973]. If, in addition, also their warping rigidity is zero (as with the V-shaped cross section), then lateral buckling comes about as a pure torsional deformation, the buckling length of which is undetermined (as in the case of pure torsional buckling described in Chapter 7, Eq. (7-22). Hence the shell-beam loaded by (uniformly) distributed forces will laterally buckle with a very short buckling length developing at the place of the maximum bending moment, so that its critical load will be equal to that of a shell-beam with a constant bending moment all along its length (Eq. (9-8)). Thus in the case of a shell-beam with a V-shaped cross section the expression of the critical load simplifies to:

$$q_{cr}^{fork} = -\frac{8GI_t}{\beta_1 l^2} \tag{9-34}$$

Case b). The beam is subjected to equal and opposite distributed forces acting at point P and at the shear centre T (*Fig. 9-6*). We have to determine the work done by the forces acting at P, as a consequence of their vertical displacements due to the rotations of the cross sections around T occurring during the deformation of case a). We have thus to assume that the beam rests on fork-like supports and is supported along the shear centre axis by continuous hinges, so that the cross sections can only rotate around T ($u_T = 0$). We assume a sine function (9-5b) for Φ and introduce it into the expressions (9-9) and (9-13). Equating these we arrive at the critical load:

$$q_{cr}^{ecc} = \frac{\pi^2}{tl^2}\left(GI_t + \frac{\pi^2}{l^2}EI_\omega\right). \tag{9-35}$$

It should be remarked that the same result can be obtained by the exact method [solving the differential equations (9-1)].

Case c). The beam, deformed already in case a), rotates as a rigid body around the axis passing through the points of suspension F. Assuming again the functions (9-5a,b) for the deformations, and introducing them into (9-24) and (9-25c), we obtain:

$$q_{cr}^{susp} = \frac{12.176f}{\dfrac{l^4}{\pi^2 EI_y} + \dfrac{t^2 l^2}{GI_t + \dfrac{\pi^2}{l^2}EI_\omega}} \tag{9-36}$$

It is worth while to treat some special cases. If $t = 0$, i.e. the load acts at the shear centre (as e.g. in the case of a beam with a rectangular cross section under its own weight), then:

$$q_{cr}^{susp} = 120\frac{f}{l^4}EI_y. \tag{9-37}$$

If the lateral bending stiffness EI_y can be taken as infinitely great, as in the case of shell-beams with cross sections according to *Fig. 8-18a*, then:

$$q_{cr}^{susp} = 12.176 \frac{f}{l^2 i^2} \left(GI_t + \frac{\pi^2}{l^2} EI_\omega \right). \tag{9-38}$$

Case d). The compressive force N is assumed to act at the centroid of the cross section. Since in the general case this does not coincide with the shear centre, we have to resort to the theory of coupled bending-torsional buckling, see e.g. in [TIMOSHENKO and GERE, 1961]. (The differential equations to be found here can also be obtained from Eqs (9-1a,b) by setting $M_x = 0$.) We can use the equilibrium method instead of the energy method.

Assuming again the functions (9-5a,b) for the deformations, and introducing them into Eqs (9-1a,b), we obtain the following equation of the second degree for the critical value N_{cr} of the compressive force:

$$\frac{i_p^2}{i_{pT}^2} N_{cr}^2 - \left(N_{y,cr} + N_{\Phi,cr} \right) N_{cr} + N_{y,cr} N_{\Phi,cr} = 0, \tag{9-39}$$

where i_{pT}^2 is defined by Eq. (9-2), furthermore

$$N_{y,cr} = \frac{\pi^2 EI_y}{l^2} \tag{9-40a}$$

and

$$N_{\Phi,cr} = \frac{GI_t + \frac{\pi^2}{l^2} EI}{i_{pT}^2} \tag{9-40b}$$

are the critical loads of the *Euler* buckling around the axis y and of the pure torsional buckling, respectively. Of the two roots of Eq. (9-39) we have to take the smaller one.

The critical compressive force can be transformed into critical load according to (9-19):

$$q_{cr}^{(N)} = \frac{2N_{cr}}{l \cot \alpha}. \tag{9-41}$$

Case e). The eccentricity of the compressive force causes a constant bending moment $M_N = aN$ along the beam axis (*Fig. 9-8*). Its critical value can be obtained from Eq. (9-6), taking the positive root as valid. The critical load q_{cr}, corresponding to the critical bending moment, is, making use also of (9-19):

$$q_{cr}^{(M_N)} = \frac{2M_{cr}}{al \cot \alpha}. \tag{9-42}$$

The critical load q_{cr}^* can be obtained by summing up the inverse values of the critical partial loads (9-32), (9-35), (9-36), (9-41) and (9-42), according to the relation (9-31b). However, we have to observe that if any of the partial critical loads has a negative sign (as e.g. that of case b) if the load acts under the shear centre), then – according to what has been said in Chapter 2 – we have to set this partial load infinitely great in Eq. (9-31b) in order to ensure that we be on the safe side by using the critical load given by (9-31b).

We have to make one more remark on the problem of suspended beams. In the case of the arrangement of *Fig. 9-8* the distance a has, when keeping the intensity of the load constant, not only a lower critical value, but also an upper one. In the first case the points of suspension lie so low that the beam can easily turn around its axis, while in the second case the bending moment M_N grows to such a value that it causes lateral buckling [LISKA, 1986], see also Eq. (9-42).

In the following we briefly survey the literature on lateral buckling of suspended beams.

DE VRIES [1947] dealt with the lateral stability of I-beams. In their Discussion to this paper, D. B. HALL and H. D. HUSSEY derived the critical load of a beam suspended at both ends on vertical cables and obtained the relation (9-37), taking the torsional rigidity of the beam as infinitely great.

In Hungary, after an accident with a prefabricated suspended bridge girder, several researchers worked on clarifying the related problems. Unfortunately, due to post-war difficulties, the Hungarian researchers were not informed on the results achieved in the West and, vice versa, but partly also due to the language barrier, the results of the Hungarians did not become known in the West. We first mention the three papers of CSONKA [1954a,b,c]. In the first paper he solved the problem of the lateral stability of a beam with rectangular cross section suspended at both ends. He also showed that the relation (9-37), derived earlier by BÖLCSKEI [1954] with the assumption $GI_t = \infty$, yields acceptable results as long as the vertical distance a between the suspension points and the centroid of the cross section remains smaller than the height of the cross section. (BÖLCSKEI, using the approximate assumption $GI_t = \infty$, also showed that if the beam is suspended at points being at a distance $0.225l$ from the ends, the critical load becomes the 67-fold of that valid for a beam suspended at the ends.) In his second paper CSONKA dealt with the stability problem of a beam suspended at the middle cross section, and in the third paper he determined the critical load of a suspended beam stiffened by cables like the masts of sailing boats. LOVASS-NAGY [1953] showed that the critical load of a beam with a rectangular cross section becomes infinitely great if $I_x = I_y$ (as shown later by TRAHAIR and WOOLCOCK [1973]). BÖRÖCZ [1956] systematically investigated the influence of the possible imperfections on the critical load, using the assumption $GI_t = \infty$. KORDA [1963] solved the problem of a beam suspended by inclined cables, again with the same assumption.

Beams with cross sections which distort in their own planes (and, consequently, have no torsional rigidity GI_t) consist a special problem. Such cross sections are shown in *Fig. 9-9*, but also trusses can be considered as belonging to this group if their bracing is not too rigid. DULÁCSKA and TARNAI [1979] dealt with this problem. TARNAI [1977, 1978] generalized the method to beams with variable cross section of the chords. The point of application of the load may also change along the length, and the support conditions of the two chords can be different.

Fig. 9-9 Distorting thin-walled sections

Of the thin-walled girders, YEGIAN [1956], who was obviously not aware of the results of the Hungarians, determined the critical load of a beam with a simply symmetric I-shaped cross section, suspended at both ends, and at two intermediate points. LEBELLE [1959] treated the problem of a beam with a simply symmetric I-shaped cross section, supported at two intermediate points (see also in [BECK and SCHACK, 1972]). KOLLÁR and GÁRDONYI [1967] solved the problem of shell-beams suspended at the ends, and GÁRDONYI [1979] that of shell-beams suspended at intermediate points.

PETTERSSON [1960] extended the investigation to beams with variable cross section.

The results of the theory of elastic stability have been generalized to reinforced concrete girders by RAFLA [1969, 1973] who took the influence of cracks, reinforcement and nonlinear $\sigma(\varepsilon)$ diagram into account, and by LISKA [1986] who performed also experiments with suspended reinforced concrete girders.

9.2 LATERAL STABILITY OF THE NODES OF PLANE TRUSSES

It is well known that the nodes of the compressed chords of trusses have to be supported against lateral displacement. However, it is less well known that in some cases also the nodes of the *chords in tension* need such a support. In the following we will present the criterion of KIRSTE [1950], which allow us to decide whether such a support is necessary or not.

We assume that the nodes of the truss have spherical hinges. Let us give a virtual displacement v to one of the nodes, denoted by C (*Fig. 9-10*). Supposing that all neighbouring nodes are rigidly supported against lateral displacement, the restoring force V acting on the node C is given by the expression

$$V = v \sum_i \frac{N_i}{l_i}, \tag{9-43}$$

where N_i is the bar force (positive if tension) acting in the ith bar joining the node C, and l_i is the length of the ith bar.

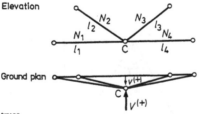

Fig. 9-10 A node of a truss

The original position of the node is stable if $\sum_i \frac{N_i}{l_i}$ has a positive sign, since in this case V becomes a restoring force. If this sum is equal to zero, then the position of the node is indifferent, and if the sum has a negative sign, then the node is unstable since V pushes it further in the direction of the displacement.

In [KIRSTE, 1950] we also find the method to deal with the case if the neighbouring nodes are not rigidly supported laterally.

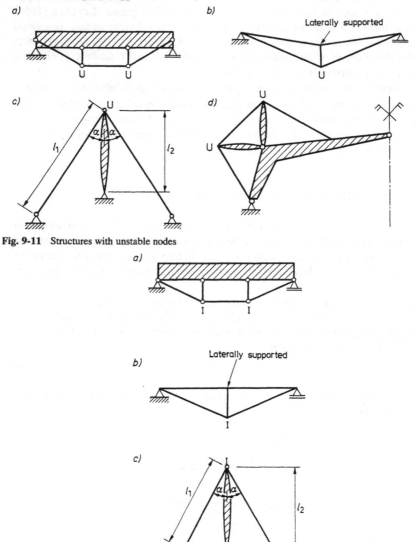

Fig. 9-11   Structures with unstable nodes

Fig. 9-12   Structures with indifferent nodes

The nodes of the actual trusses have no spherical hinges, nevertheless the criterion of *Kirste* is a good basis to decide whether the nodes need a lateral support or not. To illustrate the practical importance of this instability phenomenon, we show some examples.

In *Fig. 9-11* structural arrangements with unstable nodes (U), in *Fig. 9-12* with indifferent ones (I), and in *Fig. 9-13* with stable ones (S) are depicted. Where we marked two nodes with U, there the two nodes displace simultaneously in lateral direction.

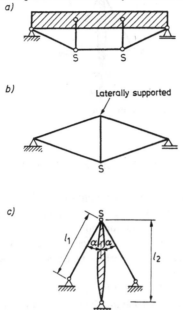

a)

b)

Laterally supported

S

c)

l_1 α α l_2

Fig. 9-13 Structures with stable nodes

To show the application of the method let us consider the structure of *Fig. 9-12c*, whose variants with longer and shorter prestressing cables are depicted in *Figs 9-11c* and *9-13c* respectively. After prestressing the cables, the vertical equilibrium of the upper node yields $N_2 = -2N_1 \cos\alpha$, where N_1 and N_2 are the forces arising in the inclined cables and in the vertical post, respectively. The sum appearing in Eq. (9-43) becomes:

$$\sum_{i=1}^{2} \frac{N_i}{l_i} = 2\frac{N_1}{l_1} + \frac{N_2}{l_2} = 2\frac{N_1}{l_1} - 2\frac{N_1}{l_2/\cos\alpha}.$$

This sum is positive if $l_1 < l_2/\cos\alpha$, i.e. in the case of *Fig. 9-13c*; negative if $l_1 > l_2/\cos\alpha$, i.e. in the case of *Fig. 9-11c*; and equal to zero in the case of *Fig. 9-12c*.

In practice it is not always enough to know that the node is stable, but we may also ask: how stable is it? That is, if the stable state of equilibrium is close to the indifferent (neutral) state, then the node may be considered almost as 'unstable' as if it were indeed in an unstable state.

TOMKA [1997] presented a very simple method to establish the 'measure of stability' of a node of a truss or cable structure. He applies two equal but opposite auxiliary forces S on the compressed bar(s) joining the node (*Fig. 9-14*), and determines their value necessary to bring the node into an indifferent state of equilibrium. If the necessary forces S cause compression, the structure is stable, and the ratio of S to the actual compressive force acting in the bar yields the measure of stability.

Fig. 9-14 The auxiliary forces S acting on the structure of *Fig. 9-13c*

9.3 SNAPPING THROUGH OF SHELL-BEAMS IN THE PLANE OF BENDING

A peculiarity of the behaviour of shell-beams shown in *Fig. 8-18* – as contrasted to thin-walled beams with 'ordinary' cross sections (*Fig. 9-1*) – is that, when subjected to bending in the vertical plane, they are prone to snapping through in the same plane. This phenomenon comes about essentially due to the fact that they bow out in the vertical plane due to bending, thus they become shell-arches which, as explained in Section 8.1.4, may snap through in the plane of the arch.

We shall give the values of the critical bending moment causing snapping through for two cases: for *individual shell-beams*, having the cross sections of *Fig. 8-18* on the basis of [KOLLÁR, 1973]; and for *shell-beam rows* with the cross sections of *Fig. 9-15*, according to [KEREK, 1981]. These latter are supported in such a way that the individual shell-beams can 'slide apart' in transverse direction, what is ensured e.g. by the arrangement shown in *Fig. 9-16*: the columns are flexible enough to allow this 'sliding'.

Fig. 9-15 Cross sections of shell-beam rows

We give the values of the critical bending moment M_{cr} causing snapping through in *Table 9-1* for individual shell-beams and for shell-beam rows as well. The notations are to be seen in *Fig. 9-2*. The values for the individual beams are equally valid for the cross sections of *Fig. 8-18a* and for the 'upside down' ones of *Fig. 8-18b*. For shell-beam rows the expression 'upside down' cross sections is meaningless, since from the wave- and V-shaped cross sections one can make only one row cross section each (see the two upper sketches in *Fig. 9-15*), and the parabolic and wing-shaped cross sections are the reversed versions of each other. (This is the reason why the critical bending moments of the rows made up of these two cross sections are equal to each other.)

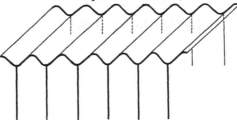

Fig. 9-16 A shell-beam row as a roof

Table 9-1 The factors λ of the bending moments $M_{cr} = \lambda \frac{EI_x v}{\sqrt{3b^2}}$ causing snapping through of shell-beams

Shape of the cross section	parabola	V	wave	wing
Individual shell-beam	0.912	0.931	0.935	1.064
Shell-beam row	1.697	1.620	1.603	1.697

The values of the critical moments of individual shell-beams coincide with those valid for shell-arches in the case of $\beta_0 b = 0$ in *Fig. 8-23*.

Finally it should be remarked that value of the critical moment causing snapping through is practically independent of the shape of the bending moment diagram: if the maximum bending moment reaches the value of M_{cr}, the beam will snap through.

10

Stability of viscoelastic structures

GYÖRGY IJJAS·

10.1 INTRODUCTION

10.1.1 General remarks

The problem of viscoelastic stability, or as it was called 'creep buckling', has emerged at the end of 1940s. It was observed that some structures collapsed after some time had elapsed after loading. That period was called 'critical time'. The researchers developed methods to calculate the critical time for various structures and investigated whether there was any connection between the elastic and the viscoelastic stability. Many studies have been published on that topic (the first reliable summary was published in 1962 [KEMPNER]), and some of them gave real impetus to research. These studies and their results will be treated in this chapter. Although the term 'creep buckling' is widely used in the literature, the expression 'viscoelastic stability' will be preferred in this chapter, because it indicates that the material of the structure has an instantaneous elastic response to the loading and the expression 'stability' is somewhat more general than the word 'buckling'.

10.1.2 Material properties

In this chapter only models made from linear viscoelastic material will be investigated. Elastic, because these materials have an initial modulus of elasticity which is constant, so it is linearly elastic and they have an elastic response. (The only exception is the *Kelvin* solid.) Viscous, because it creeps. Linear, because their constitutive equations are linear differential equations. This simply means that if the load is doubled then both the elastic response and the creep deformation is doubled too.

In *Table 10-1* the most important features of the models of the materials which will be used in the following are shown. A more detailed description of these models can be found in several books, e.g. in [FLÜGGE, 1975]. In addition, the so-called *Dischinger* material [DISCHINGER, 1937] will be treated. It was created to describe the creep of concrete.

Dischinger established his model without realizing that it is a linear elastic spring and an aging dashpot connected in series. Later BAŽANT and CEDOLIN [1991] called the materials of a *Dischinger* type when the characteristic of the spring is aging too. However, in this chapter the expression will be confined to the case only when the behaviour of the spring is time-independent. The models of materials in *Table 10-1* are classified as fluid and solid. (The *Dischinger* material is of the solid type.) The basic feature of the fluid-type material is that it creeps without any limit until it is loaded. The basic feature of the solid type material is that its deformation is limited and stops after some time if the load does not change (e.g. if the load is constant).

10.2 VARIOUS KINDS OF CREEP BUCKLING

Let the model shown in *Fig. 10-1b* be investigated. This model consists of a *Maxwell* fluid $(A - B)$ and a rigid bar $(B - C)$. Two cases are treated. In the first case the model is loaded by a conservative force P of constant value, while in the second case an inert mass m is placed at point B. The weight of the inert mass, in the second case, is equal to the value of the force in the first case. The change of the central angle in the first case is described by a first order, while in the second case by a third order differential equation. The parameters of the governing equations and the initial values are given. β_{0g} is the initial imperfection and β_0 is the elastic deformation just after loading. The parameters of the governing equations of the structures are different and the structure with inert mass shows two qualitatively different behaviours, while the curves of the structure without inert mass differ only quantitavely. The solution of the third order differential equation (the case of the structure with inert mass) is shown by a full line, while the solution of the first order differential equation (the structure loaded by a force) is given by a dashed line. The difference is evident: the continuous line describes more or less the behaviour of a real structure, while the dashed line only approximates it. It can be seen that the dashed line has a vertical tangent in both cases, while the full lines does not. Accordingly, although both cases show 'some kind of collapse', until now it is not possible to define, in the case of the inert mass, which part of the solution can be called stable and which part cannot. So in this chapter it will not be dealt with such structures any longer.

On the other hand, the case of the structure without inert mass is much simpler. The time-rotation function has a vertical tangent. This is equivalent to the fact that the velocity of the rotation is infinite and it can be considered as the collapse of the structure. The time interval between the beginning of the loading and the infinite velocity of the rotation is called 'critical time.' To our knowledge the notion of the critical time was introduced by HOFF [1954].

The example showed earlier was given only to compare the solution of the governing equations of structures with inert mass and without it. In this chapter only the behaviour of structures without inert mass will be investigated.

A contradiction, however, has to be mentioned. In the engineering practice, creep is supposed to be a slow deformation, so the inert mass of the structure can be neglected. On the other hand, this fact gives the possibility of infinite velocity of deformation. However,

TABLE 10-1	LINEAR ELASTIC SPRING	NEWTONIAN FLUID	DISCHINGER DASHPOT
MODELS AND CONSTITUVE EQUATIONS	$Q = kq_k$ k = spring constant (force/elongation)	$\frac{dq}{dt} = \frac{Q}{v}$ v = viscosity coefficient (force·time/elongation)	$\frac{dq}{dt} = \frac{d\varphi}{dt}\frac{Q}{k}$ $\varphi = \varphi_{max}(1-e^{-\lambda t})$ $\frac{dq}{dt} = \varphi_{max}\lambda e^{-\lambda t}$ k (force/elongation) λ (1/time) φ_{max} is dimensionless

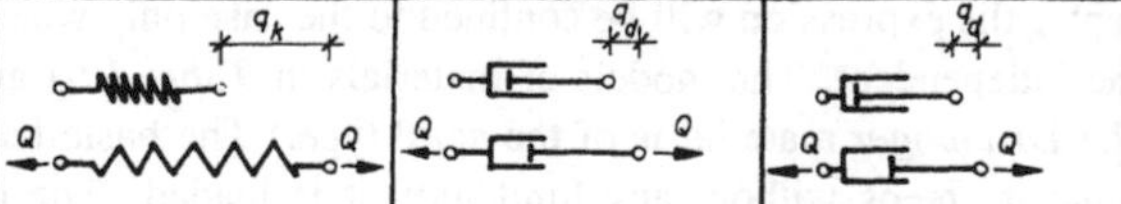

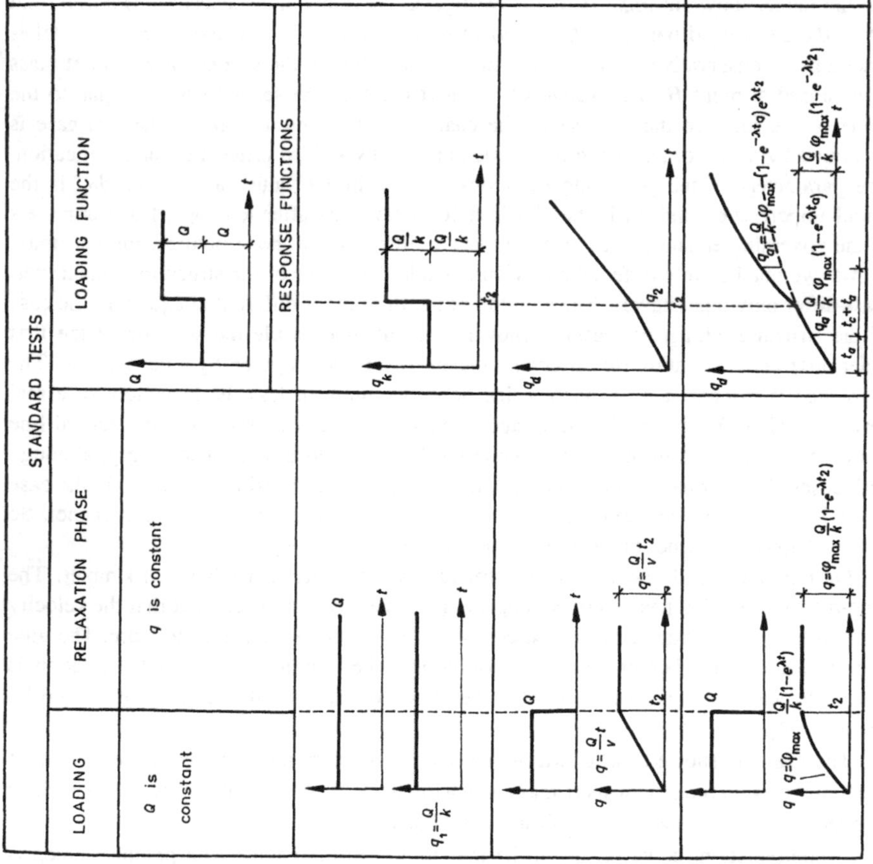

STANDARD TESTS
LOADING FUNCTION
RESPONSE FUNCTIONS
RELAXATION PHASE
q is constant
LOADING
Q is constant
$q_1 = \frac{Q}{k}$
$q = \frac{Q}{v}t$
$q = \frac{Q}{v}t_2$
$q = \varphi_{max}\frac{Q}{k}(1-e^{-\lambda t})$
$q = \varphi_{max}\frac{Q}{k}(1-e^{-\lambda t_2})$
$q_a = \frac{Q}{k}\varphi_{max}(1-e^{-\lambda t_a})$
$q_{a1} = \frac{Q}{k}\varphi_{max}(1-e^{-\lambda t_a})e^{\lambda t_2}$
$\frac{Q}{k}\varphi_{max}(1-e^{-\lambda t_2})$
t_a
$t_c + t_a$
t_2

MAXWELL FLUID	KELVIN SOLID	DISCHINGER MATERIAL	THREE PARAMETER SOLID

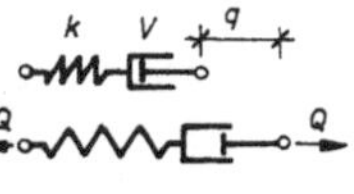

MAXWELL FLUID

$$\frac{dq}{dt} = \frac{1}{k}\frac{dQ}{dt} + \frac{Q}{v}$$

KELVIN SOLID

$$Q = kq + v\frac{dq}{dt}$$

DISCHINGER MATERIAL

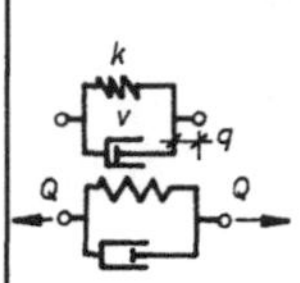

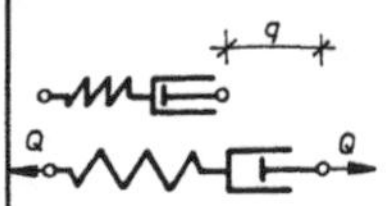

$$\frac{dq}{dt} = \frac{1}{k}\frac{dQ}{dt} + \frac{Q}{k}\frac{d\varphi}{dt}$$

or

$$\frac{dq}{d\varphi} = \frac{1}{k}\frac{dQ}{d\varphi} + \frac{Q}{k}$$

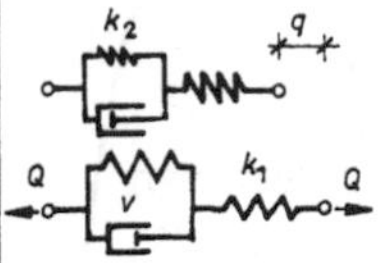

THREE PARAMETER SOLID

$$(k_1+k_2)\,Q + \frac{dQ}{dt} = k_1 k\, q + v k_1 \frac{dq}{dt}$$

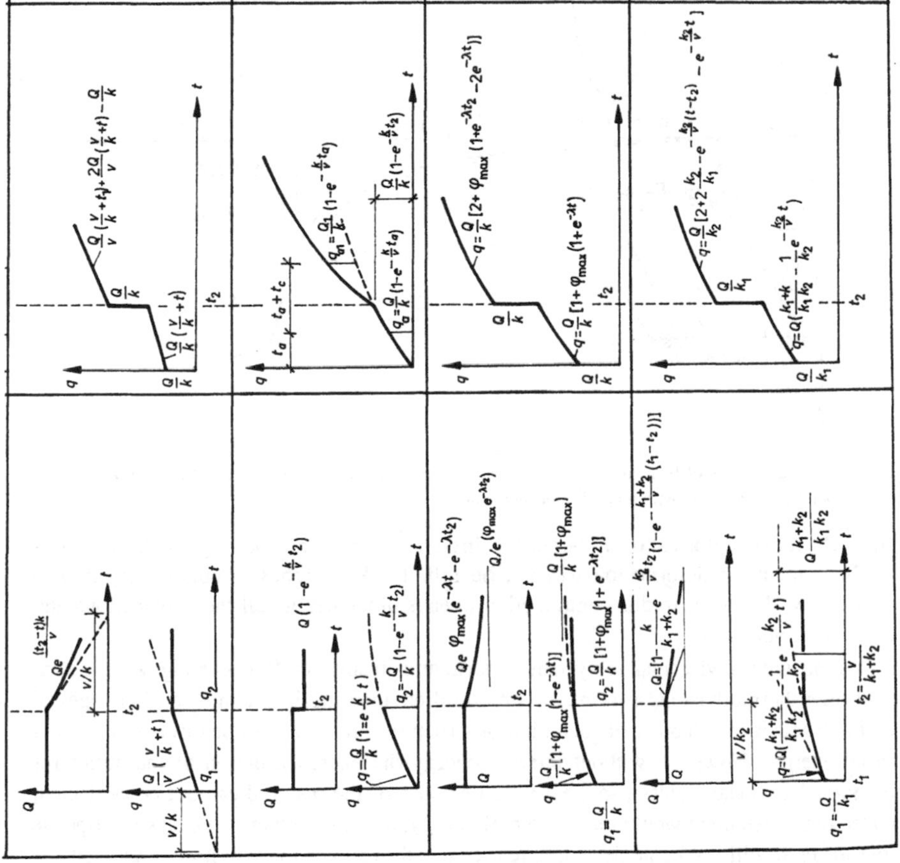

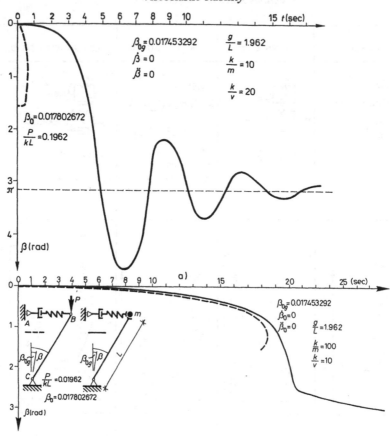

Fig. 10-1 Various kinds of behaviour of a simple structure made of *Maxwell*-fluid in the case of different parameters, furthermore with and without inert mass

if the velocity of deformation is large, the inert mass must not be neglected, so in this case the velocity of deformation will not be infinite. Researches in engineering mostly do not follow this way, so this contradiction always hides behind all the explanations and derivations given later.

Dealing with a viscoelastic system or structure means, in fact, dealing with a non-equilibrium thermodynamic system. In structural engineering it is almost a philosophical question whether the stability of such kind of structure can or can not be discussed without internal energy. However, without internal energy the energy criterion of the structures does not exists (meaningless), so the elastic response of the material can not be neglected, because otherwise there would be no internal energy. So in this chapter the elastic response of the material will not be neglected, hereafter the above question will not be relevant.

The purposes of this chapter are, first, to show the stability behaviour of different viscoelastic structures, second, to extend the notion of the equilibrium path to the pseudo-equilibrium surface in the case of viscoelastic structures and, third, to show the link between the critical time and the vanishing minimum of the total potential in the subspace of external co-ordinates (to be defined later). The expression 'pseudo-equilibrium' will be circumstantially explained in Section 10.3.2.

To give a short summary of the topic of this chapter first it has to be mentioned, that if there is no initial imperfection (the rigid bar in *Fig. 10-1b* is vertical), the structure will not creep, so that this case is not interesting for viscoelastic stability.

As it was shown in *Fig. 10-1* structures made of fluid type material with symmetric unstable post-critical behaviour always deform until they buckle or some external constraint hinders their further deformation. Naturally the *Maxwell* fluid ($A - B$) in the structure of *Fig. 10-1* can be replaced by a three-parameter solid or a *Dischinger* material. In these cases the behaviour of the model is different. If the load is higher than a well-defined one in the case of a three-parameter solid, or the dashpot ages so slowly that the deformation reaches the point of buckling in the case of a *Dischinger* material, the structure will exhibit the same behaviour as the structure without inert mass did in *Fig. 10-1*. Otherwise the structure will, as a consequence of the solid-type of the materials, not buckle. These cases belong to the group of the structures exhibiting symmetric unstable post-critical behaviour.

In the case of structures exhibiting asymmetric post-critical behaviour the situation is similar on the side of unstable behaviour.

There are structures exhibiting stable symmetric post-critical behaviour. If their material is fluid-type, they will deform until some external restraint does not hinder their deformation. If the material is solid-type, two different cases are possible. First, the structure deforms and after some time the deformation stops. Second, the structure deforms until some external constraint hinders its deformation. In the case of asymmetric post-critical behaviour, on the stable side the situation is similar. It should be remarked that the following statement is valid in all aforementioned cases: if the load exceeds the elastic critical load of the structure determined by the initial modulus of elasticity, then elastic buckling occurs without creep.

Because creep rearranges the stresses or internal forces in a statically indeterminate structure, this may change the buckling behaviour of the structure. This problem will not be discussed here. However, the basic statement, namely the connection between the critical time and the vanishing minimum of the total potential energy in the subspace of external co-ordinates is true in every cases. A useful survey on the various kinds of creep buckling can be found in [DULÁCSKA, 1984].

10.3 STRUCTURES MADE OF FLUID-TYPE MATERIAL, EXHIBITING SYMMETRIC UNSTABLE POST-CRITICAL BEHAVIOUR

10.3.1 Description of the phenomenon

Let the structure shown in *Fig. 10-1* be investigated in detail. The governing differential equation of this structure will be derived and solved. The geometric equation of the

Viscoelastic stability

structure is

$$q = q_M + q_{0g} = L\sin\beta, \tag{10-1}$$

where

q is the distance between the points A and B, called external parameter,
q_M is the elongation of the *Maxwell* fluid,
q_{0g} is the initial imperfection (the distance of points A and B before loading).

The equilibrium equation is

$$-QL\cos\beta + PL\sin\beta = 0 \tag{10-2}$$

which expresses the moment equilibrium about the point C. The constitutive equation of the *Maxwell* fluid can be found in *Table 10-1*. After some mathematical operations the result is:

$$\dot{\beta} = \frac{kP\sin\beta\cos\beta}{\nu\left(kL\cos^3\beta - P\right)} = \frac{P\sin\beta\cos\beta}{\nu L\left(\cos^3\beta - \dfrac{P}{kL}\right)}, \tag{10-3}$$

where the dot means differentation with respect to time. Before solving Eq. (10-3), let its denominator be investigated. It can be seen, that the denominator is equal to zero if

$$\beta = \arccos\sqrt[3]{\frac{P}{kL}}. \tag{10-4}$$

If the numerator of Eq. (10-3) at this value of angle β is different from zero, the rate of rotation (velocity) becomes infinite and the structure collapses. It was HULT [1962] who first investigated *von Mises* trusses and low arches with two or three hinges made of linearly viscoelastic material. He derived the governing equation of these structures without solving them. He recognized the phenomenon shown above, namely that if the denominator becomes zero, the 'velocity of deformation' will become infinite. He also expressed: 'If the elastic part of deformation is neglected, the resulting equations will not predict such a finite jump.' And that is the question. Namely: 'May one speak about buckling in the case if the material has no elastic response?'. At the end of this chapter a simple example will be given to illustrate this problem.

The solution of the differential equation (10-3) becomes after separation of the variables:

$$\frac{\nu}{k}\int_{\beta_0}^{\beta}\frac{kL\cos^3\beta - P}{P\sin\beta\cos\beta}d\beta = \frac{\nu}{k}\left[\frac{kL}{P}\cos\beta - \ln\frac{\left(\tan\dfrac{\beta}{2}\right)^{\frac{kL}{P}}}{\tan\beta}\right]_{\beta_0}^{\beta} = t, \tag{10-5}$$

where β_0 is the solution of the governing equation of the same structure if the *Maxwell* fluid is replaced by a spring:

$$P\sin\beta = kL\left(\sin\beta_0 - \sin\beta_{0g}\right)\cos\beta. \tag{10-6}$$

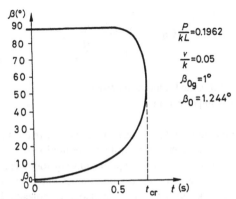

$$\frac{P}{kL}=0.1962$$

$$\frac{v}{k}=0.05$$

$$\beta_{0g}=1°$$

$$\beta_0=1.244°$$

Fig. 10-2 The function $\beta(t)$ given by Eq. (10-5)

The function (10-5) can be seen in *Fig. 10-2* in the case of parameters given in the same figure. The curve shows that the solution is, on the one hand, an approximation of the solution of a 'real' structure and, on the other hand, if the time passes the critical one, the solution becomes physically meaningless.

10.3.2 The pseudo-equilibrium surface

In the following the picture of the structure in *Fig. 10-1* is drawn again in *Fig. 10-3*. Now the motion of point *B* will be investigated. The equation of the pseudo-equilibrium surface

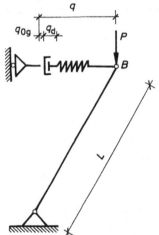

Fig. 10-3 The structure investigated in Sections 10.3.1, 10.3.2 and 10.3.3

of the structure is:

$$k \left(q - q_d - q_{0g} \right) - P \frac{q/L}{\sqrt{1 - \left(\frac{q}{L} \right)^2}} = 0, \tag{10-7}$$

where

q_d is the creep of the dashpot, called internal parameter,

q_{0g} corresponds to the initial imperfection.

The prefix 'pseudo' is used because Eq. (10-7) does not express a real equilibrium. Equilibrium in Newtonian mechanics means rest or a straight-line motion with constant velocity. In this case and later in the chapter the static equilibrium equations are always fullfilled (see e.g. Eq. (10-2)). However, because the structure creeps due to the change of q_d, so there is no rest, and the phenomenon belongs to the scope of nonequilibrium thermodynamics. In thermodynamics, equilibrium corresponds to the condition that entropy has a stationary point, and stable equilibrium corresponds to the maximum of entropy. Creep increases entropy, so that the system reaches equilibrium when entropy production stops. So from the point of view of thermodynamics, creep is a nonequilibrium phenomenon. On the other hand, it is a question whether one can speak in this case about a real motion or not because the inert mass is neglected.

The other equation, necessary to describe the behaviour of the structure, is:

$$b\dot{q}_d - k \left(q - q_d - q_{0g} \right) = 0. \tag{10-8}$$

Eqs (10-7) and (10-8) together are the so-called *Biot* equations of the linear viscoelastic material, which is based on the principles of nonequilibrium thermodynamics. It was published first without derivation in 1940 [KÁRMÁN and BIOT, 1940], and later it was derived by [BIOT 1954, 1955] and in various other ways by several authors too, e.g. [SCHAPERY, 1964].

Let P/k be expressed from Eq. (10-7):

$$\frac{P}{k} = \frac{\sqrt{1 - \left(\frac{q}{L} \right)^2} \left(q - q_d - q_{0g} \right)}{q/L}. \tag{10-9}$$

Taking into account that

$$q = L \sin \beta,$$

$$q_d + q_{0g} = L \sin \left(\beta_d + \beta_{0g} \right),$$

the following expression can be obtained:

$$\frac{P}{kL} = \frac{\sin \beta - \sin \left(\beta_d + \beta_{0g} \right)}{\tan \beta}. \tag{10-10}$$

This equation is very similar to the equation of equilibrium paths of an elastic structure. In fact, if β_d is taken equal to zero, the equilibrium paths of an elastic structure with an unstable symmetric post-buckling behaviour will be obtained for different values of β_{0g}. If both β_d and β_{0g} are zero, then the result will be the unstable bifurcating path.

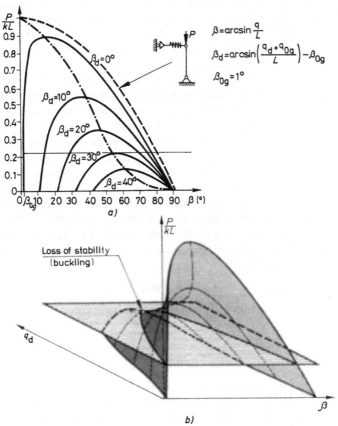

Fig. 10-4 *a*) The intersection lines of the pseudo-equilibrium surface of the structure shown in *Fig. 10-3* with planes parallel to the $(P/kL, \beta)$ plane, *b*) view of the pseudo-equilibrium surface

In *Fig. 10-4* the function given by Eq. (10-10) is shown for positive values of β. If different values of β_d are taken as initial imperfections, then the curves in *Fig. 10-4a* become the equilibrium paths. HAYMAN, who investigated the stability of viscoelastic structures [HAYMAN, 1974a,b, 1978a,b, 1980], tried to find a link between the equilibrium paths of an elastic structure and the stability of the corresponding viscoelastic structure. Although he clearly recognized that the loss of stability corresponded to the maximum point of an equilibrium path, he did not notice that the path was not an equilibrium path of an elastic structure but the intersection curve of a pseudo-equilibrium surface with a plane parallel to the co-ordinate plane $P/kL, \beta$. *Fig. 10-4a* gives these intersection lines. The dash-dotted line (locus of critical points – after HAYMAN) connects the maximum points of the intersections. However HAYMAN's observations are very important because he confirmed that buckling is inherently an elastic phenomenon.

He also did not notice that the maximum point of an equilibrium path (the maximum point of an intersection line of the pseudo-equilibrium surface with a plane parallel to the $P/kL,\beta$ plane) corresponded to the zero value of the denominator of Eq. (10-3).

It has to be mentioned that in the case of conformal mapping the mathematicians call a point to be critical if the first derivative of the mapping function, $w = f(z)$, with respect to z is equal to zero $\left(f'(z) = 0\right)$, [KORN and KORN, 1975]. However, it seems reasonable to use HAYMAN's definition because it expresses the physical substance of the phenomenon.

In *Fig. 10-4b* the pseudo-equilibrium surface is shown. Some curves of intersections with the planes parallel to the plane $P/kL,\beta$ are given too.

The explanation of the viscoelastic buckling can be easily given using the pseudo-equilibrium surface. For the sake of simplicity let the force P be supposed constant. It corresponds to a plane parallel to the co-ordinate plane, β, β_{0g}. This plane intersects the pseudo-equilibrium surface and when one point of this intersection curve coincides with a maximum point of an intersection curve parallel to the plane $P/kL,\beta$, then buckling occurs. It is clear that the cause of buckling is the elastic response of the structure.

For completeness let the tangent of the intersection curve of a plane parallel to β, β_d co-ordinate plane with the pseudo-equilibrium surface be defined at the point of the loss of stability. Expressing β_d from Eq. (10-10) results in:

$$\beta_d = \arcsin\left(\sin\beta - \frac{P}{kL}\tan\beta\right) - \beta_{0g}. \tag{10-11}$$

Differentiating with respect to β the following expression is obtained:

$$\frac{\mathrm{d}\beta_d}{\mathrm{d}\beta} = \frac{\dfrac{1}{kL\cos^2\beta}\left(kL\cos^3\beta - P\right)}{\sqrt{1 - \left[\sin\beta - \dfrac{P}{kL}\tan\beta\right]^2}}. \tag{10-12}$$

This expression is equal to zero if its numerator is equal to zero.

The expression $kL\cos^3\beta - P$ is the same as that in parenthesis in the denominator of the right-hand side of Eq. (10-3) or the coefficient in parenthesis on the right-hand side of Eq. (10-15). If $kL\cos^3\beta - P = 0$, then the angular velocity $\dot{\beta}$ (Eq. (10-3)) becomes infinite, and $\mathrm{d}\beta_d/\mathrm{d}\beta$ (Eq. (10-12)) and $\partial^2 V/\partial q^2$ (Eq. (10-15)) become zero. This mathematically supports the statements above.

10.3.3 The total potential energy

In the following, the behaviour of the total potential energy of the structure will be investigated. Let the total potential energy of the structure be defined, as usual:

$$V = \frac{1}{2}k\left(q - q_d - q_{0g}\right)^2 - PL\left[\sqrt{1 - \left(\frac{q_{0g}}{L}\right)^2} - \sqrt{1 - \left(\frac{q}{L}\right)^2}\right]. \tag{10-13}$$

It differs from the total potential energy of an elastic structure only that it contains the deformation parameter q_d of the dashpot. Let this equation be differentiated with respect to q. The result is:

$$\frac{\partial V}{\partial q} = k\left(q - q_d - q_{0g}\right) - P\frac{q/L}{\sqrt{1 - \left(\frac{q}{L}\right)^2}} = 0. \tag{10-14}$$

Eq. (10-14) is the same as (10-7) which was established as an equation giving the pseudo-equilibrium surface, or as one equation of *Biot*'s system of governing equations of a viscoelastic structure. This equation means that the total potential energy surface has a horizontal tangent in the intersection plane parallel to the plane V, q. Let Eq. (10-14) be differentiated with respect to q again. The result is:

$$\frac{\partial^2 V}{\partial q^2} = k - \frac{P}{L\left[1 - \left(\frac{q}{L}\right)^2\right]^{\frac{3}{2}}} = k - \frac{P}{L\cos^3\beta} = \left(kL\cos^3\beta - P\right)\frac{1}{L\cos^3\beta}. \tag{10-15}$$

In this equation the independent variable was changed from q to β. If Eq. (10-15) is compared to the denominator of Eq. (10-3) it is evident that the denominator of (10-3) is the same as the coefficient of $1/(L\cos^3\beta)$ in Eq. (10-15). This means that the necessary condition $d\beta/dt$ becoming infinite is equivalent to that assuming $\partial^2 V/\partial q^2$ taking zero value. We can also say that the local minimum of V disappears in the subspace of external parameters (β or q). This is the necessary condition of the buckling of viscoelastic structures, first established by the author [IJJAS, 1982].

It might be interesting to examine what will happen if both the numerator and the denominator of Eq. (10-3) become zero at the same time. It has to be remembered that a viscoelastic fluid always contains a spring and a dashpot connected in series. If the numerator of Eq. (10-3) becomes zero, the creep stops, i.e. the dashpot becomes forceless and the spring – being connected in series with it – becomes forceless too. This case can be ensued if some external constraint stops the process. However, if the spring becomes forceless, there is no internal energy and it is senseless to speak about the total potential energy. So in this case the energy criterion of the viscoelastic structure does not exist. This statement is also true in the case of structures with several degrees of freedom if the velocity of any or all the external parameters become zero at the same time.

However, in the case of a structure with several degrees of freedom, if only the velocity of one or more external parameters become zero at the time when the determinant of the *Hessian* matrix of the total potential energy becomes zero, then two possibilities exist. (The zero value of the determinant of the *Hessian* matrix is equivalent to the case of the zero value of the right-hand side of Eq. (10-15). The only difference is that in this chapter, for the sake of simplicity, only systems with one degree of freedom are investigated.) The first is: the velocity of some parameter(s) is rendered to zero by some external constraint which hinders the movement or creep. In this case it has to be dealt with another structure having other total potential energy because some parameters drop out from the equation. The second case is that the velocity of some or all of the parameters change their signs when the determinant of the *Hessian* matrix of the total potential energy becomes zero. In this case it is possible to decide only by a suitable transformation of co-ordinates whether the situation is really a rest or only a point of a process.

The intersections of the total potential energy surface with planes parallel to V, β are shown in *Fig. 10-5a*, while the total potential energy surface can be seen in *Fig. 10-5b*.

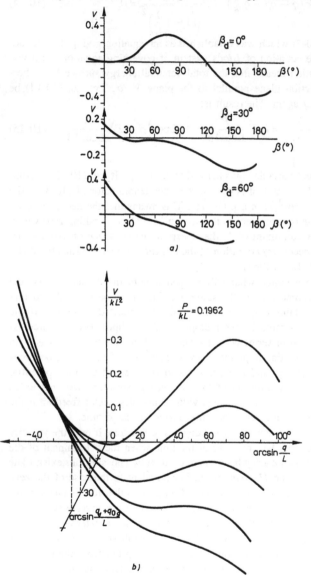

Fig. 10-5 *a*) The total potential energy function of the structure shown in *Fig. 10-3* for various constant β_d values, *b*) view of the total potential energy surface

10.3.4 Supplementary remarks

In the preceding sections the symmetric unstable behaviour of structures made of fluid type material was investigated. The phenomenon is basically the same in the case of asymmetric bifurcation on the side of unstable behaviour and in the case of structures which buckle with a limit point (*Fig. 10-6*).

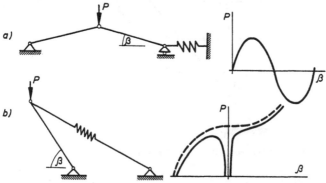

Fig. 10-6 Simple structures and their equilibrium paths. *a*) Structure with limit point, *b*) structure exhibiting asymmetric post-bucking behaviour

Here the study of [HUANG 1967] have to be mentioned. He was the first to solve the governing equation of a *von Mises* truss made of *Maxwell* fluid-type material and obtain a 'backward bending' function as a solution, like that in *Fig. 10-2*. He also stated: 'The buckling process is instantaneous and the response during buckling is elastic'. Otherwise his other study [1965], the study of HUANG and FUNK [1974] and the study of OBRECHT [1977] are worthy of mention.

10.4 STRUCTURES MADE OF SOLID-TYPE MATERIAL, EXHIBITING SYMMETRIC UNSTABLE POST-CRITICAL BEHAVIOUR

10.4.1 Description of the phenomenon

Let the structure shown in *Fig. 10-7* be investigated. It consists of a rigid bar with length L, supported at the lower end by a hinge, and supported at the upper end by a three-parameter solid. The constitutive equation of the three-parameter solid can be found in *Table 10-1*, the geometric equation is given by Eq. (10-1), and the equilibrium equation is identical with (10-2). So the governing equation of the structure is

$$\dot{\beta} = \frac{\left[\dfrac{P}{L}\tan\beta - \dfrac{k_1 k_2}{k_1 + k_2}\left(\sin\beta - \sin\beta_{0g}\right)\right]\dfrac{k_1 + k_2}{k_1 k_2}\cos^2\beta}{\dfrac{b}{k_2}\left(\cos^3\beta - \dfrac{P}{k_1 L}\right)}. \qquad (10\text{-}16)$$

The initial value of Eq. (10-16) can be defined again from Eq. (10-6), substituting k_1 for k. The solution of Eq. (10-16) can be obtained numerically, and it is shown for two different sets of parameters in *Figs. 10-8* and *10-9*. The two solutions have a basically different character, although the values of the load P differ only by 15%.

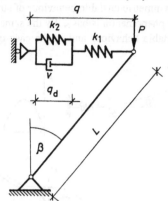

Fig. 10-7   The structure investigated in Sects. 10.4.1, 10.4.2 and 10.4.3

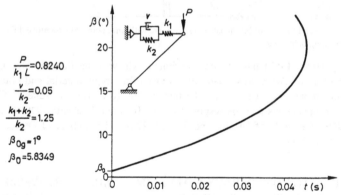

$$\frac{P}{k_1 L} = 0.8240$$

$$\frac{v}{k_2} = 0.05$$

$$\frac{k_1 + k_2}{k_2} = 1.25$$

$$\beta_{0g} = 1°$$

$$\beta_0 = 5.8349$$

Fig. 10-8   The solution of Eq. (10-16) for $P > k_1 k_2 L / (k_1 + k_2)$

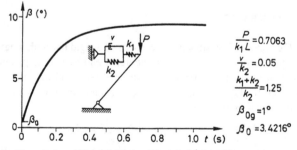

$$\frac{P}{k_1 L} = 0.7063$$

$$\frac{v}{k_2} = 0.05$$

$$\frac{k_1 + k_2}{k_2} = 1.25$$

$$\beta_{0g} = 1°$$

$$\beta_0 = 3.4216°$$

Fig. 10-9   The solution of Eq. (10-16) for $P < k_1 k_2 L / (k_1 + k_2)$

In the case of the higher load the structure behaves like the structure shown in *Fig. 10-3*, while in the case of the lower load the velocity of rotation does not become infinite. Similar results were published by BAŽANT and CEDOLIN [1991]. In *Fig. 10-8* the situation corresponds to the case when the denominator decreases to zero. In *Fig. 10-9* the case is shown when the numerator of Eq. (10-16) has become zero before the denominator did. In this case the creep stops before the structure buckles.

It is evident that if the bar is vertical then creep will not develop. So if the load exceeds the limit $k_1 L$, the primary equilibrium path becomes unstable.

10.4.2 The pseudo-equilibrium surface

Biot's equations of the structure shown in *Fig. 10-7* are

$$k_1 \left(q - q_d - q_{0g} \right) - P\frac{q/L}{\sqrt{1 - \left(\frac{q}{L}\right)^2}} = 0, \tag{10-17}$$

and

$$b\frac{dq_d}{dt} + k_2 q_d - k_1 \left(q - q_d - q_{0g} \right) = 0. \tag{10-18}$$

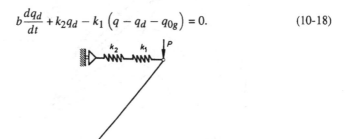

Fig. 10-10 The structure whose equilibrium path is necessary to construct the pseudo-equilibrium surface of the structure of *Fig. 10-7*

In the following, Eq. (10-18) will not be needed. However, for the examination of the structure shown in *Fig. 10-7* the equilibrium equation of the structure shown in *Fig. 10-10* is necessary. This case corresponds to the situation when the dashpot does not carry any load. So the equilibrium equation is

$$\frac{k_1 k_2}{k_1 + k_2} \left(q - q_{0g} \right) - P\frac{q/L}{\sqrt{1 - \left(\frac{q}{L}\right)^2}} = 0. \tag{10-19}$$

The left-hand side of this equation is equivalent to the numerator of the right-hand side of Eq. (10-16). The curves given by the right-hand side of Eq. (10-17) and (10-19) are shown in *Fig. 10-11* where the following relations were taken into account:

$$q = L\sin\beta, \tag{10-19a}$$

$$q_d = L\left[\sin\left(\beta_d + \beta_{0g}\right) - \sin\beta_{0g}\right], \tag{10-19b}$$

$$q_{0g} = L\sin\beta_{0g}. \tag{10-19c}$$

So Eq. (10-19) assumes the form:

$$\frac{P}{L}\tan\beta = \frac{k_1 k_2}{k_1 + k_2} L\left(\sin\beta - \sin\beta_{0g}\right).$$

(10-20)

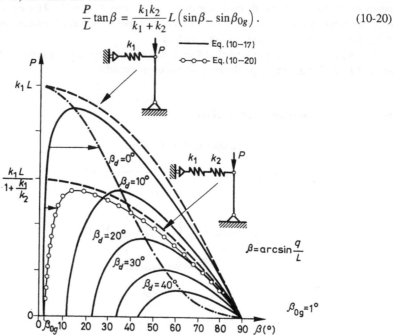

Fig. 10-11 The intersection lines of the pseudo-equilibrium surface of the structure shown in *Fig. 10-10* with planes parallel to the (P,β) plane, and the directrix of the equilibrium surface (marked by circles)

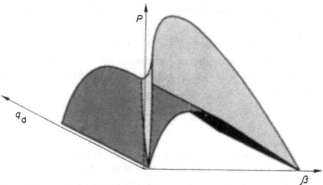

Fig. 10-12 View of the pseudo-equilibrium surface

The continuous lines in *Fig. 10-11* are given by Eq. (10-17) for different values of q_d or β_d. The curve marked with small circles is given by Eq. (10-20). The dashed lines correspond to the structures with geometries shown in the figure. The dash-dotted line connects the maximum points of the continuous lines (critical point locus). Almost the same is shown in *Fig. 10-12* in a three-dimensional co-ordinate system, only the surfaces corresponding the dashed lines are not shown for a clearer representation. The two different behaviours in *Figs. 10-8* and *10-9* can be explained with the help of the pseudo-equilibrium surface, as follows.

If the load is higher than the highest generatrix of the surface which corresponds to the intersection line shown in *Fig. 10-11* by a curve with circles, then the creep increases until the deformation reaches the critical point locus. Then the structure will snap or collapse. However, if the load is lower than the highest generatrix of the surface with generatrices perpendicular to the plane P, β, then the creep will stop after a certain time. At the end of this process the dashpot ceases to carry the load (becomes unloaded) and the two springs are able to keep the structure in equilibrium. Rest will follow.

HAYMAN [1974a, b, 1978a, b, 1980] came to the same conclusion but he did not use *Biot*'s equations. That is, he did not take into account the internal parameters which basically define the whole process. However, his use of the equilibrium paths of the structure has been a very important step in understanding viscoelastic instability. BAŽANT and CEDOLIN [1991] also published results similar to HAYMAN's without referring to him.

Let now the following question be investigated: is there any state when both the numerator and the denominator of Eq. (10-16) become equal to zero at the same time (or rather at the same central angle). For this purpose let the maximum point of the curve, given by Eq. (10-19), be determined. Taking into account Eqs (10-19a) and (10-19c), the result is

$$\frac{k_1 k_2}{k_1 + k_2} L \left(\sin\beta - \sin\beta_{0g}\right) - P \tan\beta = 0. \tag{10-21}$$

The left-hand side of this equation is equivalent to the expression in brackets in the numerator of (10-16). If Eq. (10-21) is differentiated with respect to β, the following expression will be obtained

$$\frac{k_1 k_2}{k_1 + k_2} L \cos\beta - P \frac{1}{\cos^2\beta} = 0, \tag{10-22}$$

which becomes, after some rearrangement,

$$\beta = \arccos \sqrt[3]{\frac{k_1 + k_2}{k_2} \frac{P}{k_1 L}}. \tag{10-23}$$

This equation gives the maximum point of the curve marked with small circles in *Fig. 10-11*. However, the denominator of Eq. (10-16) becomes zero if

$$\beta = \arccos \sqrt[3]{\frac{P}{k_1 L}}. \tag{10-24}$$

The two values are entirely different. This means that the structure either snaps, or reaches rest (equilibrium position) at a limited value of the deformation, so that it can not happen that after an infinite time both the numerator and the denominator become zero.

10.4.3 The total potential energy

The total potential energy of the structure shown in *Fig. 10-7* is

$$V = \frac{1}{2} k_1 \left(q - q_d - q_{0g}\right)^2 + \frac{1}{2} k_2 q_d^2 - PL \left[\sqrt{1 - \left(\frac{q_{0g}}{L}\right)^2} - \sqrt{1 - \left(\frac{q}{L}\right)^2}\right]. \tag{10-25}$$

Differentiating (10-25) two times with respect to q yields the following result:

$$\frac{\partial^2 V}{\partial q^2} = k_1 - \frac{P}{L\left[1 - \left(\frac{q}{L}\right)^2\right]^{3/2}} = \frac{k_1 L \cos^3 \beta - P}{L \cos^3 \beta} \qquad (10\text{-}26)$$

The numerator of this expression becomes zero at the same value of β as the denominator of Eq. (10-16) . This result proves the fact that the necessary condition of buckling of viscoelastic structures corresponds to the case of the vanishing minimum of the total potential energy in the subspace of external co-ordinates. This necessary condition is very similar to the stability condition of elastic structures in the case of conservative forces.

The reasoning in connection with the zero value of the numerator, presented in Section 10.3.3., remains valid also here. That is, if the dashpots are carrying any load, the velocity of external parameters do not become equal to zero. (The exceptions were discussed in Section 10.3.3.) However, if the velocities of all the external parameters become equal to zero, there are three possibilities. First, external restrains hinder the creep; second, the dashpots become free, they do not carry any load. The first case was discussed earlier. The second case corresponds to an elastic structure with an initial modulus of elasticity, and its stability has to be investigated as that of an elastic structure. In the case of the structure shown in *Fig. 10-7*, k_1 means the initial modulus of elasticity. The third case was addressed in Section 10.3.3.

10.4.4 Supplementary remarks

In the preceding sections the symmetric unstable behaviour was investigated in the case of a three-parameter solid. However, the situation is essentially the same in the case of asymmetrical bifurcation on the side of unstable behaviour or in the case of buckling with a limit point.

10.5 STRUCTURES MADE OF DISCHINGER-TYPE MATERIAL, EXHIBITING SYMMETRIC UNSTABLE POST-CRITICAL BEHAVIOUR

10.5.1 Description of the phenomenon

Let the structure shown in *Fig. 10-3* be investigated again but let the *Maxwell* fluid be replaced by a *Dischiger* material. Using the constitutive equation of the material (see *Table 10-1*) as the function of φ, the equation governing the behaviour of the structure becomes

$$\frac{d\beta}{d\varphi} = \frac{P \sin \beta \cos \beta}{k L \cos^3 \beta - P}. \qquad (10\text{-}27)$$

This equation is formally the same as (10-3). The only difference is that the independent variable is φ instead of t. The *Dishinger* material can thus be considered as a *Maxwell* fluid, whose 'time' (i.e. φ) does not increase beyond every limit, but stops at a value φ_{max}.

After integrating and substituting φ as a function of t (see *Table 10-1*) the solution becomes:

$$t = -\frac{1}{\lambda}\ln\left\{1 - \frac{1}{\varphi_{max}}\left[\frac{kL}{P}\cos\beta + \ln\frac{\left(\tan\frac{\beta}{2}\right)^{\frac{kL}{P}}}{\tan\beta}\right]_{\beta_0}^{\beta}\right\}. \qquad (10\text{-}28)$$

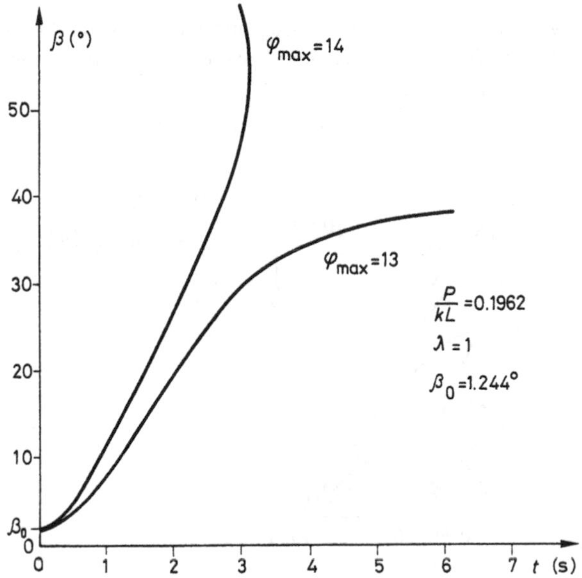

Fig. 10-13 The curves given by Eq. (10-28) for two different φ_{max} values

The results are shown for two different values of φ_{max} in *Fig 10-13*. Let the problem of the 'limit' function $\beta(t)$ be investigated again as was done in Section 10.4.2. For this purpose let Eq. (10-27) be rewritten in a form where for φ the function $\varphi = \varphi_{max}\left(1 - e^{-\lambda t}\right)$ is introduced:

$$\frac{d\beta}{dt} = \frac{\lambda\varphi_{max}\sin\beta\cos\beta\, e^{(-\lambda t)}}{\frac{kL}{P}\cos^3\beta - 1}. \qquad (10\text{-}29)$$

Inserting t from (10-28) the result is

$$\frac{d\beta}{dt} = \frac{\lambda\left\{\left[\varphi_{max} - \frac{kL}{P}\left(\cos\beta + \ln\tan\frac{\beta}{2}\right) - \ln\tan\beta\right]_{\beta_0}^{\beta}\right\}}{\frac{kL}{P}\cos^3\beta - 1}\sin\beta\cos\beta. \qquad (10\text{-}30)$$

To decide whether both the numerator and the denominator become zero at the same time, it has to be taken into account that $\beta_0 = 1.244296804$ and the denominator takes its zero value at $\beta_0 = \arccos(0.1962)^{-1/3} = 54.47373571$. Expressing φ_{max} from the numerator of (10-30), the result is 13.6701962. So there is a φ_{max} which gives a 'limit' solution. That is the structure buckles at the same time when it reaches its resting state at $t = $ infinite. This is an important difference between the behaviours of our structures made of three-parameter solid or of *Dischinger* material. It has to be mentioned that this solution is very sensitive numerically, because of its limit situation.

10.5.2 The pseudo-equilibrium surface

Because the *Dischinger* material is a linear spring and an ageing dashpot connected in series, the structure of *Fig. 10-3* can be considered again. The equation which defines the pseudo-equilibrium surface (*Biot*'s equation) is

$$kL\left[\sin\beta - \sin\left(\beta_d - \beta_{0g}\right)\right]\cos\beta - P\sin\beta = 0 \qquad (10\text{-}31)$$

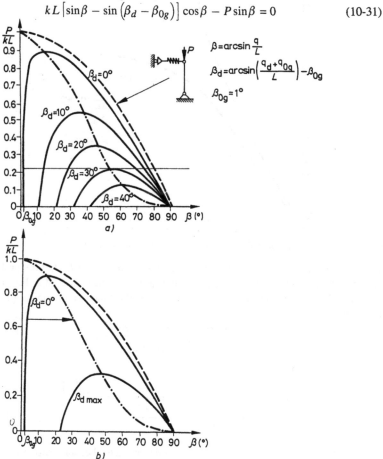

$$\beta = \arcsin\frac{q}{L}$$

$$\beta_d = \arcsin\left(\frac{q_d + q_{0q}}{L}\right) - \beta_{0g}$$

$$\beta_{0g} = 1°$$

Fig. 10-14 The intersection lines of the pseudo-equilibrium surface of the structure made of *Dischinger* material with planes parallel to the $(P/kL, \beta)$ plane

Let the two types of behaviour, shown in *Fig. 10-13*, be investigated with the help of this equation. If the time rate of aging is quicker than the velocity of creep, the deformation will stops before the structure will collapse (*Fig 10-14a*), and the structure remains in a stable equilibrium state. On the other hand, if the aging is slower than the velocity of creep, the structure will collapse (*Fig. 10-14b*).

10.5.3 The total potential energy

It can be realized with the help of Sections 10.3.3 and 10.3.4, without any detailed explanation, that the necessary condition of buckling is, also in the case of a *Dischinger* material, the vanishing minimum of the total potential energy in the subspace of external co-ordinates [IJJAS, 1994*b*].

If the numerator of the right-hand side of Eq. (10-30) becomes zero, this means that the dashpot becomes rigid. So the structure behaves from this moment on, elastically, and its stability can be determined as that of an elastic structure. In this case the value (or, more precisely, the sign) of the denominator is determinative, because this defines whether the total potential energy has a minimum or not.

If both the numerator and the denominator become zero at the same time, then it becomes an elastic structure which just has lost its stability. This is true also in the case of several degrees of freedom.

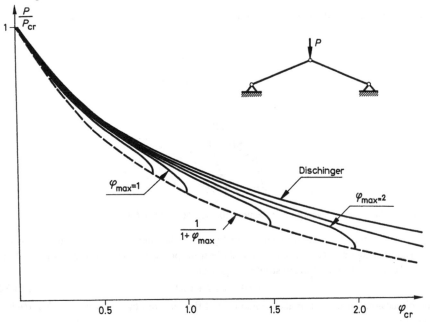

Fig. 10-15 φ_{cr} values belonging to the snap-through of a *von Mises* truss made of three-parameter solid and of *Dischinger* material.

10.5.4 Supplementary remarks

It is worth mentioning the study of SZALAI [1989] who investigated the behaviour of *von Mises* trusses made from *Dischinger* material and from a three-parameter solid. The trusses were loaded by a concentrated force of a constant value at the mid-point hinge. She used the approach $\cos(x) = 1 - x^2/2$, which means that she investigated the behaviour of the model by a second-order approximation. (In all earlier derivations there were no approximation made in the geometric equations.) Her results are shown in *Fig. 10-15*. To interpret these it must be remembered that φ_{max} is the total deformation minus the elastic deformation divided by the elastic deformation. So, in the case of a three-parameter solid, $\varphi_{max} = k_1/k_2$. The horizontal co-ordinate in *Fig. 10-15* gives the critical φ values at which the truss buckles. The vertical co-ordinate gives the proportion of the total load to the elastic critical load of the truss. From this figure it can be seen that the *Dischinger* material gives only one function as the upper limit of the load carrying capacity, and one φ_{cr} defines one P/P_{cr}, independently of φ_{max}. On the other hand, in the case of the three-parameter solid, the function $P/P_{cr}(\varphi_{cr})$ depends on the value of φ_{max}. The lower limit of the load-carrying capacity is, of course, $1/(1 + \varphi_{max})$.

10.6 STRUCTURES MADE OF FLUID-TYPE MATERIAL, EXHIBITING SYMMETRIC STABLE POST-CRITICAL BEHAVIOUR

10.6.1 Description of the phenomenon

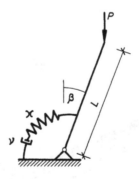

Fig. 10-16 The structure investigated in Sections 10.6.1, 10.6.2 and 10.6.3

Let the behaviour of the structure shown in *Fig. 10-16* be considered. The structure consists of a rigid bar again, free to rotate about its bottom point, and whose rotation is hindered by a *Maxwell* fluid with the constitutive equation:

$$v\kappa\dot{\beta} = \kappa M + v\dot{M} \tag{10-32}$$

The dimensions of the constants are naturally different from those given in *Table 10-1*.

That is, the dimension of κ is moment divided by angle (radian), the dimension of ν is moment multiplied by time and divided by angle (radian). The moment of the force P about the hinge is

$$M = PL\sin\beta \tag{10-33}$$

Inserting this into Eq. (10-32), after some rearrangement the following differential equation is obtained:

$$\dot\beta = \frac{\kappa}{\nu}\frac{PL\sin\beta}{\kappa - PL\cos\beta} = \frac{\kappa}{\nu}\frac{\sin\beta}{\dfrac{\kappa}{PL} - \cos\beta} \tag{10-34}$$

This is a separable equation whose solution is:

$$t = \frac{\nu}{\kappa}\int_{\beta_0}^{\beta}\frac{\kappa - PL\sin\beta}{PL\sin\beta}\,\mathrm{d}\beta = \frac{\nu}{\kappa}\left[\ln\frac{\left(\tan\frac{\beta}{2}\right)^{\frac{\kappa}{PL}}}{\sin\beta}\right]_{\beta_0}^{\beta} \tag{10-35}$$

If β is equal to zero, then β_0 is also equal to zero. So the equation yields a trivial (zero) solution, because the upper and lower bounds of the integral coincide, thus no creep occurs. If $PL/\kappa < 1$, then $\kappa/(PL) > 1$ and the sign of $\mathrm{d}\beta/\mathrm{d}t$ depends only on the numerator of Eq. (10-34) or on the sign of β. It is clear that the denominator never becomes zero, so the velocity of the central angle never becomes infinite. Consequently, the structure will not buckle. $\kappa/(PL) = 1$ is a limit state corresponding to the bifurcation point. If $(PL)/\kappa > 1$, i.e. $\kappa/(PL) < 1$ the problem no longer belongs to viscoelastic stability, but to elastic stability. A curve given by Eq. (10-35) is shown in *Fig. 10-17*. The parameters of this solution are given there, too.

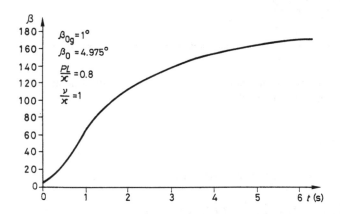

Fig. 10-17 The curve given by Eq. (10-35)

10.6.2 The pseudo-equilibrium surface

Let *Biot*'s equations of the structure shown in *Fig. 10-16* be written as

$$PL\sin\beta - \kappa\left(\beta - \beta_d - \beta_{0g}\right) = 0 \qquad (10\text{-}36a)$$

and

$$v\dot{\beta} - \kappa\left(\beta - \beta_d - \beta_{0g}\right) = 0, \qquad (10\text{-}36b)$$

where

β_{0g} is the initial imperfection

β_d is the rotation of the dashpot.

In the following only Eq. (10-36a) is necessary, which after some mathematical operation becomes:

$$\frac{PL}{\kappa} = \frac{\beta - \beta_d - \beta_{0g}}{\sin\beta}. \qquad (10\text{-}37)$$

This equation gives again a surface in the PL/κ, β, β_{0g} co-ordinate system and the intersection curves with planes parallel to the PL/κ,β co-ordinate plane are given in *Fig. 10-18*.

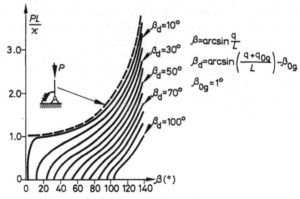

Fig. 10-18 The intersection lines of the pseudo-equilibrium surface of the structure shown in *Fig. 10-16* with planes parallel to the $(PL/\kappa,\beta)$ plane

Observing this figure it can be stated that, since the surface has no critical point locus during full rotation, the structure will not buckle.

10.6.3 The total potential energy

The total potential energy of the structure of *Fig. 10-16* is:

$$V = \frac{1}{2}\kappa\left(\beta - \beta_d - \beta_{0g}\right)^2 - PL\left(\cos\beta_{0g} - \cos\beta\right). \qquad (10\text{-}38)$$

Differentiating it two times with respect to β results in the following:

$$\frac{\partial V^2}{\partial \beta^2} = \kappa - PL\cos\beta \qquad (10\text{-}39)$$

If κ is larger than *PL*, the right-hand side of (10-39) will be positive for every value of β. That is, the total potential energy will always be a minimum, so the structure will not buckle (*Fig. 10-19*). Attention has to be called for the similarity of the right-hand side of (10-39) and the denominator of the right-hand side of (10-34).

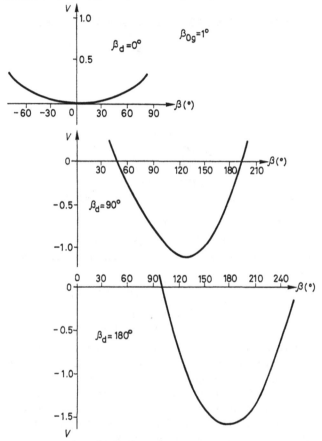

Fig. 10-19 The total potential energy for various constant β_d values

10.6.4 Supplementary remarks

A behaviour similar to the stable-symmetric one can be observed in the case of structures with asymmetric bifurcation on the side of stable behaviour. If the *Maxwell* fluid is replaced by a *Dischinger* material, the behaviour of the structure is also similar. However, if the aging of the dashpot is quicker than the velocity of the central angle of rotation, then the creep will stop before the structure reaches its lowest point. In this sense the behaviour of the structure is similar to that which will be investigated in the next section.

10.7 STRUCTURES MADE OF SOLID-TYPE MATERIAL, EXHIBITING SYMMETRIC STABLE POST-CRITICAL BEHAVIOUR

10.7.1 Description of the phenomenon

Let the structure shown in *Fig. 10-20* be investigated. In this case the constitutive equation

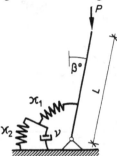

Fig. 10-20   The structure investigated in Sections 10.7.1, 10.7.2 and 10.7.3

of the three-parameter solid is

$$(\kappa_1 + \kappa_2) M + \nu \dot{M} = \kappa_1 \kappa_2 \left(\beta - \beta_{0g}\right) + \nu \kappa_2 \beta \qquad (10\text{-}40)$$

Taking into account that $M = PL\sin\beta$, after some mathematical operations the following differential equation is obtained:

$$\dot{\beta} = \frac{\dfrac{\kappa_1 + \kappa_2}{\kappa_2} \dfrac{PL}{\kappa_1} \sin\beta - \left(\beta - \beta_{0g}\right)}{\dfrac{\nu}{\kappa_2}\left(1 - \dfrac{PL}{\kappa_1}\cos\beta\right)} \qquad (10\text{-}41)$$

The investigation of Eq. (10-41) shows that the velocity of the angle of rotation β will become infinite only if the numerator is limited, and if $PL/\kappa_1 \geq 1$. If the denominator is equal to zero, then buckling occurs when the load acting on the structure is equal to the (elastic) critical load of the structure, determined with the spring having a constant κ_1. If $PL/\kappa_1 \leq 1$ the denominator will never become zero. In this case the structure never loses its stability, and if it has a given initial imperfection, after some time it will come to rest. Here the study of HUANG [1973] has to be mentioned.

10.7.2 The pseudo-equilibrium surface

Let a structure be investigated similar to the one in *Fig. 10-20* which is made of a rigid bar, loaded by a force P at the top of the rigid bar, and supported by a spring κ_1, which hinders the rotation of the rigid bar about the hinge at the bottom. Let its equilibrium be defined as:

$$\kappa_1 \left(\beta - \beta_d - \beta_{0g}\right) - PL\sin\beta = 0. \qquad (10\text{-}42)$$

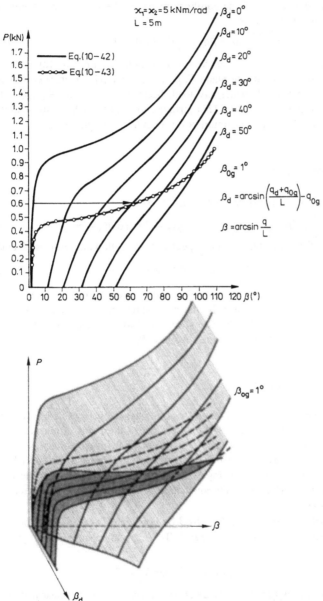

Fig. 10-21 *a*) The intersection lines of the pseudo-equilibrium surface of the structure shown in *Fig. 10-16* with planes parallel to the (P, β) plane, and the curve given by Eq. (10-43), (marked by circles), *b*) view of the pseudo-equilibrium surface.

Furthermore let the equation of the equilibrium paths of the structure of *Fig. 10-20*, neglecting the dashpot, be written as:

$$\frac{\kappa_1 \kappa_2}{\kappa_1 + \kappa_2} \left(\beta - \beta_{0g} \right) - PL \sin \beta = 0. \tag{10-43}$$

In *Figs. 10-21a,b* the intersection lines of the surfaces, given by Eqs (10-42) and (10-43), with planes parallel to the co-ordinate plane P, β, are shown. This picture clearly shows that the structure does not buckle under any load, because the function $P(\beta, \beta_d)$ always reaches the surface given by Eq. (10-43). This surface corresponds to the elastic equilibrium and, because it does not have any maximum point, buckling is impossible.

10.7.3 The total potential energy

The total potential energy of the structure of *Fig. 10-20* is

$$V = \frac{1}{2}\kappa_1 \left(\beta - \beta_d - \beta_{0g} \right)^2 - \frac{1}{2}\kappa_2 \beta_d^2 - PL \left(\cos \beta_0 - \cos \beta \right). \tag{10-44}$$

Differentiating Eq. (10-44) two times with respect to β the following result is obtained:

$$\frac{d^2 V}{d\beta^2} = \kappa_1 - PL \cos \beta. \tag{10-45}$$

The similarity of the right-hand side of Eq. (10-45) and the denominator of the right-hand side of Eq. (10-41) is clear. More precisely, the two expressions become zero at the same central angle. This means again that the necessary condition of buckling is the vanishing minimum of the total potential energy in the subspace P, β.

10.8 CREEP OF THE DASHPOT

So far the phenomenon of viscoelastic stability and its physical background have been explained. Accordingly, the creep of the dashpot was not in the focus of our attention and its value was not determined. However, in some numerical investigations, the determination of the creep of the dashpot can not be avoided. The typical solution of buckling of viscoelastic structures becomes meaningless after the critical time elapses (because the inert mass was neglected). This has a consequence in the determination of the creep of the dashpot [IJJAS, 1994a].

To show this, let the structure of *Fig. 10-3* be considered again. The function $\beta(t)$ is defined by Eq. (10-5). With the help of Eq. (10-10) and after some rearrangement the relation

$$\beta_d = \arcsin \left(\sin \beta - \frac{P}{\kappa L} \tan \beta \right) - \beta_{0g} \tag{10-46}$$

is obtained. The functions $\beta_d(\beta)$ and $\beta_d(t)$ are shown in *Fig. 10-22*. It is clear that the $\beta_d(t)$ function is physically meaningless 'after the peak'. The results, especially the function $\beta_d(t)$, show the numerical problems which may arise during solution.

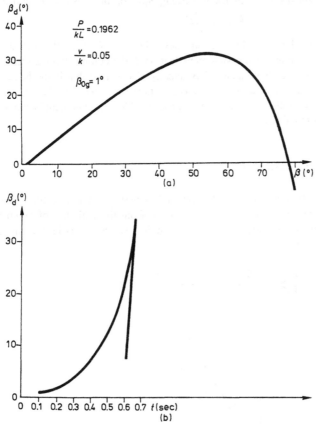

$\beta_d (^\circ)$

$\dfrac{P}{kL} = 0.1962$

$\dfrac{v}{k} = 0.05$

$\beta_{0g} = 1^\circ$

(a)

$\beta (^\circ)$

$\beta_d (^\circ)$

(b)

$t (\text{sec})$

Fig. 10-22   *a*) The function $\beta_d(\beta)$ in the case of the structure shown in *Fig. 10-3, b*) the function $\beta_d(t)$ of the creep of the dashpot in dependence of time

10.9 TWO REMARKS ABOUT THE PROBLEMS APPEARING IN THE LITERATURE

10.9.1 Importance of the degree of approximations

During the investigation it is very important to decide how the behaviour of the structure can be approximated to obtain acceptable results. Let the structure shown in *Fig. 10-3* be considered again. Confining our investigation to small displacements, $\sin\beta$ is approximated by β and $\cos\beta$ by 1. Thus the geometric equation becomes:

$$q = L\beta \qquad (10\text{-}47)$$

and the equilibrium equation is:

$$-QL + PL\beta = 0 \qquad (10\text{-}48)$$

After the usual mathematical operations the governing equation of the structure will be:

$$\dot{\beta} = \frac{kP\beta}{b(kL - P)} \tag{10-49}$$

From Eq. (10-49) it is evident that the structure never buckles if $kL > P$. Such an approximate investigation resulted in the statement that an *Euler* column with a small initial imperfection never buckles if its material is linearly viscoelastic, but if the behaviour of the dashpot is nonlinear, the *Euler* column has a critical time (see e.g. [HOFF, 1958]). ŻYCZKOWSKI [1962] showed that this statement is unacceptable. His statement is supported by the paper of VINOGRADOV [1985].

10.9.2 Importance of the elastic behaviour

In the literature the elastic behaviour of the material is very often neglected, i.e. it is assumed that the material is viscous. It is well known from the definiton of creep that infinite velocity of creep occurs only if the load is infinite. To show this let the structure in *Fig. 10-23* be considered.

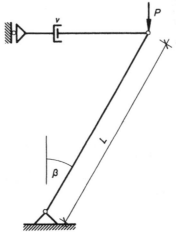

Fig. 10-23 The structure investigated in Section 10.9.2

In this case the rigid bar with length L is supported by a Newtonian fluid. Taking into account the constitutive equation of the Newtonian fluid [see *Table 10-1* and Eqs (10-1) and (10-2)], after the usual mathematical operations, the governing equation of the structure shown in *Fig. 10-23* takes the form:

$$\dot{\beta} = \frac{P}{bL}\frac{\sin\beta}{\cos^2\beta} \tag{10-50}$$

This shows that the velocity of the rotation of the central angle becomes infinite only if $\beta = \pi/2$. In the earlier investigations it was shown that creep buckling occurs at a smaller central angle if the elastic response of the material is taken into account. So neglecting the elastic response overestimates the critical time.

11

Buckling under dynamic loading

Lajos KOLLÁR

11.1 DESCRIPTION OF THE DYNAMIC LOADING PROCESS

Loading is called dynamic if it is not applied 'infinitely slowly' (as assumed in the case of static loading) but suddenly. Such a loading may be caused by impact, caused e.g. by a mass dropped onto the bar. It can be seen by simple engineering consideration that, on the one hand, dynamic loading is more detrimental to structures than static loading because a dynamic load causes greater internal forces than a static load of the same magnitude. On the other hand, the fact that the dynamic load acts only for a limited time may be advantageous, because it may exceed the static critical load without destroying the structure.

To investigate the effects of dynamic loading we have to take the inert mass of the structure into account, i.e. we have to use the equations of movement. So far researchers have investigated mostly the simple bar and dealt with two basic problems: they considered the variation in time of either the force or the displacement of the bar end. The first case represents impact loading, and the second the behaviour of a bar tested in the compression machine. They assumed in both cases that the bar has an initial crookedness. By so doing, they avoided both the eigenvalue problem and the difficulties of transition from the straight to the curved bar [KONING and TAUB, 1933; TAUB, 1933; MEIER, 1945; HOFF, 1951; DAVIDSON, 1953; SEVIN, 1960; HOFF, 1965; HOLZER and EUBANKS, 1969; ELISHAKOFF, 1978].

The movement equations of the dynamically loaded, initially curved elastic bar (*Fig. 11-1a*) can be written as follows [HOFF, 1951; DAVIDSON, 1953; SEVIN, 1960].

Equilibrium in the direction y perpendicular to the bar axis:

$$EI\eta'''' + \left[N\left(y' + \eta'\right)\right]' + \rho A\ddot{\eta} = 0. \tag{11-1a}$$

Equilibrium parallel to the bar axis:

$$-N' = \rho A\ddot{\zeta}. \tag{11-1b}$$

Geometric relation between the displacements and the elongation [cf. *Fig. 11-1b* and Eq. (11-3)]:

$$\varepsilon_z = -\frac{N}{EA} = \zeta' + \frac{1}{2}\left[\left(y' + \eta'\right)^2 - \left(y'\right)^2\right] = \zeta' + y'\eta' + \frac{1}{2}\left(\eta'\right)^2. \tag{11-1c}$$

The notations are (see also *Fig. 11-1a*):

The notations are (see also *Fig. 11-1a*):

$y(z)$ initial shape of the bar

$\eta(z,t)$ additional displacements of the bar in direction y, due to the dynamic load

$\zeta(z,t)$ displacements of the bar cross sections in direction z

$N(z,t)$ compressive force acting in the bar (positive if compression)

A area of cross section of the bar

I moment of inertia of the cross section of the bar

E modulus of elasticity

ρ density of the bar material

$' = \partial/\partial z$ partial derivative with respect to z

$\cdot = \partial/\partial t$ partial derivative with respect to time.

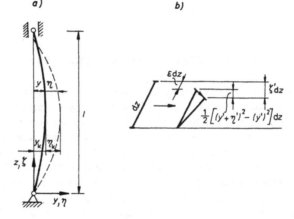

Fig. 11-1 Notations

These three equations – together with the initial and boundary conditions – determine the three unknown functions: $\eta(z,t)$, $\zeta(z,t)$ and $N(z,t)$. The solution can be obtained numerically [DAVIDSON, 1953; SEVIN, 1960].

From the results of DAVIDSON and SEVIN, we can draw some basic conclusions. Of these, the most important is that the longitudinal shock wave of compression travels much faster in the bar than the bending vibration connected with the lateral displacements. Hence we can assume that the compressive force is constant along the bar length and varies only with time. The investigations of SEVIN also showed that, as a rule, the shortening of the bar can also be neglected in comparison to the bending deformation. The consequence of this assumption is that we cannot investigate the straight bar but – as mentioned before – we have to give the bar a certain initial curvature. Hence the result of our investigation will also not be a critical load but a relation by which the maximum lateral deflection and bending moment caused by the impact can be computed. The task to be solved thus ceases to be a stability problem, but becomes that of strength of materials.

The above assumptions greatly simplify the otherwise complicated phenomenon and enable us to solve the problem not only in knowledge of the variation in time of the dynamic load, but also if we merely know the mass and velocity of the falling body, but not the variation in time of the impact force. However, we still need some additional assumptions. First, we have to make the process of impact univocal. To this purpose we need the *coefficient of resilience* (see Section 11.3), which is different for every material, even for every body. Hence, for the sake of simplicity, we first assume that the impact is perfectly plastic, i.e. the falling body does not rebound from the bar but adheres to it and moves further as attached to the bar. Later we will generalize our solution to impacts which are not perfectly plastic.

Second, we have to assume that the impact process is terminated before the movement (bending deformation) of the bar begins.

We will neglect the influence of the weight of the bar on its bending deformation.

The problem of buckling of a bar under a falling load has been investigated by UNGER [1984], but he treated only perfectly elastic impacts.

11.2 BUCKLING OF AN INITIALLY CURVED BAR UNDER A FALLING LOAD

Let us investigate the buckling of an initially curved, hinged-hinged bar with a continuous mass distribution, shown in *Fig. 11-2*, under a falling load with the mass m_0 and the arriving velocity v_0. The total mass of the bar be m_{bar}, its initial shape be a half sine wave:

$$y = y_m \sin \frac{\pi}{l} z. \tag{11-2a}$$

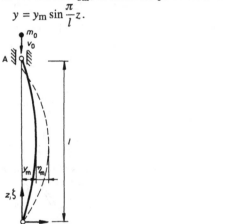

Fig. 11-2 The bar model with continuous mass distribution

According to our assumptions the compressive force can be considered as constant along the bar length, so that the additional deformation of the bar has also the shape of a sine curve:

$$\eta = \eta_m \sin \frac{\pi}{l} z. \tag{11-2b}$$

For the perfectly plastic impact we have to apply the impulse theorem (conservation of momentum), since a part of the energy gets lost and thus the conservation of energy theorem is not valid. To this purpose we need the vertical and horizontal velocities of the points of the bar axis after impact, as functions of the vertical velocity v_{zA} of the upper bar end A.

These velocities are proportional to the displacements ζ and η (assumed as small) of these points.

Considering the assumed inextensional deformation of the bar axis, the following relation between the horizontal displacements η and the vertical displacements ζ of the points of the bar axis can be established, taking into account Eqs (3-13) and (3-14):

$$\zeta' = \frac{1}{2}\left[(y' + \eta')^2 - (y')^2\right] = y'\eta' + \frac{1}{2}(\eta')^2. \tag{11-3}$$

Introducing here (11-2a) and (11-2b):

$$\zeta' = \frac{\pi^2}{l^2}\left(y_m\eta_m + \frac{1}{2}\eta_m^2\right)\cos^2\frac{\pi}{l}z, \tag{11-4}$$

$$\zeta(z) = \int_0^z \zeta'\,dz = \frac{\pi^2}{2l}\left(y_m + \frac{1}{2}\eta_m\right)\eta_m\left(\frac{z}{l} + \frac{1}{2\pi}\sin\frac{2\pi}{l}z\right). \tag{11-5}$$

Confining our investigation to small displacements we have $\eta \ll y$ and thus $y_m + \frac{1}{2}\eta_m \approx y_m$. Since we are interested now in the *initial* velocities after impact, the assumption of small displacements is justified. So we have:

$$\zeta(z) = \frac{\pi^2}{2}\frac{y_m}{l}\eta_m\left(\frac{z}{l} + \frac{1}{2\pi}\sin\frac{2\pi}{l}z\right), \tag{11-6}$$

and since v_z and v_y are proportional to ζ and η, respectively:

$$v_z(z) = \frac{\pi^2}{2}\frac{y_m}{l}v_{ym}\left(\frac{z}{l} + \frac{1}{2\pi}\sin\frac{2\pi}{l}z\right). \tag{11-7}$$

At $z = l$:

$$v_{zA} = \frac{\pi^2}{2}\frac{y_m}{l}v_{ym}. \tag{11-8}$$

Introducing this into (11-7):

$$v_z(z) = v_{zA}\left(\frac{z}{l} + \frac{1}{2\pi}\sin\frac{2\pi}{l}z\right). \tag{11-9}$$

The distribution of these vertical velocities is shown in *Fig. 11-3*.

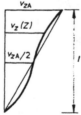

Fig. 11-3 The distribution of the vertical velocities

The horizontal velocities are proportional to η (Eq. (11-2b)). Taking into account also (11-8) we obtain:

$$v_y(z) = v_{ym}\sin\frac{\pi}{l}z = \frac{2l}{\pi^2 y_m}v_{zA}\sin\frac{\pi}{l}z. \tag{11-10}$$

The duration of the impact be τ, and, for the sake of simplicity, we will write for the impulse $N\tau$, instead of $\int N \mathrm{d}\tau$. After this preparation we can write the impulse theorem for the impact, see *Fig. 11-4*.

In the *vertical direction*:

$$\frac{m_{bar}}{l} \int_0^l v_z(z)\mathrm{d}z + m_0 v_z A - m_0 v_0 = -N\tau. \qquad (11\text{-}11a)$$

According to *Fig. 11-3*, the value of the integral is $\frac{v_z A l}{2}$, thus (11-11a) can be written as

$$\left(\frac{m_{bar}}{2} + m_0\right) v_z A - m_0 v_0 = -N\tau. \qquad (11\text{-}11b)$$

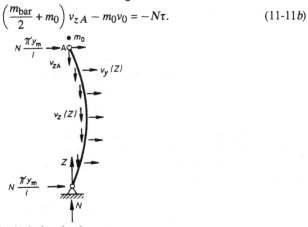

Fig. 11-4 The quantities appearing in the impulse theorem

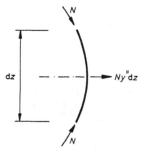

Fig. 11-5 The horizontal force resulting from the compressive force due to the curvature of the bar

In the *horizontal direction*: According to *Fig. 11-5*, a horizontal force $Ny''\mathrm{d}z$ acts on every bar section of the length $\mathrm{d}z$. From Eq. (11-2a) we have:

$$y'' = -y_m \frac{\pi^2}{l^2} \sin\frac{\pi}{l}z. \qquad (11\text{-}12)$$

(The minus sign has to be omitted, since N is considered positive when compression.)

Thus we can write the impulse theorem for the section $\mathrm{d}z$:

$$\frac{m_{bar}}{l}\mathrm{d}z\, v_y(z) = N\tau\, y''\mathrm{d}z, \qquad (11\text{-}13a)$$

that is, with the aid of Eq. (11-10) and (11-12):

$$\frac{m_{bar}}{l}dz\frac{2l}{\pi^2 y_m}v_{zA}\sin\frac{\pi}{l}z = N\tau\frac{\pi^2}{l^2}y_m\sin\frac{\pi}{l}z\,dz. \qquad (11\text{-}13b)$$

We obtain the impulse theorem for the whole bar by integration:

$$m_{bar}v_{zA}\frac{4l}{\pi^3 y_m} = 2N\tau\frac{\pi y_m}{l}. \qquad (11\text{-}13c)$$

As can be seen, at both ends of the bar a horizontal force of the value $N\pi y_m/l$ has to act for a time τ (*Fig. 11-4*); these give, together with the vertical force N, a resultant falling in the tangent of the bar axis. We can express $N\tau$ from Eq. (11-13) and introduce it into (11-11b):

$$\left(\frac{m_{bar}}{2} + m_0 + m_{bar}\frac{2}{\pi^4}\frac{l^2}{y_m^2}\right)v_{zA} - m_0 v_0 = 0. \qquad (11\text{-}14a)$$

From this we obtain the vertical velocity of point A after impact:

$$v_{zA} = \frac{m_0 v_0}{m_0 + \dfrac{m_{bar}}{2}\left(1 + \dfrac{4l^2}{\pi^4 y_m^2}\right)}. \qquad (11\text{-}14b)$$

In order to obtain the internal forces in the bar we have to use the theorem of conservation of energy, since we assume that the deformation after impact is perfectly elastic, so that no energy gets lost. We want to determine the maximum bending moment of the bar. The kinetic energy of the bar is:

$$E = \frac{m_0}{2}v_{zA}^2 + \frac{m_{bar}}{2l}\int_0^l v_z^2 dz + \frac{m_{bar}}{2l}\int_0^l v_y^2 dz. \qquad (11\text{-}15)$$

The first integral becomes, using (11-9):

$$\int_0^l v_z^2 dz = v_{zA}^2 l\left(\frac{1}{3} - \frac{3}{8\pi^2}\right) = 0.2954 v_{zA}^2 l,$$

and the second integral is, on the basis of (11-10):

$$\int_0^l v_y^2 dz = \frac{4l^2}{\pi^4 y_m^2}v_{zA}^2\frac{l}{2}.$$

We thus have:

$$E = \left(\frac{m_0}{2} + m_{bar}0.1477 + m_{bar}\frac{l^2}{\pi^4 y_m^2}\right)v_{zA}^2. \qquad (11\text{-}16)$$

The strain energy of bending of the bar is:

$$U = \frac{1}{2EI}\int_0^l M^2 dz. \qquad (11\text{-}17a)$$

Since the bending moment varies according to the relation $M = M_m \sin\frac{\pi}{l}z$, thus

$$U = \frac{M_m^2 l}{4EI}. \qquad (11\text{-}17b)$$

Setting $E = U$:

$$M_m = 2v_{zA}\sqrt{\frac{EI}{l}\left[\frac{m_0}{2} + m_{bar}\left(0.1477 + m_{bar}\frac{l^2}{\pi^4 y_m^2}\right)\right]}. \qquad (11\text{-}18a)$$

Introducing the value of v_{zA} from (11-14b):

$$M_m = \frac{2m_0 v_0}{m_0 + \frac{m_{bar}}{2}\left(1 + \frac{4l^2}{\pi^4 y_m^2}\right)}\sqrt{\frac{EI}{l}\left[\frac{m_0}{2} + m_{bar}\left(0.1477 + \frac{l^2}{\pi^4 y_m^2}\right)\right]}, \qquad (11\text{-}18b)$$

and the maximum deflection of the middle cross section of the bar is:

$$\eta_m = \frac{l^2}{\pi^2}\frac{M_m}{EI} \qquad (11\text{-}19)$$

As can be seen, the bending moment increases if we increase either the bending rigidity EI/l of the bar or the relative eccentricity y_m/l. The assumption of neglecting the shortening of the bar axis becomes less and less justified when the value y_m/l approaches zero.

11.3 GENERALIZATIONS

If the impact is *not perfectly plastic*, then it can be characterized by the *coefficient of resilience* Ψ which defines the ratio of the impulse parts coming about in the phases of compression and resilience, respectively:

$$\Psi = \frac{I_2}{I_1}, \qquad (11\text{-}20)$$

where

$$I_2 = \int_{\tau_1}^{\tau_2} P dt$$

and

$$I_1 = \int_{\tau_0}^{\tau_1} P dt$$

are the impulses acting in the phase of resilience and compression, respectively. With a perfectly plastic impact $\Psi = 0$, and with a perfectly elastic one $\Psi = 1$.

That is, with a not perfectly plastic impact, the bar is acted upon by the impulse part I_1 of the compression phase and, in addition, by the impulse part $I_2 = \Psi I_1$ of the resilience phase, during which the mass m_0 acquires a velocity of a sign opposite to the arriving one. Since all dynamic quantities are caused by the impulse acting during impact (see Eqs (11-11) and (11-13)), and the impulse is linearly proportional to the change of velocities, i.e. in the present case to the velocities of the bar after impact, so if $\Psi > 0$, we have to multiply all velocities derived in Section 11.2 by $(1 + \Psi)$. However, after an imperfectly plastic impact, the mass m_0 detaches from the bar, so that its kinetic energy should be subtracted from the expression (11-16), i.e. we have to omit the term $m_0 v_{zA}^2 / 2$. (We assume that the rebouncing mass m_0 falls back to the bar later than the maximum bending moment develops, so that its kinetic energy need not be taken into account when computing the maximum bending moment.) Hence Eq. (11-18b) takes the following modified form, taking into account that the bending moment is also linearly proportional to the velocities

$$M_m = \frac{2m_0 v_0 (1 + \Psi)}{m_0 + \dfrac{m_{bar}}{2} \left(1 + \dfrac{4l^2}{\pi^4 y_m^2}\right)} \sqrt{\frac{EI}{l} m_{bar} \left(0.1477 + \frac{l^2}{\pi^4 y_m^2}\right)}. \tag{11-21}$$

The coefficient of resilience can be determined by the well-known drop test (*Fig. 11-6*)

$$\Psi = \sqrt{\frac{h_2}{h_1}}. \tag{11-22}$$

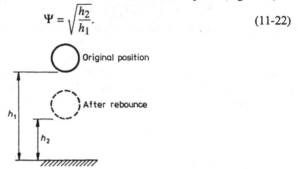

Fig. 11-6 Determination of the coefficient of resilience by the drop test

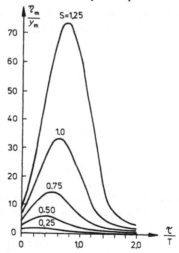

Fig. 11-7 The maximum deflection as a function of the duration of impact in the cases of various impulse magnitudes

We explained in Section 11.1 why we can take the *compressive force constant* along the bar. The *incompressibility of the bar*, however, is not an entirely justified assumption: its validity decreases with the decreasing ratio y_m / l. Nevertheless we have to observe that with this assumption we are on the safe side, since the bar is considered as stiffer than in reality and thus the impulse of the impact becomes greater.

Finally we have to assess the assumption that the *process of impact ends before the movement of the bar starts*. The validity of this statement has been investigated by KONING and TAUB [1933]. We reproduce one of their diagrams in *Fig. 11-7*, which gives the maximum deflection η_m as a function of the impact time τ, due to various impulses. The period of vibration of the bar is T, the impulse is measured by the quantity $S = \frac{N}{N_E}\frac{\tau}{T}$, where N_E is the *Euler* critical force of the bar. They considered the compressive force in the bar as constant. The curves show that as long as the maximum deflection η_m does not exceed the four- to fivefold of the initial curvature amplitude y_m, the deflection (and the bending moment) depends, in fact, only on the magnitude of the impulse, and is independent of the duration of the impact.

12

Stability paradoxes

Lajos KOLLÁR

There are several problems of stability, the solution of which does not comply with sound engineering sense. Some of these problems cease to be paradoxes if we thoroughly study them and consider points of view hitherto not included. Other problems, however, cannot be explained in a visual way, at least not yet.

We shall treat the problems in two groups. In the first group are those structures whose stability behaviour differs from common engineering sense. In the second group we gathered structures whose critical load increases when softened and decreases when stiffened.

12.1 STRUCTURES BEHAVING DIFFERENTLY FROM COMMON ENGINEERING SENSE

12.1.1 Instability of blown-up rubber balloons

The surprising feature of this kind of instability is that it is caused by tension.

The phenomenon can be shown in the simplest way on a complete sphere. Let us assume that the material of the sphere is perfectly elastic and has an unlimited capability to elongate. Let us apply an overpressure p in the interior of the sphere, which causes a tension

$$N_m = N_h = \frac{pR}{2} \tag{12-1}$$

in the shell wall [CSONKA, 1987], with N_m and N_h as the meridional and hoop stresses, respectively, and R as the actual radius of the sphere (*Fig. 12-1*).

These tensile forces cause an elongation

$$\varepsilon_m = \varepsilon_h = \varepsilon = \frac{N_m}{Et_0} - v\frac{N_h}{Et_0} = \frac{N_m}{Et_0}(1-v) \tag{12-2}$$

where t_0 is the original thickness of the shell wall, and v is *Poisson*'s ratio. Hence the original radius R_0 increases to

$$R = R_0(1 + \varepsilon). \tag{12-3}$$

Expressing ε from (12-3) and introducing it into (12-2) yields:

$$N = \frac{Et_0}{1-v}\left(\frac{R}{R_0} - 1\right). \tag{12-4}$$

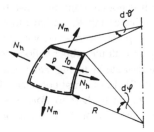

Fig. 12-1 The forces acting on an elementary part of the shell

If we set this equal to (12-1) we obtain the actual radius R as a function of p:

$$R = \frac{1}{\dfrac{1}{R_0} - \dfrac{1-\nu}{2Et_0}p}. \tag{12-5}$$

Hence if

$$\frac{1}{R_0} = \frac{1-\nu}{2Et_0}p,$$

that is if:

$$p = p_{cr} = \frac{2Et_0}{R_0(1-\nu)}, \tag{12-6}$$

then $R \to \infty$, so that the balloon cannot remain in equilibrium, but its radius increases to infinity [TROSTEL, 1962].

The phenomenon can be visualized in *Fig. 12-2*. The tensile force N_p due to the overpressure p, and the tensile force N_m taken by the membrane material, are given by Eqs (12-1) and (12-4) respectively, both as functions of R.

Plotting these in the co-ordinate system (N, R) we obtain two inclined straight lines; the inclination of N_p depends on the magnitude of p. At $R = R_0$, N_m is zero, thus N_p is larger. If the straight line of N_p is flatter than that of N_m, then they intersect each other and the shell comes to a stable equilibrium state; if, however, the line of N_p is steeper than that of N_m, then no such equilibrium state exists (*Fig. 12-2*). The line of N_p belonging to the critical overpressure p_{cr} is parallel to that of N_m.

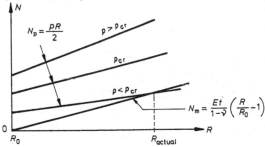

Fig. 12-2 The behaviour of the linearly elastic balloon

The explanation of this surprising instability phenomenon is that, although it comes about by tension, the total load causing it is not constant, but steadily increasing: in order to keep the overpressure p constant with increasing R, we have to pump more and more air into the balloon. The critical value of p can be reached without tearing the material only if it has a small E and a great σ_{ult}, as with the rubber balloons of children.

Let us render the investigation more accurate by taking into account the thinning of the shell wall due to tension [DULÁCSKA, 1972]. Let us first consider a bar with the original cross section A_0 and apply a tensile force on it. Due to *Poisson*'s ratio v a transverse shortening ε_t comes about, which causes the area of the cross section to decrease to the value

$$A = A_0(1 - \varepsilon_t)^2 = A_0(1 - v\varepsilon)^2 . \tag{12-7}$$

Assuming that the modulus of elasticity E of the material remains constant, i.e. the relation

$$\varepsilon = \frac{\sigma}{E}, \tag{12-8}$$

as referred to the stress σ computed with the actual cross section, holds, then, by applying a tensile force F on the bar, the relation

$$\sigma = \frac{F}{A} = \frac{F}{A_0(1 - v\varepsilon)^2} = E\varepsilon \tag{12-9}$$

can be written. Expressing the 'nominal' stress F/A_0, referred to the original cross section A_0, we obtain the equation

$$\frac{F}{A_0} = E\varepsilon(1 - v\varepsilon)^2 , \tag{12-10}$$

plotted in *Fig. 12-3*. As can be seen, this curve corresponds to the $\sigma(\varepsilon)$ diagram of a nonlinear, stress-softening material.

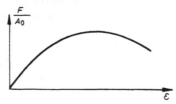

Fig. 12-3 The $\sigma(\varepsilon)$ diagram of a softening material

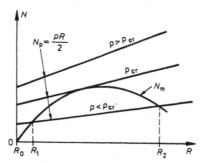

Fig. 12-4 The behaviour of a balloon made of softening material

We can plot the diagram corresponding to *Fig. 12-2* also for softening materials (*Fig. 12-4*). We again find that at values of p greater than a certain value p_{cr} there is no equilibrium: the balloon expands beyond every limit. However, for pressure values smaller than p_{cr}, two equilibrium positions exist, so that the balloon can be in equilibrium at two values of R (indicated as R_1 and R_2 in the figure).

The existence of two equilibrium positions has an interesting consequence [DULÁCSKA, 1972]. To explore this we first have to know how the pressure p varies with the radius R.

The behaviour of soft rubber materials is characterized by the fact that these materials keep their volume constant during deformation. We can take this approximately into account by setting $\nu = 0.5$ in Eq. (12-10). However, it is simpler to write the equality of volumes of the original and the deformed body. Using the notations of *Fig. 12-1*, and denoting the actual thickness of the shell by t, we can express this as:

$$t_0 R_0^2 d\varphi\, d\vartheta = t R^2 d\varphi\, d\vartheta, \tag{12-11a}$$

or

$$t = t_0 \frac{R_0^2}{R^2}. \tag{12-11b}$$

The actual stress is

$$\sigma = \frac{N}{t}, \tag{12-12}$$

and the elongation:

$$\varepsilon = \frac{R}{R_0} - 1. \tag{12-13}$$

If we by introduce these into (12-8) and observe also (12-1), we arrive at the relation

$$p = \frac{2Et_0}{R_0} \left(\frac{R_0^2}{R^2} - \frac{R_0^3}{R^3} \right), \tag{12-14}$$

plotted in *Fig. 12-5*.

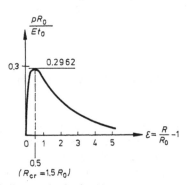

Fig. 12-5 The behaviour of the balloon made of soft rubber

The maximum (critical) value of the pressure p is:

$$p_{cr} = 0.2962 \frac{Et_0}{R_0}, \tag{12-15a}$$

to which the radius

$$R_{cr} = 1.5 R_0 \tag{12-15b}$$

belongs.

At first sight it would appear as if at values $R > R_{cr}$ the equilibrium state were unstable, since to increasing R a decreasing p belongs. However we have to consider that if we do not increase the quantity of air contained in the balloon, then with increasing R the pressure p decreases because, according to the general law of gases, the product of pressure and volume of the gas remains constant. It is easy to show that, if the balloon expands, p decreases faster than the value of p necessary to maintain equilibrium (*Fig. 12-5*). That is, starting from an arbitrary state characterized by p_1 and R_1, we can write the general law of gases as $p_1 R_1^3 = p_2 R^3$, or

$$p = p_1 \frac{R_1^3}{R^3},$$ (12-16)

which decreases much faster than that given by (12-14).

Consequently, the equilibrium state of a *closed* balloon is stable even at pressure values $p > p_{cr}$.

Let us now connect two balloons of different radii of curvature in such a way that the air can freely flow between them. We may ask the question whether it is possible to inflate this pair of balloons? (We consider them to be inflated if their radii of curvature are at least 2 or 3 times their original values.)

To inflate the balloons their critical state has to be overcome, i.e. the value p_{cr} defined by Eq. (12-15a) has to be reached. However, since the initial radii of curvature R_0 of the two balloons are not equal, and p_{cr} is inversely proportional to R_0, the value p_{cr} of the greater balloon is lower than that of the smaller one. Hence after reaching the p_{cr} of the greater balloon, the value of the pressure that can be taken by the greater balloon decreases, and we never can reach the higher p_{cr} of the smaller balloon, although this would be necessary to inflate it. Consequently, the smaller balloon can never be inflated.

This explains why rubber bunnies with great ears can be inflated only if we squeeze the bunny's body first, inflate the ears, and we let loose and begin to inflate the body only after the pressure has decreased in the ears.

12.1.2 The buckling length in the case of a load of varying direction, passing through a fixed point

If the force acting on the structure changes its direction during buckling, then it is not conservative, and its critical value can be determined, as a rule, only by the kinetic method, see e.g. in [TIMOSHENKO and GERE, 1961]. There are, however, cases in which it is possible to determine the buckling length (and from it the critical value of the load) by simple engineering considerations.

A very simple case is to be seen in *Fig. 12-6a*: the force P points always to the bottom point of the (clamped) bar. This is the static model of the columns used in the so-called *lift-slab* building method: at the beginning of lifting, the floors to be lifted prevent the lower point of the lifting cable from moving away horizontally from the bottom of the column.

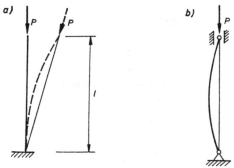

Fig. 12-6 A column loaded by a force which changes its direction

Since the force acting on the buckled column passes through its clamped bottom cross section, no bending moment arises here. Hence this cross section acts as if it were hinged. Consequently, the column buckles as a hinged-hinged no-sway bar (*Fig. 12-6b*), its buckling length is thus $l_0 = l$.

The solution of similar cases are to be found in [TIMOSHENKO and GERE, 1961] and in [PETERSEN, 1980].

12.1.3 Instability of a bar in tension

Let us consider the bar having the bending rigidity EI, depicted in *Fig. 12-7*, which is subjected to tension through the two infinitely rigid stumps attached to its ends [BIEZENO and GRAMMEL, 1953].

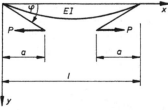

Fig. 12-7 Buckling of a bar in tension

If the bar is slightly curved, the bending moment of the forces P bends the bar further, resisted only by its bending rigidity. The equilibrium of the bar can be written as

$$P(a \sin\varphi - y) = -EI\frac{d^2y}{dx^2}. \tag{12-17}$$

Making use of the relation valid for small deformations:

$$\sin\varphi \approx \varphi \approx \left(\frac{dy}{dx}\right)_{x=0} = -\left(\frac{dy}{dx}\right)_{x=l}, \tag{12-18}$$

we arrive at the differential equation

$$EI\frac{dy^2}{dx^2} - Py = -Pa\varphi \tag{12-19}$$

with the solution:

$$y = A \cosh \mu x + B \sinh \mu x + a\varphi, \qquad (12\text{-}20)$$

where

$$\mu = \sqrt{\frac{P}{EI}}. \qquad (12\text{-}21)$$

For the two constants A and B we use the boundary conditions:

$$\left.\begin{array}{c} x = 0 \\ x = l \end{array}\right\} y = 0 \qquad (12\text{-}22)$$

which yield the characteristic equation:

$$\frac{l}{2a} = \frac{\mu l}{2} \tanh \frac{\mu l}{2}. \qquad (12\text{-}23)$$

Its smallest root μ furnishes the critical load P_{cr} according to (12-21). If $l \gg a$, then the argument of the function tanh in (12-23) becomes very large, so that the value of the function tanh can be taken equal to unity, thus

$$P_{cr} \approx \frac{\pi^2 EI}{(\pi a)^2}, \qquad (12\text{-}24)$$

i.e. the buckling length is approximately πa.

The explanation of this rather surprising phenomenon is that it is, in fact, not the bar in tension which loses its stability, but the two compressed stumps which are elastically clamped by the bar in tension.

Similar problems are treated in [ZIEGLER, 1968].

12.1.4 Structures with infinitely great critical forces

The critical loads of certain idealized structures become infinitely great at special geometric arrangements, but with other arrangements they remains finite [KOLLÁR, 1990]. Such is, e.g., the model consisting of two infinitely rigid, weightless bars and a spring with the constant c, shown in *Fig. 12-8*, the critical load of which becomes infinitely great at $\alpha = 0$, but remains finite at all other values of α.

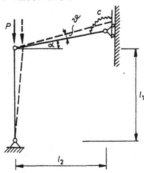

Fig. 12-8 The model having an infinitely great critical load at $\alpha = 0$

The explanation of this unusual behaviour is that the structure – omitting the spring – can be considered as a mechanism which has a dead point at $\alpha = 0$ under the driving effect of the force P.

The value of the angle α can thus be regarded as an 'initial imperfection'. Its change after buckling will be denoted by ϑ (*Fig. 12-8*). The variation of P with ϑ yields the 'post-critical load-bearing capacity', while its variation with α gives the 'imperfection sensitivity'. All these are plotted in *Fig. 12-9*.

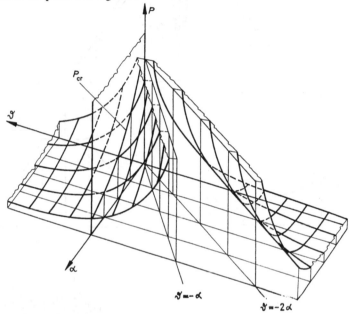

Fig. 12-9 The behaviour of the model of *Fig. 12-8*

12.1.5 Structures with abruptly changing rigidity characteristics

Let us investigate the plane structure shown in *Fig. 12-10*, consisting of an infinitely rigid mast and two inclined cables with a finite rigidity EA, loaded by a concentrated force P [HEGEDŰS, 1988].

If the structure is unprestressed, the tilting mast is stiffened only by the cable undergoing tension. Since the behaviour of the bar stiffened by a single inclined cable is 'asymmetric' (cf. what has been said in Section 1.1.3 in connection with *Fig. 1-3*), the post-critical behaviour of the structure of *Fig. 12-10* is characterized by *Fig. 12-11a*, consisting of two curves, each describing an asymmetric behaviour. The critical load of the structure is:

$$P_{\text{cr}} = \frac{EA}{2\sqrt{2}}. \tag{12-25}$$

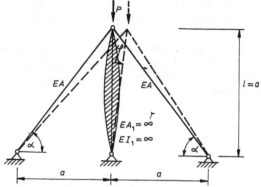

Fig. 12-10 A mast stiffened by two cables

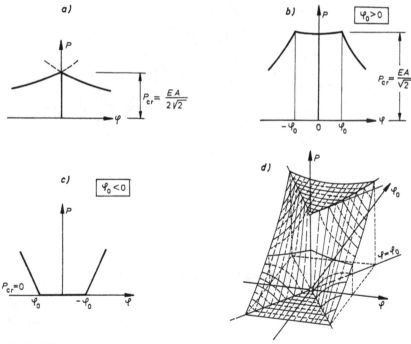

Fig. 12-11 The behaviour of the model of *Fig. 12-10*

If the structure is prestressed, the cable on the compression side slackens only when the mast has tilted by a finite angle $\varphi = \varphi_0$. This has two consequences. On the one hand, the critical load increases to the twofold of (12-25). On the other hand, the post-buckling load-bearing behaviour becomes symmetric until the tilting φ_0, in particular: stable if the initial inclination of the cables α is between α_0 and α_1:

$$\alpha_0 < \alpha < \alpha_1, \tag{12-26}$$

with

$$\alpha_0 = \frac{1}{2} \arcsin \sqrt{\frac{4}{5}} \approx 31.72°, \qquad (12\text{-}27a)$$

$$\alpha_1 = 90° - \alpha_0; \qquad (12\text{-}27b)$$

and unstable if α is outside this domain. After reaching φ_0 the cable on the compressed side slackens and ceases to stiffen the mast, so that from then on the asymmetric behaviour becomes valid. We plotted the diagram of the post-critical behaviour in *Fig. 12-11b* for the stable case.

If φ_0 is negative, i.e. if both cables are initially slack, then only after a tilting of the mast by φ_0 becomes one of them tightened. Consequently, the critical load is zero at the beginning, and increases only after a tilting φ_0 (*Fig. 12-11c*).

All these cases are shown in *Fig. 12-11d*, using the co-ordinate system $[P, \varphi, \varphi_0]$.

12.2 DESTABILIZING BY STIFFENING AND STABILIZING BY SOFTENING

12.2.1 Stabilizing by increasing the length

ZASLAVSKY [1979] presents several structures in which the increase of the length of the bar or plate increases the critical load. The examples of ZASLAVSKY can be classified into three groups: elastically supported bars, structures with a festoon curve and, finally, those loaded by nonconservative forces.

The basic type of *elastically supported bars* is the infinitely rigid bar supported by a spring, shown in *Fig. 12-12* (see also *Fig. 1-2a*). Its critical load is:

$$P_{\text{cr}} = cl, \qquad (12\text{-}28)$$

which is proportional to the bar length. This surprising result can be explained by the fact that in the moment equilibrium equation, written with respect to the point 0 for the bar tilted by an angle φ, the moment of the external force is $P(l\varphi)$, while that of the spring is $c(l\varphi)l$. Thus the restoring moment grows faster with l than that of the external load, which explains the result.

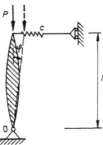

Fig. 12-12 An infinitely rigid bar supported by a spring

ZASLAVSKY [1979] enumerates many examples belonging to this category, of which we show some in *Fig. 12-13*. The infinitely rigid bar sections are marked by hatching. The bar lengths which should be increased in order to increase the critical load are denoted by l. (In the cases of Figs *12-13a,b* this phenomenon occurs only in the cases of buckling shapes drawn by dashed lines, i.e. as long as the bar of the length l does not buckle itself.)

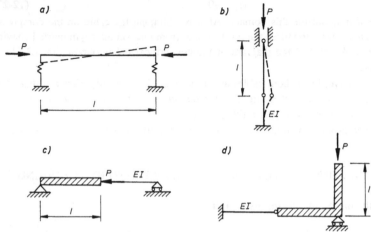

Fig. 12-13   Elastically supported bars

In the second group belong structures the critical load of which is described by a *festoon curve* as a function of the length of the bar (or plate), *(Fig. 12-14)*. With these structures if we increase the length l_1, determining the minimum critical force $P_{cr,1}$, up to l_2, the critical force also increases to $P_{cr,2}$. Here the next buckling mode prevails, and further increasing l the critical force decreases.

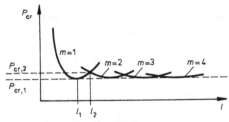

Fig. 12-14   The festoon curve describing the critical load

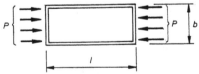

Fig. 12-15   The plate compressed in one direction

The well-known example is the simply supported plate compressed in one direction *(Fig. 12-15)*, having a constant width b, whose length l is varied. P_{cr} reaches its minimum value $P_{cr,1}$ at $l = l_1 = b$ *(Fig. 12-14)*. Increasing l to $l_2 = \sqrt{2}b$, P_{cr} increases to $P_{cr,2}$, while the plate buckles in one half wave in the longitudinal direction ($m = 1$). From here on two half waves develop ($m = 2$), and so on.

Fig. 12-16 The beam on elastic foundation

A similar phenomenon can be observed with the beam on elastic foundation (*Fig. 12-16*): increasing its length we obtain a festoon curve similar to that of *Fig. 12-14*.

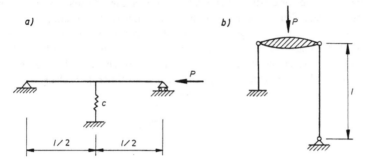

Fig. 12-17 Structures behaving according to *Fig. 12-14*

In *Fig. 12-17* we show two other structures belonging to this group.

In the third group some structures belong which are *loaded by nonconservative forces*. One of these is shown in *Fig. 12-18*. The lower end of the bar is restrained against rotation by a spring; the load P always passes through the point C. It is interesting to notice that P_{cr} is a *tensile* force.

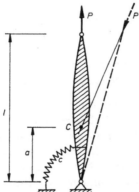

Fig. 12-18 The bar loaded by a force passing through a given point

12.2.2 The destabilizing effect of an additional support

According to our engineering sense, if we stiffen a structure, its critical load cannot decrease; in fact, it should increase.

To our knowledge it was NEAL and MANSELL [1963] who first showed an example to the contrary: if we replace the hinged support of the lower bar of the structure of *Fig. 12-19a* by a clamping (*Fig. 12-19b*), the load P causing its elastic-plastic buckling decreases. The phenomenon can essentially be explained by the fact that the clamping causes second-order stresses which promote plastification.

BARTA [1970] showed for elastic structures that the increase of some stiffnesses may decrease the critical load. Thus it can be seen that if we increase the bending stiffness of the lower beam of the structure of *Fig. 12-20*, then the compressive force in the vertical bar increases, so that it buckles at a lower value of P. This phenomenon can be explained by observing that we stiffened the structure at an 'improper place'. It is thus necessary to clarify: when do we stiffen a structure at a 'proper' and at an 'improper' place.

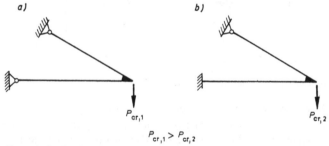

$$P_{cr,1} > P_{cr,2}$$

Fig. 12-19 Example for the destabilizing effect of an additional restraint

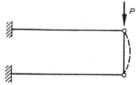

Fig. 12-20 BARTA's example for the destabilizing effect of a stiffening

This problem has been treated by TARNAI [1980] who showed that the stiffening can reduce the critical load only if it hinders displacements which do not lie in the 'direction' of the buckling deformation, i.e. which do not constitute a part of the buckling deformation. Let us consider e.g. the beam of *Fig. 12-21a*, resting on fork-like supports [TARNAI and SZALAI, 1978; TARNAI, 1980]. If we support it at the middle by a vertical bar (*Fig. 12-21b*), its critical load causing lateral buckling will diminish. The vertical bar prevents, in fact, only the deflection of the beam in the vertical plane, which does not appear in the deformation of lateral buckling.

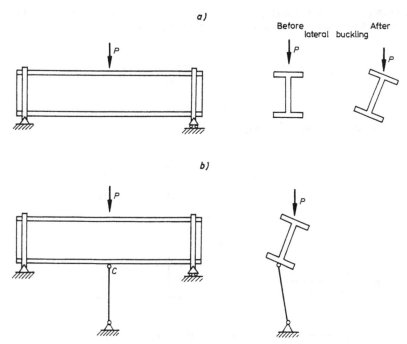

Fig. 12-21 The destabilizing effect of the additional support on the lateral buckling of a beam

This somewhat surprising result can be visually explained by observing that the additional support is applied at an 'improper' place: on the one hand, it supports the beam at the lower flange and, on the other hand, it does not prevent the rotation of the middle cross section. Hence it represents a more disadvantageous constraint than the original supports shown in *Fig. 12-21a*, and since it takes over a greater part of the load, the beam comes into a worse situation than before.

It follows from all these that if we raise the joining point C of the vertical bar to the beam, we can reach a height at which the destabilizing effect of the additional support disappears.

TARNAI [1980] also shows how *Dunkerley's* theorem can be used to determine the upper and lower bound of the critical load modified due to the additional support.

12.2.3 Paradoxes with torsional buckling

It is a well-known fact that the critical load of pure torsional buckling of a bar with a cross-shaped cross section is the greater, the shorter the protruding 'wings' of the cross section are (*Fig. 12-22*), cf. also what has been said in Section 1.5 in connection with *Fig. 1-15*. This phenomenon can be explained by considering that the pure torsional buckling of such a bar consists, in fact, of the buckling of the four protruding plates having free edges, and the critical compressive stress of such plates is inversely proportional to the square of the width b_1, so that their critical load is inversely proportional to b_1.

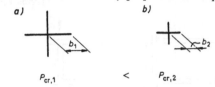

Fig. 12-22   The paradox of pure torsional buckling

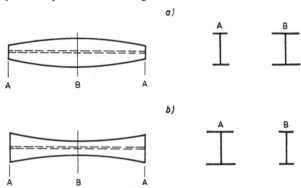

Fig. 12-23   The paradox of torsional buckling of a bar with varying I-shaped cross section

There is, however, another paradox. CYWIŃSKI and KOLLBRUNNER [1971] have shown that the critical load of torsional buckling of a bar with an I-shaped cross section, having flanges with parabolically varying width (*Fig. 12-23a*) is greater than that of a bar with constant cross section, taking either the minimum or the maximum cross section of the original bar as a basis. CYWIŃSKI and KOLLBRUNNER [1982] also investigated a bar with inversely varying cross section (*Fig. 12-23b*) and found that its critical force is smaller than that of either of the two bars with constant cross section as described above [CYWIŃSKI, 1986].

      This paradox has been clarified by KANOK-NUKULCHAI and SUSUMPOW [1993]. They found that if the effect of the in-plane shear stresses (neglected in the usual theory of torsion of thin-walled bars, used in the investigations of CYWIŃSKI and KOLLBRUNNER) is also considered, the paradox disappears: the critical load of the bar with constant cross section, taking the minimum and the maximum cross section as a basis, turns out to be smaller and greater, respectively, than that of the bar with varying cross section, in accordance with simple engineering considerations.

12.2.4   The destabilizing effect of damping in the case of nonconservative forces.

It is well known that the critical values of nonconservative forces can only be determined by the kinetic method [TIMOSHENKO and GERE, 1961]: we have to look for the greatest load intensity at which the root of the characteristic equation of the vibrating bar is still a pure imaginary number. That is, the movement of the bar is described in this case by trigono-metric functions, so that its vibration amplitude does not increase in time. If, however, by increasing the load the root will have a positive real part too, the vibration amplitude will have a part exponentially increasing in time, which means instability.

Fig. 12-24 The *Beck* bar

Let us investigate the so-called *Beck* bar, clamped at the bottom and having a free upper end, loaded by a force which is always tangential to the deflection curve of the bar (*Fig. 12-24*). Since the bar performs merely bending deformation, the damping (which is always present) can be related to the bending deformation only. Let us assume that the damping has a viscous character, i.e. it is in every cross section proportional to the first derivative of the curvature with respect to the time. The question is: how does this damping influence the value of the critical force?

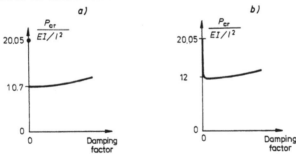

Fig. 12-25 The destabilizing effect of damping

We know from the theory of vibrations that the viscous damping modifies the eigenfre-quency of the structure, but if the damping is small, this modification is hardly perceptible. However, ZIEGLER [1952] found, when approximating the bar of *Fig. 12-24* by a two-member model, that the critical force of the undamped case changes by a finite value even if the damping is very small (see also [ZIEGLER, 1968]). In *Fig. 12-25a* we show the results of the computation of POPPER [1976] referring to the *Beck* bar.

The so-called *Ziegler* jump of *Fig. 12-25a* has taken up the attention of researchers for a long time. LOTTATI and ELISHAKOFF [1987] tried to investigate the vicinity of this jump by a more refined numerical calculation and found that the critical load continuously varies with increasing damping, so that no such jump exists (*Fig. 12-25b*). Their calculation, however, cannot be considered a proof, since the computer cannot exactly compute certain values 'in limit' (e.g. those of the type $0 \times \infty$). Thus it seems probable that the model of the damped *Beck* bar allows this jump. It is conceivable that a more refined model would yield a continuous transition, but this needs further research. In any case that part of the paradox will remain that even a very small damping reduces the critical load to about half its original value.

References

REFERENCES TO CHAPTER 1

BRUSH, D. O. and ALMROTH, B. O. 1975 Buckling of Bars, Plates, and Shells. McGraw-Hill, New York.

CROLL, J. G. A. and WALKER, A. C. 1972 Elements of Structural Stability. Macmillan, London.

GÁSPÁR, ZS. 1984 Buckling model for a degenerated case. *Newsletter*, **2**, No. 4, 5–8. Technical University of Budapest.

KOITER, W. T. 1945 Over de stabiliteit van het elastisch evenwicht. (On the stability of the elastic equilibrium.) PhD Thesis, Delft. H. J. Paris, Amsterdam.

KOITER, W. T. 1963 Elastic stability and post-buckling behaviour. In: Nonlinear Problems (Proceedings of a Symposium 1962, R. E. Langer, Ed.). The University of Wisconsin Press, Madison. 257–275.

KOITER, W. T. 1967 Post-buckling analysis of a simple two-bar frame. In: Recent Progress in Applied Mechanics (B. Broberg, J. Hult, F. Niordson, Eds.). Almquist and Wiksell, Stockholm. 337–354.

KOLLÁR, L. 1972 Some practical considerations on the postcritical behaviour of structures. In: Preliminary Report, 8th Congress of the International Association for Bridge and Structural Engineering, Amsterdam.

KOUNADIS, A. N., GIRI, J. and SIMITSES, G. J. 1977 Nonlinear stability analysis of an eccentrically loaded two-bar frame. *Journ. Appl. Mech.*, **44**, 701–706.

SÄTTELE, J. M., RAMM, E. and FISCHER, M. 1981 Traglastkurven einachsig gedrückter Rechteckplatten mit Seitenverhältnissen $\alpha \leq 1$ bei vorgegebenen geometrischen Imperfektionen. *Der Stahlbau*, **50**, 205–210.

TIMOSHENKO, S. P. and GERE, J. M. 1961 Theory of Elastic Stability. McGraw-Hill, New York.

TIMOSHENKO, S. P. and WOINOWSKY-KRIEGER, S. 1959 Theory of Plates and Shells. McGraw-Hill, New York.

THOMPSON J. M. T. and HUNT, G. W. 1973 A General Thery of Elastic Stability. Wiley, London.

REFERENCES TO CHAPTER 2

AKHIEZER, N. I. and GLAZMAN, I. M. 1977 Theory of Linear Operators in Hilbert Space (in Russian). Vishcha Skola, Khar'kov.

BLEICH, F. 1952 Buckling Strength of Metal Structures. McGraw-Hill, New York.

DUNKERLEY, S. 1894 On the whirling and vibration of shafts. *Phil. Trans. Roy. Soc. Lond.* A **185**, 279–360.

FÖPPL, L. 1933 Über das Ausknicken von Gittermasten, insbesondere von hohen Funk tü rmen. *ZAMM* **13**, 1–10.

HORNE, M. R. 1995 The Rankine-Merchant load and its application. In: Summation Theorems in Structural Stability (TARNAI, T. Ed.), CISM Courses and Lectures No. 354. Springer, Wien, 111–139.

KÁRMÁN, T. and BIOT, M. A. 1940 Mathematical Methods in Engineering. McGraw-Hill, New York.

KOLLÁR, L. 1971 Recent results in the theory of stability through the eyes of a designer (in Hungarian). *Magyar Építőipar*, **20**, 333–337.

KOLLÁR, L. (Editor) 1991 Special Problems of Structural Stability (in Hungarian). Akadémiai Kiadó, Budapest.

MELAN, H. 1917 Kritische Drehzahlen von Wellen mit Langsbelastung. *Zeitschrift der Österr. Ingenieur- und Architekten-Vereines*, **69**, 610–612, 619–621.

MIKHLIN, S. G. 1970 Variational Methods in Mathematical Physics (in Russian). Nauka, Moscow.

PAPKOVICH, P. F. 1963 Works on Structural Mechanics of Ships, Vol. 4, Stability of bars, roofs and plates (in Russian). Sudpromgiz, Leningrad.

PLANTEMA, F. J. 1952 Theory and Experiments on the Elastic Overall Instability of Flat Sandwich Plates. PhD Thesis, Delft.

RANKINE, W. J. M. 1863 A Manual of Civil Engineering, 2nd Ed. Charles Griffin, London.

SOUTHWELL, R. V. 1922 On the free transverse vibrations of a uniform circular disc clamped at its centre; and on the effects of rotation. *Proc. Roy. Soc. Lond.* A, **101**, 133–153.

STRIGL, G. 1955 Das nicht lineare Überlagerungsgesetz für die Lösungen von zusammengesetzten Stabilitätsproblemen mit Verzweigungspunkt. *Der Stahlbau*, **24**, 33–39, 51–61.

TARNAI, T. 1980 Generalization of Southwell's and Dunkerley's theorems for quadratic eigenvalue problems. *Acta Techn. Hung.*, **91**, 201–221.

TARNAI, T. 1995 The Southwell and the Dunkerley theorems. In: Summation Theorems in Structural Stability (Tarnai, T., Ed.), CISM Courses and Lectures No. 354. Springer, Wien. 141–185.

TIMOSHENKO, S. P. and GERE, J. M. 1961 Theory of Elastic Stability. McGraw-Hill, New York.

WEINBERGER, H. 1974 Variational Methods for Eigenvalue Approximation. SIAM, Philadelphia, Pennsylvania.

WEINBERGER, H. 1995 Some mathematical aspects of buckling. In: Summation Theorems in Structural Stability (TARNAI, T., Ed.), CISM Courses and Lectures No. 354. Springer, Wien. 1–37.

REFERENCES TO CHAPTER 3

BUDIANSKY, B. and HUTCHINSON, J. 1964 Dynamic Buckling of Imperfection-Sensitive Structures. *Proc. 11th Int. Congr. Appl. Mech.* Springer, Berlin, 1966.

BÜRGERMEISTER, G., STEUP, H. and KRETSCHMAR, H. 1963 Stabilitätstheorie. Teil 2. Akademie-Verlag, Berlin.

FLÜGGE, W. 1973 Stresses in Shells, Springer, Berlin.

FOK, W. C., RHODES, J. and WALKER, A. C. 1976 Local buckling of outstands in stiffened plates. *Aeronautical Quarterly*, **27**, 277–291.

FOK, W. C., RHODES, J. and WALKER, A. C. 1977 Buckling of locally imperfect stiffeners in plates, *Proc. ASCE, Journ. Eng. Mech. Div.*, **103**, EM5, 895–291.

HUTCHINSON, J. 1965 Axial buckling of pressurized imperfect cylindrical shells. *AIAA Journ.*, **3**, 1461–66.

HUTCHINSON, J. 1967 Imperfection sensitivity of externally pressurized spherical shells. *Journ. Appl. Mech.*, **34**, 49–55.

KOITER, W. T. 1945 Over de stabiliteit van het elastisch evenwicht. (On the stability of the elastic equilibrium.) PhD Thesis. Delft, H. J. Paris, Amsterdam.

KOITER, W. T. 1963*a* Elastic stability and post-buckling behaviour, In: Nonlinear Problems (Proceedings of a Syposium 1962, R. E. Langer, Ed.). The University of Wisconsin Press, Madison, 257–275.

KOITER, W. T. 1963*b* The effect of axisymmetric imperfections on the buckling of cylindrical shells under axial compression. *Proc. Royal Netherlands Academy of Sciences, Amsterdam, Series B*, **66**, 265–279.

KOLLÁR, L. and DULÁCSKA, E. 1984 Buckling of Shells for Engineers, Wiley, Chichester, and Akadémiai Kiadó, Budapest.

NEUT, A. VAN DER 1969 The interaction of local buckling and column failure of thin-walled compression members. *Proc. 12th Int. Congr. Appl. Mech.* (M. Hetényi, W. G. Vincenti, Eds). Springer, Berlin.

PFLÜGER, A. 1964 Stabilitätsprobleme der Elastostatik. 2. Aufl. Springer, Berlin.

REIS, J. and ROORDA, J. 1979 Post-buckling behaviour under mode interaction. *Proc. ASCE, Journ Eng. Mech. Div.*, **105**, EM4, 609–621.

SVENSSON, S. E. and CROLL, J. G. A. 1975 Interaction between local and overall buckling. *Int. Journ. Mech. Phys. Solids*, **17**, 307–321.

THOMPSON, J. T. M. and HUNT, G. W. 1973 A General Theory of Elastic Stability. Wiley, London.

THOMPSON, J. T. M. and LEWIS, G. M. 1972 On the optimum design of thin-walled compression members. *Journ. Mech. Phys. Solids*, **20**, 101–109.

TIMOSHENKO, S. P. and GERE, J. M. 1961 Theory of Elastic Stability. McGraw-Hill, New York.

TULK, J. D. and WALKER, A. C. 1976 Model studies of the elastic buckling of a stiffened plate. *Journ. Strain Analysis*, **11**, 137–143.

WALKER, A. C. 1975 Interactive buckling of structural components. *Sci. Prog., Oxf*, **62**, 579–597.

WILLIAMS, D. G. and WALKER, A. C. 1975 Explicit solutions for the design of initially deformed plates subject to compression. *Proc. ICE*, Part 2, **59**, 763–787.

REFERENCES TO CHAPTER 4

ARNOL'D, V. U. 1972 Normal forms for functions near degenerate critical points, the Weyl groups of A_k, D_k, and E_k, and Lagrangian singularities. *Functional Anal. Appl.*, **6**, 254–272.

BOLOTIN, V. V. 1965 Statisticheskie Metody v Stroitelnoï Mekhanike. Izd. Literatury po stroitelstvu, Moscow.

BOLOTIN, V. V. 1982 Metody Teorii Veronatnochesteï i Teorii Nadezhnosti v Raschetakh Sooruzheniï. Stroïzdat, Moscow.

ELISHAKOFF, I. 1983 How to introduce the imperfection-sensitivity concept into design. *Proc. IUTAM Symposium on Collapse, London.* (J. M. T. THOMPSON, G. W. HUNT, Eds). 345–357.

GÁSPÁR, ZS. 1977 Buckling models for higher catastrophes. *J. Struct. Mech.*, **5**, 357–368.

GÁSPÁR, ZS. 1982 Critical imperfection territory. *J. Struct. Mech.*, **11**, 297–325.

GÁSPÁR, ZS. 1983a Probability of the instability of imperfect structures. *ZAMM*, **63**, T 48–T 49.

GÁSPÁR, ZS. 1983b Computation of imperfection-sensitivity at two-fold branching points. *ZAMM*, **63**, 359–370.

GÁSPÁR, ZS. 1985 Imperfection-sensitivity at near-coincidence of two critical points. *J. Struct. Mech.*, **13**, 45–65.

GÁSPÁR, ZS. and DOMOKOS, G. 1991 Global description of the equilibrium paths of a simple mechanical model. In: Stability of Steel Structures 1990, Budapest (M. IVÁNYI Ed.). Akadémiai Kiadó, Budapest. 79–86.

GÁSPÁR, ZS. and MLADENOV, K. A. 1995 Post-critical behavior of a column loaded by a polar force. *Proc. Int. Coll. on Stability of Steel Structures* 1995, Budapest (IVÁNYI, M. VERŐCI B. Eds). Vol. II., 275–281.

GILMORE, R. 1981 Catastrophe Theory for Scientists and Engineers. Wiley, New York.

GIONCU, V. and IVAN, M. 1984 Teoria comportárii critice si postcritice a structurilor elastice. Editura Academiei rep. Soc. Romania, Bucuresti.

HACKL, K. 1990 Ausbreitung von Instabilitäten in einem Knickmodell von Thompson und Gáspár. *ZAMM*, **70**, T 189 T 192.

HUNT, G. W. 1978 Imperfections and near-coincidence for semi-symmetric bifurcations, *Annals of the New York Academy of Sciences*, **316**, 572–589.

HUNT, G. W., REAY, N. A. and YOSHIMURA, T. 1979 Local diffeomorphism in the bifurcational manifestations of the umbilic catastrophes. *Proc. Roy. Soc. Lond.* A, **369**, 47–65.

PAJUNEN, S. and GÁSPÁR, ZS. 1996 Study of an interactive buckling model – From local to global approach. *Proc. 2nd Int. Conf. on Coupled Instabilities in Metal Struct. CIMS'96* (J. Rondal, D. Dubina, V. Gioncu, Eds). 35–42.

POSTON, T. and STEWART, I. N. 1978 Catastrophe Theory and its Applications. Pitman, London.

THOM, R. 1972 Stabilité Structurelle et Morphogenèse. Benjamin, New York.

THOMPSON, J. M. T. 1982 Instabilities and Catastrophes in Science and Engineering. Wiley, London.

THOMPSON, J. M. T. and GÁSPÁR, ZS. 1977 A buckling model for the set of umbilic catastrophes. *Math. Proc. Camb. Phil. Soc.*, **82**, 497–507.

THOMPSON, J. M. T. and HUNT, G. W. 1973 A General Theory of Elastic Stability. Wiley, London.

THOMPSON, J. M. T. and HUNT, G. W. 1984 Elastic Instability Phenomena. Wiley, Chichester.

ZEEMAN, E. C. 1976 The umbilic bracelet and the double-cusp catastrophe. *Lecture Notes in Mathematics*, **525**, 328–366.

ZEEMAN, E. C. 1977 Catastrophe Theory: Selected Papers (1972-1977). Addison-Wesley, Reading, Massachusetts.

REFERENCES TO SECTION 5.1

AGENT, R. 1970 Considerarea conlucrarii spatiale in calculul sarcinii critice de flambaj a stilpilor la hale industriale cu poduri rulante. *Constructii si Materiale de Constructii*, No. 5, 256.

APPELTAUER, J. and BARTA, TH. 1960 Stabilitatea cadrelor. Metode si formule practice pentru determinarea lungimilor de flambaj la stilpii de cadru. Partea I-a. *Bul. de Studii si Cerc. INCERC-ISCAS*, No. 3, 96.

APPELTAUER, J. and BARTA, TH. 1961 Stabilitatea cadrelor. Metode si formule practice pentru determinarea lungimilor de flambaj la stilpii de cadru. Partea II-a. *Bul. de studii si Cerc. INCERC-ISCAS*, No. 1, 3.

APPELTAUER, J. 1961 Näherungsverfahren zur Knicklängenberechnung bei mehrgeschossigen und mehrfeldrigen Rahmen. *Revue de Mécanique Appliquée, Bucuresti*, **VI**, No. 4, 483.

APPELTAUER, J. and BARTA, TH. 1962 Stabilitatea cadrelor. Calculul de stabilitate al cadrelor de otel. Partea I-a. Bara izolata. *Bul. de Studii si Cerc. INCERC-ISCAS*, No. 1, 31. Partea II-a. Cadre. No. 2, 39. Partea a III-a. Verificare sectiunilor cu considerarea flambajului prin incovoiere-rasucire al barelor cu pereti subtiri. No. 4, 41.

APPELTAUER, J. and BARTA, TH. 1964 Critical loads of plane frames. *Constructional Eng.*, No. 8, 293.

APPELTAUER, J., GIONCU, E. and CUTEANU, E. 1966 Metoda pentru calculul la stabilitate al cadrelor de otel. Studii si cerc. INCERC. *Mecanica constructiilor, Bucuresti*, No. 1.

APPELTAUER, J. 1970*a* Contributii la calculul de stabilitate al cadrelor de otel. PhD Thesis, Timisoara.

APPELTAUER, J. 1970*b* Zum Stabilitätsnachweis von Hallenrahmen mit gestuften Stützen infolge Kranbelastung. *European Civil Eng.*, No. 6, 276.

BASLER, K. 1956 Der exzentrisch gedrückte und querbelastete, prismatische Druckstab. *Schweizerische Bauzeitung*, **74**, 587, 627.

BECK, C. F. and ZAR, M. 1963 Steel column design for multistory rigid frames. *Proc. ASCE, Journ. Struct. Div.*, **89**, ST4, 537.

BEEDLE, L. S. and TALL, L. 1960 Basic column strength. *Proc. ASCE, Journ. of the Struct. Div.*, **86**, ST7, 139.

BEER, H. 1966 Beitrag zur Stabilitätsuntersuchung von Stabwerken mit Imperfektionen. *Publ. IABSE*, **25**, 43.

BERNARDINIS, M. D. 1968 Instabilità globale dei telai multipiani. *Costruzzioni Metalliche*, **20**, No. 5, 43.

BOLTON, A. 1955*a* A quick approximation to the critical load of rigidly jointed trusses. *Struct. Eng.*, **33**, No. 3, 90.

BOLTON, A. 1955*b* The critical load of portal frames when sidesway is permitted. *Struct. Eng.*, **33**, No. 8, 229.

BOWLES, R. E. and MERCHANT, W. 1958 Critical Loads of Tall Building Frames. *Struct. Eng.*, **36**, No. 6, 819.

BRITVEC, S. J. and CHILVER, A. H. 1963 Elastic buckling of rigidly-jointed frames. *Proc. ASCE, Journ. Eng. Mech. Div.*, **89**, EM6, 217.

BÜRGERMEISTER, G. and STEUP, H. 1957 Zur iterativen Lösung von Problemen der Verformungstheorie. *Stahlbau*, **26**, No. 7, 183.

BÜRGERMEISTER, G., STEUP, H. and KRETZSCHMAR, H. 1963 Stabilitätstheorie mit Erläuterungen zu den Knick- und Beulvorschriften, Teil II. Akademie–Verlag, Berlin.

CAMPUS, F. and MASSONET, C. 1955 Rechereches sur le flambement de colonnes en acier 37, à profil en double té, socillicitées obliquement. *Bull. CERES*, **7**, 119.

CASSENS, J. 1961 Biegung und Längkraft im Stahlbau. *Bautechnik*, **38**, No. 1, 22.

CHILVER, A. H. 1956 Buckling of a simple portal frame. *J. Mec. Phys. Solids*, **5**, 18.

CHU, K. H. and PABARCIUS, A. 1964 Elastic and inelastic buckling of portal frames. *Proc. ASCE, Journ. Eng. Mech. Div.*, **90**, EM5, 221.

CHU, K. H. and CHOW, H. L. 1969 Effective column length in unsymmetrical frames. *Publ. IABSE*, **29**, 1.

CHWALLA, E. 1938 Die Stabilität lotrecht belasteter Rechteckrahmen. *Bauingenieur*, **19**, No. 5–6, 69.

CHWALLA, E. and JOKISCH, FR. 1941 Über das ebene Knickproblem des Stockwerkrahmens. *Stahlbau*, **14**, No. 8–9, 33; 10–11, 47.

CHWALLA, E. 1959 Die neuen Hilfstafeln zur Berechnung von Spannungsproblemen der Theorie zweiter Ordnung und von Knickproblemen. *Bauingenieur*, **34**, No. 4, 6, and 8, 128, 240, 299.

CSELLÁR, Ö. and HALÁSZ, O. 1961 Die Stabilitätsuntersuchung gedrückter dünnwandiger Stäbe. *Acta Techn. Acad. Sci. Hung.*, **35–36**, 546.

DULÁCSKA, E. and KOLLÁR, L. 1960 Angenäherte Berechnung des Momentenzuwaches und der Stabilität von gedrückten Rahmenstielen. *Bautechnik*, **37**, No. 3, 98.

DUTHEIL, J. 1957 La stabilité des colonnes en acier soumises au flambement-déversement. *Ann. ITBTP*, **118**, No. 10, 679.

DUTHEIL, J. 1961 La prise en compte des imperfections inévitables dans la determination des systèmes hyperstatiques en acier sollicités au flambement. *Publ. IABSE*, **21**, 61.

DUTHEIL, J. 1966 Vérification des pièces comprimées. Principes fondamentaux. *Constr. Mét.*, **3**, No. 2, 3.

FEY, T. 1966 Vereinfachte Berechnung von Rahmensystemen des Stahlbetonbaus nach der Theorie 2. Ordnung. *Bauingenieur*, **41**, No. 6, 231.

FREIHART, I. and FREIHART, G. 1968 Näherungsweise Ermittlung der Knickhöhe von verschieblichen Rahmenstielen. *Bautechnik*, **45**, No. 7, 235.

GALAMBOS, TH. V. 1965a Subassemblages and Restrained Columns without Sway. Plastic Design of Multi-Storey Frames. Lecture Notes (PDMSF-LN). Lehigh Univ., Depart. of Civil Eng. Fritz Eng. Lab., Bethlehem, Pennsylvania.

GALAMBOS, TH. V. 1965b Column Deflection Curves. PDMSF-LN. Lehigh Univ., Depart. of Civ. Eng. Fritz Eng. Lab., Bethlehem, Pennsylvania.

GALAMBOS, TH. V. 1965c Restrained Columns Free to Sway. PDMSF-LN. Lehigh Univ., Depart. of Civ. Eng. Fritz Eng. Lab., Bethlehem, Pennsylvania.

GALAMBOS, TH. V., LU, L. W. and DRISCOLL, G. C. JR. 1968 Research on Plastic Design of Multi-Storey Frames at Lehigh University. IABSE, Eighth Congress, New York.

GENT, A. R. 1962 The design of frame structures considering strength, stability and deflections. *Proc. ICE*, **23**, No. 11, 337.

GIONCU, V. 1982 New Conceptions, Trends and Perspectives in the Theory of Postcritical Behaviour of Structures. *Proc. of the 3rd Int. Coll. on Stab.* First Sess., Timisoara.

GIONCU, V. and IVAN, M. 1983 Bazele Calcului Structurilor la Stabilitate. Ed. Facla, Timisoara.

GODLEY, M. H. R. and CHILVER, A. H. 1967 The elastic post-buckling behaviour of unbraced plane frames. *Int. Journ. Mech. Sci.*, **9**, No. 5, 323.

GOLDBERG, J. E. 1960 On the Lateral Buckling of Multi-Storey Building Frames with Shear Bracing. IABSE, Sixth Congress, Stockholm, Final report, 231.

GOLDBERG, J. E. 1968 Lateral buckling of braced multi-storey frames. *Proc. ASCE, Journ. Struct. Div.*, **94**, ST12, 2963.

HABEL, A. 1958 Die Knicklänge der Rahmenstiele aus Stahlbeton. *Beton- und Stahlbetonbau*, **53**, No. 11, 273.

HALÁSZ, O. 1967 Adoption of the theory of plasticity to steel structures. *Acta Techn. Acad. Sci. Hung.*, **59**, No. 1–2, 57.

HALÁSZ, O. 1969 Design of elastic-plastic frames under primary bending moments. *Periodica Polytechnica*, Budapest, **13**, No. 3–4, 95.

HALÁSZ, O. 1972 Theorems for a Simplified Second Order Limit Analysis of Elastic-Plastic Frames. IABSE, Ninth Congress, Amsterdam, Prelim. Report 17.

HALÁSZ, O. 1977 Effect of Initial Imperfections on the Elastic-Plastic Failure Load of Simple Frames. *Proc. Stability of Steel Structures*, Liège, Prelim. Rep., 13.

HALÁSZ, O. and IVÁNYI, M. 1977 Stability problems in Hungarian specifications and recommendations for steel structures. *Proc. Stability of Steel Structures*, Budapest–Balatonfüred, 25–35.

HALÁSZ, O. and IVÁNYI, M. 1979 Tests with simple elastic-plastic frames. *Periodica Polytechnica*, Budapest, **23**, 151–182.

HALLDORSSON, O. P. and WANG, CH. K. 1968 Stability analysis of frameworks by matrix methods. *Proc. ASCE, Journ. Struct. Div.*, **94**, ST7, 1745.

HANGAN, M. D. 1967 Poduri industriale din beton armat. Edit. Technica, Bucuresti.

HERBER, K. H. 1956 Vereinfachte, einheitliche Berechnung von Knickproblemen für Stäbe mit starrer Lagerung und Rahmentragwerke. *Bautechnik*, **33**, No. 5, 157, No. 9, 326, No. 10, 358, and No. 12, 435.

HOLMES, M. 1961 Steel frames with brickwork and concrete infilling. *Proc. ICE*, **19**, No. 8, 473.

HOLMES, M. and GANDHI, S. N. 1965 Ultimate load design of tall building frames allowing for instability. *Proc. ICE*, **30**, No. 1, 147.

HORNE, M. R. 1961 The stability of elastic-plastic structures. In: Sneddon, I. N., Hill, R. Progress in Solid Mechanics, **2**, Chap. VII.

HORNE, M. R. 1962 The effect of finite deformation in the elastic stability of plane frames, *Proc. Roy. Soc. Series A. Math. and Phys. Sci.*, **266**, No. 1324, 47.

HORNE, M. R. 1963a Generalized approximate method of assessing the effect of deformations on failure loads. *Publ. IABSE*, **23**, 205.

HORNE, M. R. 1963b Elastic-plastic failure loads of plane frames. *Proc. Roy. Soc. Series A. Math. and Phys. Sci*, **274**, No. 1358, 343.

HORNE, M. R. and MERCHANT, W. 1965 The Stability of Frames. Pergamon Press, Oxford.

HUDDLESTON, J. V. 1967 Nonlinear buckling and snap-over of a two-member frame. *Int. Journ. Solids and Struct.*, **3**, 1023.

JENNINGS, A. 1968 Frame analysis including change of geometry. *Proc. ASCE, Journ. Struct. Div.*, **94**, ST3, 627.

JOHNSON, D. E. 1960 Lateral stability of frames by energy method. *Proc. ASCE, Journ. Eng. Mech. Div.*, **86**, EM4, 23.

JUNG, F. W. 1961 Zur näherungsweisen Berechnung von einfachen, gelenkig gelagerten Rahmen und Stäben nach Theorie II. Ordnung. *Stahlbau*, **30**, 146.

KÄRCHER, H. 1968 Eine Konvergenzverbesserung der näherungsweisen Traglastberechnung von Rahmen. *Stahlbau*, **37**, No. 8, 250.

KAZAZAEV, V. N. 1960 O granitsakh primenimosti deformatsionnykh raschetov odnoproletnoi nesimmetrichnoi rami s zashchemlennimi stoikami. *Stroitel'naia Mekh. i Raschet Sooruzh.*, No. 2, 27.

KERKHOFS, W. 1965 Flambement d'emsemble d'un bâtiment à étages. *La Techn. des Trav.*, **41**, No. 3–4, 123; No. 5–6, 185; No. 7–8, 249; No. 9–10, 313.

KETTER, R. L. 1961 Further studies of the strength of beam-columns. *Proc. ASCE, Journ. Struct. Div.*, **87**, ST6, 135.

KIRSTE, L. 1956 Simplified calculus of the stability of multistorey frames. *Publ. IABSE*, **16**, 295.

KLÖPPEL, K. and GODER, W. 1957 Näherungsweise Berechnung der Biegemomente nach Spannungstheorie II. Ordnung zur Bemessung von außermittig gedrückten Stäben nach DIN 4114 Ri 10.2. *Stahlbau*, **26**, No. 7, 188.

KLÖPPEL, K. and FRIEMANN, H. 1964 Übersicht über Berechnungsverfahren für Theorie II. Ordnung. *Stahlbau*, **33**, No. 9, 270.

KLÖPPEL, K. and UHLMANN, W. 1968 Die Berechnung der Traglasten beliebig gelagerter Einfeldrahmen mit beliebiger Querschnittsform unter Berücksichtigung der plastischen Zonen in Stablängsrichtung mit Hilfe elektronischer Rechenautomaten. *Stahlbau*, **37**, No. 3, 65 and No. 5, 145.

KNOTHE, K. 1963 Vergleichende Darstellung der Näherungsmethoden zur Bestimmung der Traglast eines biegesteifen Stahlstabwerkes. *Stahlbau*, **32**, No. 11, 330.

KOLLBRUNNER, C. F. and MEISTER, M. 1955 Knicken. Theorie und Berechnungen von Knickstäben. Knickvorschriften. Springer, Berlin.

KORNOUKHOV, N. V. 1948 Prochnosti i Ustoichivosti Sterzhnevykh Sistem. Stroiizdat, Moskva.

LIGTENBERG, I. F. K. 1965 Stability and Plastic Design. *Heron*. English Edit., No. 3, 1 and 13.

LIPTÁK, L. 1963 Stabilitätsnachweis bei Hallenrahmen mit Kranbahn. *Bautechnik*, **40**, No. 4, 132.

LOW, M. W. 1959 Some model tests on multi-storey rigid steel frames. *Proc. ICE*, **13**, No. 7, 287.

LU, L. W. 1965*a* Frame Buckling. PDMSF-LN. Lehigh Univ. Fritz Eng. Lab., Bethlehem, Pennsylvania.

LU, L. W. 1965*b* Design of Columns in Unbraced Frames. PDMSF-LN. Lehigh Univ. Fritz Eng. Lab., Bethlehem, Pennsylvania.

LU, L. W. 1965*c* Inelastic buckling of steel frames. *Proc. ASCE, Journ. Struct. Div.*, **91**, ST6, 185.

MAJID, K. J. 1967 An evaluation of the elastic critical load and the Rankine load of frames. *Proc. ICE*, **36**, No. 3, 579.

MAJID, K. J. and ANDERSON, D. 1968 Elastic-plastic design of sway frames by computer. *Proc. ICE*, **41**, No. 12, 705.

MASSONNET, CH. 1976 Die europäischen Empfehlungen (EKS) für die plastische Bemessung von Stahltragwerken. *Acier-Stahl-Steel*, No. 4, 146.

MATEESCU, D., APPELTAUER, J. and CUTEANU, E. 1980 Stabilitatea la Compresiune a Structurilor din Bare de Otel. Edit. Acad., Bucuresti.

MERCHANT, W. and SALEM, A. H. 1960 The Use of Stability Functions in the Analysis of Rigid Frames. IABSE, Sixth Congress, Stockholm, Prelim. Report.

MOSES, F. 1964 Inelastic frame buckling. *Proc. ASCE, Journ. Struct. Div.*, **90**, ST6, 105.

MURASEW, 1958 See Beton si Beton Armat. Comentarii STAS. Edit. Tehnica, Bucuresti.

OPLADEN, K. 1959 Rahmenformeln für symmetrische Rechteckrahmen mit elastisch eingespannten Stielfüßen. *Deutsche Bauzeitschr.*, No. 12, 1463.

ORAN, C. 1967 Complementary Energy Method for Buckling. *Proc. ASCE, Journ. Eng. Mech. Div.*, **93**, EM1, 57.

OSTAPENKO, A. 1965 Behaviour of Unbraced Frames. PDMSF-LN. Lehigh Univ. Fritz Eng. Lab., Bethlehem, Pennsylvania.

OXFORT, J. 1961 Über die Begrenzung der Traglast eines statisch unbestimmten biegesteifen Stabwerkes aus Baustahl durch das Instabilwerden des Gleichgewichtes. *Stahlbau*, **30**, No. 2, 33.

OXFORT, J. 1963 Die Verfahren zur Stabilitätasberechnung statisch unbestimmter biegesteifer Stahlstabwerke, verglichen an einem Untersuchungsbeispiel. *Stahlbau*, **32**, No. 2, 42.

PENELIS, G. 1968 Knickung räumlicher einstöckiger Rahmenträger. *Bauingenieur*, **43**, No. 7, 250.

PETERS, P. 1963 Beiwerte β zur Ermittlung der Knicklängen $S_K = \beta h$ freistehender Rechteckrahmen. *Bauplanung u. Bautechnik*, **17**, No. 2, 96.

POPOV, I. G. 1959 Ob ustoichivosti ploskikh simmetrichnik mnogoetazhnikh ram. Nauchnie doklavi vyschei shkoli. *Stroitel'stvo*, No. 2, 79.

PUWEIN, M. G. 1936-1938 Die Knickfestigkeit des Stockwerkrahmens. *Stahlbau*, **9**, 201; **10**, 7; **11**, 119.

RAEVSKI, A. N. 1964 Opredelenie ratsional'nykh otnoshenii zhiostkostei stoiek iz uslovia ih ravnoustoichivosti. *Stroitel'naia Mekh. i Raschet Sooruzh.*, **10**, No. 1, 23.

RESINGER, F. and STEINER, H. 1959 Näherungslösung von statischen Problemen der Theorie II. Ordnung und der Wölbkrafttorsion mit Hilfe der Deformationsmethode. *Österreichische Ing.-Zeitschr.*, **2**, No. 5, 172.

ROORDA, J. 1965 Stability of structures with small imperfection. *Proc. ASCE, Journ. Eng. Mech. Div.*, **91**, EM1, 87.

ROSMAN, R. 1961 Näherugsverfahren zur Lösung des Spannungsproblems freistehender Stockwerkrahmen nach der Theorie II. Ordnung. *Österreichische Ing.-Zeitschr.*, **4**, No. 11, 377.

ROSMAN, R. 1962 Beitrag zur Berechnung von Stockwerkrahmen nach der Spannungstheorie II. und III. Ordnung mit Hilfe des stellvertretenden Kragträgers. *Bautechnik*, **39**, No. 9, 275.

ROZENBLAT, G. I. 1962 Raschet na ustoichivosti ram so stupentsatimi stoikami. *Stroitel'naia Mekh. i Raschet Sooruzh*, No. 1, 23.

RÓZSA, M. 1967 A new iterative method for the finite deflection analysis of frameworks *Acta Techn. Acad. Sci. Hung.*, **57**, No. 1–2, 35.

SAAFAN, S. A. 1963 Nonlinear behaviour of structural plane frames. *Proc. ASCE, Journ. Struct. Div.*, **89**, ST4, 557.

SAHMEL, P. 1955 Näherungsweise Berechnung der Knicklängen von Stockwerkrahmen. *Stahlbau*, **24**, No. 4, 89.

SATTLER, K. 1953 Das Durchbiegungsverfahren zur Lösung von Stabilitätsproblemen. *Bautechnik*, **30**, No. 10, 288 and No. 11, 326.

SAWKO, F. and WILDE, A. M. B. 1967 Automatic analysis of strain hardening structures. *Proc. ICE*, **37**, No. 5, 195.

SCARLAT, A. 1969 Stabilitatea si calculul de ordinul II al structurilor si stabilitatea structurilor. Probleme speciale. Edit. Techn., Bucuresti.

SCHABER, E. 1960 Näherungsweise Berechnung der Schnittlasten von Stockwerkrahmen unter Berücksichtigung der Verformungen. *Bauingenieur*, **36**, No. 1, 7.

SCHEER, J. 1966 Zur Netzgleichung des auf Theorie II. Ordnung erweiterten Formänderungsgrößenverfahrens. *Stahlbau*, **35**, No. 7, 211.

SCHINEIS, M. 1960 Die Stabilität des Dreigelenk-Rechteckrahmens bei Riegelbelastung. *Bautechnik*, **37**, No. 12, 453.

SLAVIN, A. 1950 Stability studies of structural frames. *Transact. New York Acad. of Sci.* Series II, **12**, 82–93.

SMITH, B. S. 1962 Lateral stiffness of infilled frames. *Proc. ASCE, Journ. Struct. Div.*, **88**, ST6, 183.

SNITKO, N. K. 1952 Ustoichivost' sterzhnevykh sistem. *Gos. Izd. Lit. po Stroit. i. Arkh.* Moskva.

SNITKO, N. K. 1963 Ustoichivost' mnogoproletnykh mnogoetazhnykh ram. Izvestiia vyschikh uchebnikh zavedenii. *Stroitelstvo i architektura*. Novosibirsk, No. 9.

STEINHARDT, O. and BEER, H. 1968 Hochhäuser. Plastizitätstheorie. IABSE Eighth Congres, New York, Preliminary Report, 209.

STEVENS, L. K. 1964a Correction of virtual work relationships for the effect of axial load. *Struct. Eng.*, **42**, No. 5, 153.

STEVENS, L. K. 1964b Control of stability by limitation of deformations. *Proc. ICE*, **28**, No. 7, 57.

STEVENS, L. K. 1967 Elastic stability of practical multi-storey frames. *Proc. ICE*, **36**, No. 1, 99.

SWITZKY, H. and WANG, P. CH. 1969 Design and analysis of frames for stability. *Proc. ASCE, Journ. Struct. Div.*, **95**, ST4, 695.

TALL, S., BEEDLE, L. S. and GALAMBOS, TH. V. 1964 Structural Steel Design. Ronald Press, New York.

THOMPSON, J. M. T. and HUNT, G. W. 1973 A General Theory of Elastic Stability. Wiley, London.

TIMOSHENKO S. P. and GERE, J. M. 1961 Theory of Elastic Stability. McGraw-Hill, New York.

VOGEL, U. 1963 Über die Traglast biegesteifer Stahltragwerke. *Stahlbau*, **32**, No. 4, 106.

VOGEL, U. 1965 Die Traglastberechnung stählerner Rahmentragwerke nach der Plastizitätstheorie II. Ordnung. Forschungshefte aus dem Gebiete des Stahlbaues. Heft 15, Stahlbau-Verlag, Köln.

WOOD, R. H. 1958 The stability of tall buildings. *Proc. ICE*, **11**, No. 9, 69, and Discussion of the paper, *Proc. ICE*, **12**, 1959, No. 4, 502.

WRIGHT, E. W. and GAYLORD, E. H. 1968 Analysis of unbraced multi-storey steel rigid frames. *Proc. ASCE, Journ. Struct. Div.*, **94**, ST5, 1143.

YEN, Y. CH., LU, L. W. and DRISCOLL, JR. G. C. 1962 Tests on the stability of welded steel frames. *Welding Research Council Bull.*, No. 81, Sept.

426   References

ZALKA, K. A. and ARMER, G. S. T. 1992 Stability of Large Structures. Butterworth-Heinemann, Oxford.

ZALKA, K. A. 1992 Characteristic Deformations and the Stability of Large Frameworks. Building Research Establishment, N214/92. Watford, England, 25.

ZALKA, K. A. 1998 Equivalent Wall for Frameworks for the Global Stability Analysis. Building Research Establishment, N33/98. Watford, England.

Discussion of the paper: DAVIES, J. M. Frame instability and strain hardening in plastic theory. *Proc. ASCE, Journ. Struct. Div.*, 92, ST3 (1966), 1 and *Proc. ASCE, Journ. Struct. Div.*, 93, ST1, (1967), 590.

AISC 1978 Specifications for the Design, Fabrication and Erection of Structural Steel for Buildings. American Institute of Steel Construction, Chicago, and Commentary.

DIN 4114 Stahlbau. Stabilitätsfälle. (Knippung, Kippung, Beulung). Berechnungsgrundlagen. Blatt 1 Vorschriften (1952), Blatt 2 Richtlinien (1953). DIN 18800 Teil 2 Stahlbauten. Stabilitätsfälle. Knicken von Stäben und Stabwerken. Entwurf.

L'Institut technique du bâtiment et des travaux publics. Le centre technique industriel de la construction métallique. Règles de calcul des constructions en acier. Société de diffusion des techniques du bâtiment et des travaux publiques, Paris, 1966.

REFERENCES TO SECTION 5.2

BLEICH, FR. 1952 Buckling Strength of Metal Structures. McGraw-Hill, New York.

DULÁCSKA, E. and KOLLÁR, L. 1960 Angenäherte Berechnung des Momentenzuwaches und der Stabilität von gedrückten Rahmenstielen. *Bautechnik*, 37, No. 3, 98–109.

DULÁCSKA, E. 1987 Stabilitási útmutató és példatár. Tervezési segédlet S-23 (Guide-book and examples of stability. Work-help for design.) Tervezésfejlesztési és Technikai Intézet, Budapest.

KORONDI, D. 1974 Acél keretszerkezetek közelítő stabilitásvizsgálata. Műszaki doktori értekezés (Approximate stability analysis of steel frames. PhD thesis.) TU Budapest.

LIPTÁK, L. 1973 Talajtörés vagy helyzeti állékonyság? (Soil failure or overturning?) *Mélyépítéstudományi Szemle*, 23, 229–240.

LIPTÁK, L. 1960 Az egyenestengelyű, prizmatikus rúd rugalmas kihajlása tetszőleges befogási viszonyok esetén (Elastic buckling of straight bars of constant cross section with arbitrary support conditions). *Építés- és Közlekedéstudományi Közlemények*, 4, 621–626.

NEMESTÓTHY, É. 1972 Rugalmasan befogott oszlopok kihajlási hossza. (Buckling lengths of elastically restained columns). *Magyar Építőipar*, 576–580.

ZALKA, K. A. and ARMER, G. S. T. 1992 Stability of Large Structures. Butterworth-Heinemann, Oxford.

REFERENCES TO CHAPTER 6

ALLEN, H. G. 1969 Analysis and Design of Structural Sandwich Panels. Pergamon Press, Oxford.

ASZTALOS, Z. 1972 Stability analysis of multi-storey frames by the continuum method. (In Hungarian: Kétlábú sokszintes keretszerkezetek kihajlásvizsgálata a kontinuummodellel, az oszlopok alakváltozásainak figyelembevételével.) *Magyar Építőipar*, **21**, 471–474.

BECK, H. 1956 Ein neues Berechnungsverfahren für gegliederte Scheiben, dargestellt am Beispiel des Vierendeel–Trägers. *Bauingenieur*, **31**, 436–443.

BECK, H. 1959 Ein Beitrag zur Berücksichtigung der Dehnungsverformungen bei Rahmen mit schlanken und gedrungenen Konstruktionsgliedern. *Die Bautechnik*, **36**, 178–184.

BENSON, A. and MAYERS, I. 1967 General instability and face wrinkling of sandwich plates, unified theory and application. *AIAA Journ.*, **5**(4), 729–739.

BLEICH, F. and MELAN, E. 1927 Die gewöhnlichen und partiellen Differenzengleichungen der Baustatik. Julius Springer, Berlin.

BOLOTIN, V. V. 1965 Osvedenyi trojmernih zadach teorii uprugoy ustoychivosty odnomernim i dvuhmernim zadacham. In: Problemi Ustoychivosty v Stroitelnoy Mechanyke. (Eds: Bolotin, V. V., Rabinovich, I. M. and Smirnov, A. F.) 166–179. Stroiizdat, Moskva.

BRAND, L. 1966 Differential and Difference Equations. Wiley, New York.

CHONG, K. P. and HARTSOCK, J. A. 1969 Flexural wrinkling in foam-filled sandwich panels. *Proc. ASCE, J. Eng. Mech. Div.*, **95**, 585–610.

CSONKA, P. 1956 Über proportionierte Rahmen. *Die Bautechnik*, **33**, 19–20.

CSONKA, P. 1961a Buckling of bars elastically built in along their entire length. *Acta Techn. Hung.*, **32**, 424–427.

CSONKA, P. 1961b Calculation of auto-portant railway carriage frames. *Acta Techn. Hung.*, **33**, 143–164.

CSONKA, P. 1962 Beitrag zur Berechnung waagerecht belasteter Stock-Werkrahmen. *Die Bautechnik*, **39**, 237–240.

CSONKA, P. 1965a Simplified analysis of multi-storey frames subjected to wind loads. (In Hungarian: Egyszerűsített eljárás szélerőkkel terhelt emeletes keretek számítására). Az *MTA VI. Oszt. Közl.*, **35**, 209–219.

CSONKA, P. 1965b Bending moments of multi-storey frames subjected to wind loads. (In Hungarian: Szélerőkkel terhelt sokemeletes derékszögű keretek legnagyobb nyomatékai.) *MTA VI. Oszt. Közl.*, **35**, 271–275.

GOSCHY, B. 1970 Räumliche Stabilität von Großtafelbauten. *Die Bautechnik*, **47**, 416–425.

HEGEDŰS, I. 1979 Buckling of axially compressed cylindrical sandwich shells. *Acta Techn. Hung.*, **89**, 377–387.

HEGEDŰS, I. and KOLLÁR, L. P. 1984a Buckling of sandwich columns with thin faces under distributed normal loads. *Acta Techn. Hung.*, **97**, 111–122.

HEGEDŰS, I. and KOLLÁR, L. P. 1984b Buckling of sandwich columns with thick faces subjected to axial loads of arbitrary distribution. *Acta Techn. Hung.*, **97**, 123–132.

428　References

HEGEDŰS, I. and KOLLÁR, L. P. 1987 Stabilitätsuntersuchung von Rahmen und Wandscheiben mit der Sandwichtheorie. *Bautechnik*, **64**, 420–425.

HEGEDŰS, I. and KOLLÁR, L. P. 1988*a* Buckling of sandwich plates with thick faces. (In Hungarian: Vastag héjalású szendvicslemez horpadása). *Mélyépítéstudományi Szemle*, **38**, 173–176.

HEGEDŰS, I. and KOLLÁR, L. P. 1988*b* Stability analysis of bars elastically restrained from rotation along their entire length. *Acta Techn. Hung.*, **101**, 57–65.

HEGEDŰS, I. and KOLLÁR, L. P. 1988*c* Generalized bar models and their physical interpretation. *Acta Techn. Hung.*, **101**, 67–93.

HEGEDŰS, I. and KOLLÁR, L. P. 1989 Wrinkling of faces of compressed and bent sandwich bars and its interaction with overall instability. *Acta Technica Hung. Civil Eng.*, **102**, 49–63.

HEGEDŰS, I. and KOLLÁR, L. P. 1990 Buckling of mirror symmetrical plane structures under nonsymmetrical loads. *Mechanics of Structures and Machines*, **18**(4).

HEGEDŰS, I. and KOLLÁR, L. P. 1999 Buckling analysis of laced columns using difference equations and the continuum method. *Acta Techn. Hung. Civil Eng.* (In press).

JORDAN, C. (K.) 1947 Calculus of Finite Differences. Chelsea, New York.

KÁRMÁN, T. and BIOT, M. A. 1940 Mathematical Methods in Engineering. McGraw-Hill, New York.

KOLLÁR, L. (ed.) 1991 Special Problems of Structural Stability (In Hungarian: A mérnöki stabilitáselmélet különleges problémái). Akadémiai Kiadó, Budapest.

KOLLÁR, L. and HEGEDŰS, I. 1985 Analysis and Design of Space Frames by the Continuum Method. Elsevier, Amsterdam and Akadémiai Kiadó, Budapest.

KOLLÁR, L. P. 1986*a* Buckling analysis of coupled shear walls by multi-layer sandwich model. *Acta Techn. Hung.*, **99**, 317–332.

KOLLÁR, L. P. 1986*b* Stability Analysis of Frames and Shar Walls by the Method of Difference Equtions and by the Continuum Method. PhD Thesis. (In Hungarian: Keretek és merevítőfalak stabilitásvizsgálata a differenciaegyenletek módszerével és a kontinuum módszerrel. Kandidátusi értekezés). Hungarian Academy of Sciences, Budapest.

KOLLÁR, L. P. 1990 Buckling of generally anisotropic shallow sandwich shells. *Journ. Reinforced Plastics and Composites*, **9**, 549–568.

KOLLÁR, L. P. 1994 Buckling of anisotropic cylinders. *Journ. Reinforced Plastics and Composites*, **13**, 954–975.

LIGETI, R. 1974 Analysis of the Stiffening System of Buildings (In Hungarian: Épületek merevítő falainak és falrendszereinek számítási módszerei). BVTV.

PLANTEMA, F. J. 1961 Sandwich Construction. The Bending and Buckling of Sandwich Beams, Plates and Shells. Wiley, New York.

POMÁZI, L. 1980 Stability of rectangular sandwich plates with constructionally orthotropic hard layers. *Periodica Polytechnica, Mech. Eng., Part. I*: **24**, 203–221.

RATZERSDORFER, J. 1936 Die Knickfestigkeit von Staben und Stabwerken. Julius Springer, Berlin.

ROSMAN, R. 1965 Die statische Berechnung von Hochhauswänden mit Öffnungsreihen. Bauingenieur-Praxis. Heft 65, Ernst, Berlin.

ROSMAN, R. 1968 Statik und Dynamik der Scheibensysteme des Hochbaues. Springer, Berlin.

ROSMAN, R. 1974 Stability and dynamics of shear-wall frame structures. *Build. Sci.*, **9**, 55–63. Pergamon Press, Oxford.

ROSMAN, R. 1980 Stabilität im Grundriss unsymmetrischer Stützen- und Wandscheiben-Systeme. *Die Bautechnik*, **57**, 21–33.

STAFFORD SMITH, B. and CROWE, E. 1986 Estimating of periods of vibration of tall buildings. *J. Struct. Engr.*, **112**, 1005–1019.

SZERÉMI, L. 1975 Analysis of the Stiffening System of High Rise Buildings by the Continuum Method (In Hungarian: Magasházak merevítő rendszerének számítása a kontinuum modell alkalmazásával). BME Építőanyagok Tanszék, *Tudományos Közlemények*, **21**, *Beton és Vasbeton*, 123–140.

SZERÉMI, L. 1978 Stiffening system of multi-storey buildings by the continuum modell. *Periodica Polytechnica Civ. Eng.*, **22**, 205–218.

SZERÉMI, L. 1984 Stiffening Systems of Building Structures. (In Hungarian: Épületmerevítések számítása). In: Mérnöki Kézikönyv II. Red.: Palotás László. Műszaki Könyvkiadó, Budapest.

TIMOSHENKO, S. and GERE, J. 1961 Theory of Elastic Stability. McGraw-Hill, New York.

WLASSOW, W. S. 1965 Dünnwandige elastische Stäbe. VEB Verlag für Bauwesen, Berlin.

ZALKA, K. 1976 Stability investigation of frameworks attacked on top by concentrated forces with the aid of the method of continuum. *Acta Techn. Hung.*, **83**, 357–374.

ZALKA, K. 1977 Stability investigation of frameworks subjected to distributed normal loads with the aid of the method of continuum. *Acta Techn. Hung.*, **84**, 43–59.

ZALKA, K. 1979 Buckling of a cantilever subjected to distributed normal load, taking shearing deformation into account. *Acta Techn. Hung.*, **89**, 497–508.

ZALKA, K. 1980 Torsional buckling of a cantilever subjected to distributed normal loads. *Acta Techn. Hung.*, **90**, 91–108.

ZALKA, K. and ARMER, G. S. T. 1992 Stability of Large Structures. Butterworth-Heinemann, Oxford.

ZALKA, K. 1987 Stabillity of large frameworks on hinged supports. *Acta Techn. Hung.*, **100**, 355–383.

REFERENCES TO CHAPTER 7

AKESSON, B. 1980 Overall buckling in bending and torsion of rack structures. In: Thin-walled Structures. Ed. J Rhodes and A. C. Walker. Chatto and Windus, London. 127–144.

BARTA, T. A. 1967 On the torsional-flexural buckling of thin-walled elastic bars with monosymmetric open cross section. In: Thin-walled Structures. Ed. A. H. Chilver. Chatto and Windus, London, 60–86.

BECK, H. and SCHÄFER, H. 1969 Berechnung von Hochhäusern durch Zusammenfassung aller aussteifenden Bauteile zu einem Balken. *Der Bauingenieur*, **44**, 80–87.

BORNSCHEUER, F. W. 1952 Systematische Darstellung des Biege- und Verdrehvorganges unter besonderer Berücksicht. *Der Stahlbau*, **21**, Heft 1, 1–9.

BRANDT, B., SCHÄFER, H. and REEH, H. 1975 Zum Stabilitätsnachweis von Hochhäusern. *Beton- und Stahlbetonbau*, **70**, 211–223.

DANAY, A., GELLÉRT, M. and GLÜCK, J. 1975 Continuum method for overall stability of tall asymmetric buildings. *Proc. ASCE, Journ. Struct. Div.*, **101**, ST12, 2505–2521.

DULÁCSKA, E. 1966 Structural design of reinforced concrete shear wall structures (in Hungarian). *Mélyépítéstudományi Szemle*, **16**, 466–475.

FLACHSBART, O. 1932 Messungen an ebenen und gewölbten Platten. Ergebnisse der Aerodynamischen Versuchsanstelt zu Göttingen. (Herausgegeben von L. Prandtl und A. Betz). Oldenbourg, München-Berlin. IV. Lieferung.

GLÜCK, A. and GELLÉRT, M. 1972 Buckling of lateral restrained thin-walled cantilevers of open cross section. *Proc. ASCE, Journ. Struct. Div.*, **98**, 2031–2042.

HEGEDŰS, I., and KOLLÁR, L. P. 1987 Stabilitätsuntersuchung von Rahmen und Wandscheiben mit der Sandwichtheorie. *Die Bautechnik*, **64**, 420–425.

IRWIN, A. W. 1984 Design of shear wall buildings. CIRIA Report 102. Construction Industry Research and Information Association, London.

KÓKAI, T. 1983 Axis of twist for the bracing system of tall buildings (in Hungarian). University doctoral thesis. Department of Reinforced Concrete Structures, Budapest Technical University.

KOLLÁR, L. 1977 Stiffening of buildings against torsional buckling. *Regional Colloquium on Stability of Steel Structures. Budapest-Balatonfüred. 19–21 October 1977*. Ed. O. Halász, M. Iványi. 411–422.

KOVÁCS, O. and FABER, G. 1963 Handbook of Elastic Stability (in Hungarian). Műszaki Kiadó, Budapest.

LIPTÁK, L. 1973a Soil failure or overturning? (in Hungarian). *Mélyépítéstudományi Szemle*, **23**, 229–240.

LIPTÁK, L. 1973b Design against overturning – a new approach (in Hungarian). *Mélyépítéstudományi Szemle*, **23**, 534–538.

LUSAS, 1995 LUSAS Finite Element System V11. User Manual. FEA Ltd, Forge House, Kingston Upon Thames, Surrey, KT1 1HN, UK.

MACLEOD, I. A., and ZALKA, K. A. 1996 The global critical load ratio approach to stability of building structures. *The Structural Engineer*, **74**, No. 15, 249–254.

PEARSON, C. E. 1956 Remarks on the centre of shear. *Z. angew. Math. Mech.*, **36**, Nr. 314, 94–96.

PFLÜGER, A. 1950 Stabilitätsprobleme der Elastostatik. Springer, Berlin.

PODOLSKI, D. M. 1970 Spatial analysis of tall buildings (in Russian) *Structural Mechanics*. **68**, 63–69.

ROSMAN, R. 1965 Die statische Berechnung von Hochhauswänden mit öffnungsreihen. Bauingenieur-Praxis, Heft **65**. Ernst, Berlin.

ROSMAN, R. 1980 Stabilität im Grundriß unsymmetrischer Stützen- und Wandscheibensysteme. *Die Bautechnik*, **57**, 21–32.

STAFFORD SMITH, B. and COULL, A. 1991 Tall Building Structures. Wiley, New York.

STÜSSI, F. 1965 Die Grenzlagen der Schubmittelpunktes bei Kastenträgern. *IVBH Publications*, **25**, 279–315.

SZMODITS, K. 1975 Handbook for the Structural Design of Panel Type Structures (in Hungarian). Építéstudományi Intézet, Budapest.

Tall Building Systems and Concepts, 1980 Monograph on Planning and Design of Tall Buildings, Vol SC, Foundation Systems. ASCE, 259–340.

TARANATH, B. S. 1988 Structural Analysis and Design of Tall Buildings. McGraw-Hill, London.

TIMOSHENKO, S.,and GERE, J. 1961 Theory of Elastic Stability. McGraw-Hill, New York.

VLASOV, V. Z. 1961 Thin-walled Elastic Beams. Israeli Program for Scientific Translation, Jerusalem.

YARIMCI, E. 1972 The equivalent-beam approach in tall building stability analysis. In: *Proc. Int. Conf. Planning and Design of Tall Buildings*. Vol. **II-16**, 553–563.

ZALKA, K. A. 1980 The torsion analysis of an I-column with rectangular openings in both flanges. Building Research Establishment, N161/80, Watford, England.

ZALKA, K. A. 1994 Mode coupling in the torsional-flexural buckling of regular multi-storey buildings. *Int. Journ. Structural Design of Tall Buildings*. **3**, 227–245.

ZALKA, K. A.,and ARMER, G. S. T. 1992 Stability of Large Structures. Butterworth-Heinemann, Oxford.

ZIENKIEWITZ, O. C., PAREKH, C. J. and TEPLY, B. 1971 Three dimensional analysis of buildings composed of floor and wall panels. *Proc. ICE*, **49**, 319–332.

REFERENCES TO SECTION 8.1

AMAZIGO, J. C. 1978 Optimal shape of shallow circular arches against snap-buckling. *Journ. Appl. Mech.*, **45**, 591–594.

AUSTIN, W. J. 1971 In-plane bending and buckling of arches. *Proc. ASCE, Journ. Struct. Div.*, , **97**, ST4, 1575–1592.

BATOZ, J.-L. 1981 Importance de la théorie non linéaire et du type de pression sur le flambement d'arcs et d'anneaux circulaires. *Journ. de Mécanique Appliquée*, **5**, 219–238.

BODNER, S. R. 1958 On the conservativeness of various distributed force systems. *Journ. Aeronaut. Sci.*, **25**, 132.

BORESI, A. P. 1955 A refinement of the theory of buckling of rings under uniform pressure. *Journ. Appl. Mech.*, **22**, 95–102.

BRAZIER, L. G. 1927 On the Flexure on thin cylindrical shells and other 'thin' sections. *Proc. Roy. Soc. London Ser. A.*, **116**, 104–114.

BRUSH, D. O. and ALMROTH, B. O. 1975 Buckling of Bars, Plates and Shells. McGraw-Hill, New York.

BUDIANSKY, B. 1974 Theory of buckling and postbuckling behaviour of elastic structures. *Adv. Applied Mechanics*, **14**, 1–65.

432 References

CARRIER, G. F. 1947 On the buckling of elastic rings. *Journ. Mathematics and Physics,* **26,** 94–103.

CHWALLA, E. and KOLLBRUNNER, C. F. 1938 Beiträge zum Knickproblem des Bogenträgers und des Rahmens. *Der Stahlbau,* **11,** 73–78; 81–84; 94–96.

DEUTSCH, E. 1940 Das Knicken von Bogenträgern bei unsymmetrischer Belastung. *Bauingenieur,* **21,** 353–360.

DICKIE, J. F. and BROUGHTON, P. 1971 Stability criteria for shallow arches. *Proc. ASCE, Journ. Mech. Div.,* **97,** EM3, 951–965.

DIN 4114 1952 Stahlbau, Stabilitätsfälle (Knickung, Kippung, Beulung).

DINNIK, A. N. 1955 Prodol'niĭ izgib. Kruchenie. Izd. Akad. Nauk SSSR, Moscow.

DISCHINGER, F. 1937 Untersuchungen über die Knicksicherheit, die elastische Verformung und das Kriechen des Betons bei Bogenbrücken. *Der Bauingenieur,* **18,** 487–520; 539–552; 595–621.

DISCHINGER, F. 1939 Elastische und plastische Verformungen der Eisenbetontragwerke und insbesondere der Bogenbrücken. *Der Bauingenieur,* **20,** 53–63 286–294; 426–437; 563–572.

DULÁCSKA, E. 1964 Stability of eccentrically compressed shell-arches. *Acta Techn. Acad. Sci. Hung.,* **45,** 351–359.

DULÁCSKA, E. and KOVÁCS, I. 1971 Vonórúddal merevített parabolaívek kihajlása (Stability of tied parabolic arches). *Ép. és Közl.-tud. Közl.,* 39–66.

DYM, C. L. 1973 Buckling and postbuckling behaviour of steep compressible arches. *Int. Journ. Solids Structures,* **9,** 129–140.

DYM, C. L. 1977 Buckling of supported arches under three pressure distributions. *Journ. Appl. Mech.,* **44,** 764–766.

EIBL, J. 1963 Zur Stabilitätsfrage des Zweigelenkbogens mit biegeweichem Zugband und schlaffen Hängestangen. Institut für Baustoffkunde und Stalhbetonbau der TH Braunschweig. Heft 3.

EL NASCHIE, M. S. 1975 The initial post-buckling of an extensional ring under external pressure. *Int. Journ. Mech. Sci.,* **17,** 387–388.

FEDERHOFER, K. 1934 Über Eigenschwingungen und Knicklasten des parabolischen Zweigelenkbogens. *Sitzungsber. d. Akad. d. Wiss. Wien,* Abt. IIa **143,** 131–150.

FIROOZBAKSH, K. and FARSHAD, M. 1976 On minimum weight funicular arches. *Mech. Res. Comm.,* **3,** 269–275.

FUNG, Y. C. and KAPLAN, A. 1952 Buckling of Low Arches or Curved Beams of Small Curvature. NACA Technical Note 2840.

HILMAN, L. 1930 Ob ustojchivosti parabolicheskikh arok pri vertikal'noi ravnomerno raspredelennoi nagruzke. Izvestiia Leningradskogo Politekhnicheskogo Instituta. 33.

HOFF, N. J. and BRUCE, V. G. 1953 Dynamic analysis of the buckling of laterally loaded flat arches. *Journ. Math. Physics,* **32,** 276–288.

HUANG, N. C. 1967 Nonlinear creep buckling of some simple structures. *Journ. Appl. Mech.* **34,** 651–658.

KÁRMÁN, TH. VON 1911 Über die Formänderung dünnwandiger Rohre, insbesondere federnder Ausgleichsrohre. *VDI-Zeitschrift,* **55,** 1889-1895.

KOLLÁR, L. 1961a Stability of centrally compressed shell-arches. *Acta Techn. Acad. Sci. Hung.*, **32**, 11–38.

KOLLÁR, L. 1961b Stability of bent shell-arches. *Acta Techn. Acad. Sci. Hung.*, **32**, 267–297.

KOLLÁR, L. 1973 Statik und Stabilität der Schalenbogen und Schalenbalken. Akadémiai Kiadó, Budapest and W. Ernst, Berlin.

KOLLBRUNNER, C. F. and MEISTER, M. 1961 Knicken, Biegedrillknicken, Kippen. Springer, Berlin.

KOVÁCS, I. 1974 Allgemeines Knicken des mit Hängestangen und Balken versteiften Parabelbogens. PhD Thesis. Technische Universität, Stuttgart.

KOVÁCS, I. 1984 Private communication.

LOKSCHIN, A. 1936 Über die Knicking eines gekrümmten Stabes. *Zeitschrift angew. Math. Mech.*, **16**, 49–55.

MASUR, E. F. and LO, D. L. C. 1972 The shallow arch – general buckling, postbuckling, and imperfection analysis. *Journ. Struct. Mech.*, **1**, 91–112.

MAYER-MITA, R. 1913 Die Knicksicherheit in sich versteifter Hängebrücken, sowie des Zwei- und Dreigelenkbogens innerhalb der Tragwandebene. *Der Eisenbau*, **4**, 423–428.

NAGY, J. 1978 Stability of flat sandwich arches. *Acta Techn. Acad. Sci. Hung.* **87**, 451–479.

ORAN, C. and BAYAZID, H. 1978 Another look at buckling of circular arches *Proc. ASCE, Journ. Eng. Mech. Div.*, **104**, EM6, 1417–1432.

PEARSON, C. E. 1956 General theory of elastic stability. *Quart. Appl. Math.*, **14**, 133–144.

PETERSEN, CHR. 1980 Statik und Stabilität der Baukonstruktionen. Fr. Vieweg, Braunschweig.

PFLÜGER, A. 1951 Ausknicken des Parbelbogens mit Versteifungsträger. *Der Stahlbau*, **20**, 117–120.

RÓZSA, M. 1964 Stability analysis of arches with vertical load. *Acta Technica Acad. Sci. Hung.*, **49**, 387–397.

SANDERS, J. L. 1963 Nonlinear theories for thin shells. *Quart. Appl. Math.*, **21**, 21–36.

SCHIBLER, W. 1948 Ebenes Knicken von Zweigelenkbogen unter berücksichtigung des Aufbaues. *Schweizerische Bauzeitung*, **66**, 482–485.

SCHMIDT, R. 1978 Initial post-buckling of hingeless circular arch. *Proc. ASCE, Journ. Eng. Mech. Div.*, **104**, EM4, 959–964.

SCHMIDT, R. 1979a Improved values of buckling pressure for one-hinged and three-hinged circular arches. *Industrial Mathematics*, **29**, 123–126.

SCHMIDT, R. 1979b A critical study of post-buckling analyses of uniformly compressed rings. *Zeitschrift angew. Math. Mech.*, **59**, 581–582.

SCHMIDT, R. 1979c Initial post-buckling behavior of circular arches with hinged ends. *Industrial Mathematics*, **29**, 27–38.

SCHMIDT, R. 1979d Initial post-buckling behavior of uniformly compressed two-hinged circular arches. *Zeitschrift angew. Math. Mech.*, **59**, 473–474.

SCHMIDT, R. 1979e Post-buckling behaviour of uniformly compressed circular arches with clamped ends. *Zeitschrift angew. Math.* , **30**, 553–556.

SCHMIDT, R. 1979*f* Initial postbuckling of three-hinged circular arch. *Journ. Appl. Mech.*, **46**, 954-955.

SCHMIDT, R. 1980*a* Buckling of rings subjected to unconventional loads. *Industrial Mathematics*, **30**, 135-143.

SCHMIDT, R. 1980*b* Critical constant-directional pressure on circular rings and hingeless arches. *Zeitschrift angew. Math. Phys.*, **31**, 776-779.

SCHMIDT, R. 1981 Critical values of centrally directed pressure acting on hingeless circular arches. *Industrial Mathematics*, **31**, 75-77.

SCHREYER, H. L. and MASUR, E. F. 1966 Buckling of shallow arches. *Proc. ASCE, Journ. Eng. Mech. Div.*, **92**, EM4. 1-19.

SCHREYER, H. L. 1972 The effect of initial imperfections on the buckling load of shallow circular arches. *Journ. Appl. Mech.* , **39**, 445-450.

SILLS, L. B. and BUDIANSKY, B. 1978 Post-buckling ring analysis. *Journ. Appl. Mech.*, **45**, 208-210.

SIMITSES, G. J. and RAPP, I. H. 1977 Snapping of low arches with nonuniform stiffness. *Proc. ASCE, Journ. Eng. Mech. Div.*, **103**, EM1, 51-65.

SINGER, J. and BABCOCK, C. D. 1970 On the buckling of rings under constant directional and centrally directed pressure. *Journ. Appl. Mech.* , **37**, 215-218.

STÜSSI, F. 1935 Aktuelle baustatische Probleme der Konstruktionspraxis. *Schweizerische Bauzeitung*, **106**, 119-122; 132-136.

TIETZE, M. 1973 Die kritische Stabilitätsfunktion NYO des Parabelbogens bei Berücksichtigung des Einflusses der Schrägstellung von Ständern bzw. Hängern, ermittelt mit Hilfe des Drehwinkelverfahrens nach Theorie 2. Ordnung. Diplomarbeit. TU, München.

TIMOSHENKO, S. P. and GERE, J. M. 1961 Theory of Elastic Stability. McGraw-Hill, New York,.

WEINEL, E. 1937 Über Biegung und Stabilität eines doppelt gekrümmten Plattenstreifens. *Zeitschrift angew. Math. Mech.*, **17**, 366-369.

WEMPNER, G. A. and KESTI, N. E. 1962 On the buckling of circular arches and rings. *Proc. 4th US Nat. Congr. Appl. Mech.* , **2**, 843-849.

REFERENCES TO SECTION 8.2

BÓDI, I. 1985 Lateral buckling of elastically restrained arches with built-in supports. *Acta Techn. Acad. Sci. Hung.*, **98**, 181-196.

GODDEN, W. G. 1954 The lateral buckling of tied arches. *Proc. Inst. Civil Engrs (London)*, **3**, 496-514.

GODDEN, W. G. and THOMPSON, J. C. 1959 An experimental study of a model tied-arch bridge. *Proc. Inst. Civil Engrs* (London) **14**, 383-394.

HAVIÁR, GY., GÁLLIK, I. and MAGYAR, Á. SZ. 1954 Vasbeton ívhíd kismintamérése. (Model test of a reinforced concrete arch bridge). *Mélyépítéstudományi Szemle*, **4**, 288-300.

KEE, C. F. 1961 Lateral inelastic buckling of tied arches. *Proc. ASCE, Journ. Struct. Div.*, **87**, ST1, 23–39.

KOLLÁR, L. and IVÁNYI, GY. 1966 Kippen und Biegedrillknicken von Schalenbogen mit Hilfe der Energiemethode. *Bautechnik-Archiv*, Heft 19, W. Ernst, Berlin.

KOLLÁR, L. 1962 Torsional buckling of thin-walled curved bars (shell-arches). *Acta Techn. Acad. Sci. Hung.*, **40**, 337–353.

KOLLÁR, L. 1964*a* Lateral buckling of thin-walled curved bars (shell-arches). *Acta Techn. Acad. Sci. Hung.*, **45**, 297–314.

KOLLÁR, L. 1964*b* Lateral buckling of shell-arches, taking into consideration nonlinear stress distribution. *Bulletin IASS*, No. 19.

KOLLÁR, L. and GÁRDONYI, Z. 1967 Kippen von Schalenbogen unter antimetrischer Belastung. *Acta Techn. Acad. Sci. Hung.*, **59**, 243–263.

KOLLÁR, L. 1973 Statik und Stabilität der Schalenbogen und Schalenbalken. Akadémiai Kiadó, Budapest. – W. Ernst, Berlin.

KOLLÁR, L. 1982 The supporting effect of the fabric of tent structures stretched onto an arch row on the lateral stability of the arches. *Acta Techn. Acad. Sci. Hung.*, **94**, 197–214.

KOLLÁR, L. and GYURKÓ, J. 1982 Lateral buckling of elastically supported arches. *Acta Techn. Acad. Sci. Hung.*, **94**, 37–45.

KOLLÁR, L. and BÓDI, I. 1982 Lateral buckling of arches with fork-like supports, elastically restrained along their entire length against lateral displacement and rotation. *Acta Techn. Acad. Sci. Hung.*, **95** 99–106.

KOVÁCS, I. 1974 Allgemeines Knicken des mit Hängestangen und Balken versteiften Parabelbogens. Dissertation. Technische Universität, Stuttgart.

ÖSTLUND, L. 1954 Lateral Stability of Bridge Arches Braced with Transverse Bars. Kungl. Tekniska Högskolans Handlingar, Stockholm. No. 84.

PETERSEN, CHR. 1980 Statik und Stabilität der Baukonstruktionen. Vieweg, Braunschweig.

SAKIMOTO, T. and KOMATSU, S. 1982 Ultimate strength of arches with bracing systems. *Proc. ASCE, Journ. Struct. Div.*, **108**, ST5, 1064–1076.

SAKIMOTO, T. and KOMATSU, S. 1983 Ultimate strength formula for steel arches. *Proc. ASCE, Journ. Struct. Div.*, **109**, 613–627.

SAKIMOTO, T. and YAMAO, T. 1983 Ultimate strength of deck-type steel arch bridges. In: *Stability of Metal Structures*. Third International Colloquium. Preliminary Report. Paris.

SHUKLA, S. and OJALVO, M. 1971 Lateral buckling of parabolic arches with tilting loads. *Proc. ASCE, Journ. Struct. Div.*, **97**, ST6, 1763–1773.

STEIN, P. 1961 Die Anwendung des 'Durchbiegungsverfahrens' zur Ermittlung der kritischen Last von Bogenträgern beim Ausweichen senkrecht zu ihrer Ebene unter Berücksichtigung von biegeweichen Hängern. *Bauingenieur*, **36**, 175–183.

STÜSSI, F. 1943-44 Kippen und Querschwingungen von Bogenträgern. *Publ. IABSE*, **7**, 327–343.

TIMOSHENKO S. P. and GERE, J. M. 1961 Theory of Elastic Stability. McGraw-Hill, New York.

TOKARZ, F. J. and SANDHU, R. S. 1972 Lateral-torsional buckling of parabolic arches *Proc. ASCE, Journ. Struct. Div.*, **98**, ST5, 1161–1179.

REFERENCES TO CHAPTER 9

BECK, H. and SCHACK, R. 1972 Bauen mit Beton- und Stahlbetonfertigteilen. *Beton-Kalender*, Bd II. 159–256.

BLEICH, F. 1952 Buckling Strength of Metal Structures. McGraw-Hill, New York.

BÖLCSKEI, E. 1954 Die Stabilität des an zwei Punkten aufgehängten geraden Balkens. *Acta Techn. Acad. Sci. Hung.*, **8**, 243–256.

BÖRÖCZ, I. 1956 Kiviteli pontatlanságok hatása a végein felfüggesztett négyszögkereszt-metszetű rúd stabilitására. (The influence of imperfections on the stability of a beam with rectangular cross section, suspended at both ends). *Mélyépítéstudományi Szemle*, **6**, 105–115.

BÜRGERMEISTER, G., STEUP, H. and KRETSCHMAR, H. 1966 Stabilitätstheorie, Teil I. Akademie-Verlag, Berlin.

CHWALLA, E. 1944 Kippung von Trägern mit einfachsymmetrischen, dünnwandigen und offenen Querschnitten. *Sitzungsberichte der Akademie der Wissenschaften in Wien.* Abt. IIa, **153**, 25–60.

CSONKA, P. 1954a Die Stabilität der an ihren Enden aufgehängten prismatischen Stäbe von rechteckigem Querschnitt. *Acta Techn. Acad. Sci. Hung.*, **8**, 79–99.

CSONKA, P. 1954b Die Stabilität des an einem Punkt aufgehängten geraden Balkens. *Acta Techn. Acad. Sci. Hung.*, **8**, 389–397.

CSONKA, P. 1954c Die Stabilität des an seinen Enden aufgehängten, an seiner seitlichen Verschiebung gehinderten Balkens. *Acta Techn. Acad. Sci. Hung.*, **10**, 31–42.

DULÁCSKA, E. and TARNAI, T. 1979 Kippen ebener Parallelgurtfachwerke. *Bauplanung – Bautechnik*, **33**, 173–175.

GÁRDONYI, Z. 1979 Felfüggesztett vékonyfalú gerendák és ívek oldalirányú stabilitása (Lateral buckling of suspended thin-walled beams and arches). PhD Thesis. Technical University of Budapest.

KEREK, A. 1981 Static and stability investigation of bent folded plates. *Acta Techn. Acad. Sci. Hung.*, **93**, 39–65.

KIRSTE, L. 1950 Das Ausknicken von Fachwerken aus ihrer Ebene. *Österr. Ing.-Archiv.*, **4**, 136–138.

KOLLÁR, L. and GÁRDONYI, Z. 1967 Déversement latéral des poutres à paroi mince sus-pendues par leurs extrémités, *Acta Techn. Acad. Sci. Hung.*, **57**, 187–210.

KOLLÁR, L. 1973 Statik und Stabilität der Schalenbogen und Schalenbalken. Akadémiai Kiadó, Budapest and W. Ernst, Berlin.

KOLLÁR, L. (ed.) 1991 A mérnöki stabilitáselmélet különleges problémái (Special problems of structural stability). Akadémiai Kiadó, Budapest.

KORDA, J. 1963 Ferde kötelekkel emelt gerenda stabilitása (The stability of a beam sus-pended by inclined cables). *Magyar Építőipar*, 283–285.

LEBELLE, P. 1959 Stabilité élastique des poutres en béton à l'égard du déversement latéral. Annales de l'Institut technique de bâtiment et des travaux publics, Série: Béton précontraint (32), **12**, No. 141, 780–831.

LISKA, A. 1986 A berepedés hatása kötelekkel emelt vasbeton tartók kifordító terhére (The influence of cracking on the lateral stability of suspended reinforced concrete beams). PhD. Thesis. Technical University, Budapest.

LOVASS-NAGY, V. 1953 Két végén szabadon felfüggesztett gerenda oldalirányú kihajlással szembeni stabilitásának vizsgálata. (Lateral stability of beams suspended at both ends). *A MTA Alk. Mat. Int. Közl.*, **2**, 33–49.

LŐRINCZ, GY. 1977 A sajátértékelmélet összegezési tételeinek alkalmazása a felfüggesztett gerendák stabilitásszámítására (Application of the summation theorems of the eigen-value theory to the stability investigation of suspended beams). Closing thesis of the course of engineering mathematics. Technical University of Budapest.

MEISSNER, F. 1955 Einige Auswertungsergebnisse der Kipptheorie einfach-symmetrischer Balkenträger. *Der Stahlbau*, **24**, 110–113.

PETTERSSON, O. 1960 Vippningsproblem vid hissning och montering av slanka balkar. *Nordisk Betong*, 231–270.

PFLÜGER, A. 1964 Stabilitätsprobleme der Elastostatik 2. Aufl. Springer, Berlin.

RAFLA, K. 1968 Beitrag zur Frage der Kippstabilität aufgehängter Träger. PhD Thesis. TU Braunschweig.

RAFLA, K. 1969 Näherungsweise Berechnung der kritischen Kipplasten von Stahlbeton-balken. *Beton- und Stahlbetonbau*, **64**, 183–187.

RAFLA, K. 1973 Hilfsdiagramme zur Vereinfachung der Kippuntersuchung von Stahlbeton-balken. *Beton- und Stahlbetonbau*, **68** 43–47.

STÜSSI, F. 1954 Vorlesungen über Baustatik, Bd. I. Birkhäuser, Basel.

TARNAI, T. 1977 Lateral buckling of plane trusses with parallel chords and hinged joints. *Acta Techn. Acad. Sci. Hung.*, **85**, 179–196.

TARNAI, T. 1978 Lateral buckling of plane trusses with parallel chords, hinged joints, simple forked support at both ends. *Acta Techn. Acad. Sci. Hung.*, **87**, 425–439.

TIMOSHENKO, S. P. and GERE, J. M. 1961 Theory of Elastic Stability. McGraw-Hill, New York.

TOMKA, P. 1997 Lateral stability of cable structures. *Int. Journ. Space Struct.*, **12**, 19–30.

TRAHAIR, N. S. 1977 Lateral buckling of beams and beam-columnns. In: Chen W. F. and Atsuta, T.: Theory of Beam-Columns. McGraw-Hill, New York. Vol. 2, 71–157.

TRAHAIR, N. S. and WOOLCOCK, S. T. 1973 Effect of major axis curvature on I-beam stability. *Proc. ASCE, Journ. Eng. Mech. Div.*, **99**, EM1, 85–98.

VRIES, K. de 1947 Strength of beams as determined by lateral buckling. Discussion. *Trans. ASCE*, **112**, 1245–1320.

YEGIAN, S. 1956 Lateral buckling of I-beams supported by cables. PhD. Thesis. University of Illionis, Urbana, Illionis.

REFERENCES TO CHAPTER 10

BAŽANT, Z. and CEDOLIN, L. 1991 Creep buckling. Chapter 9 in: Stability of Structures, Elastic, Inelastic, Fracture and Damage. Oxford University Press, Oxford, 584–631.

BIOT, M. A. 1954 Theory of stress-strain relations in anisotropic viscoelasticity and relaxation phenomena. *Journ. Applied Physics*, **25**, No. 11, 1385–1391.

BIOT, M. A. 1955 Variational principles in irreversible thermodynamics with application to viscoelasticity. *Physical Review*, **97**, No. 6, 1463–1469.

DISCHINGER, FR. 1937 Untersuchungen über die Knicksicherheit, die elastische Verfomung und das Kriechen des Betons bei Bogenbrücken. *Der Bauingenieur*, **18**, Heft 35/36, 539–552.

DULÁCSKA, E. 1984 Influence of creep. In: Buckling of Shells (L. Kollár, E. Dulácska, Eds) Wiley, Chichester and Akadémiai Kiadó, Budapest. 220–230.

FLÜGGE, W. 1975 Viscoelasticity. Springer, Berlin.

HAYMAN, B. 1974a The Influence of Post-Buckling Characteristics on Creep Buckling. University of Leicester, Engineering Department, Report 74–8, May.

HAYMAN, B. 1974b Large and Small Deflection Analyses in Creep Buckling Problems. University of Leicester, Engineering Department, Report 74-9, May.

HAYMAN, B. 1978a Aspect of creep buckling I. The influence of post-buckling characteristics. *Proc. Royal Society of London, Series A. Mathematical and Physical Sciences*, **364**, 393–414.

HAYMAN, B. 1978b The effects of small deflection approximations on predicted behaviour II. *Proc. Royal Society of London, Series A. Mathematical and Physical Sciences*, **364**, 415–433.

HAYMAN, B. 1980 Creep buckling - A general view of the phenomena. IUTAM Symposium Leicester/UK 1980. In: Creep in Structures. (A. P. S. Ponter, D. R. Hayhurst, Eds) Springer-Verlag, Berlin. 289–307.

HOFF, N. J. 1954 Buckling and Stability. *Journ. Royal Aeronautical Society*, **58**, 3–52.

HOFF, N. J. 1958 A Survey of the Theories of Creep Buckling. *Proc. of the Third U. S. National Congress of Applied Mechanics*, Brown University, Providence, Rhode Island, June 11-14. Pergamon Press, Oxford.

HUANG, N. C. 1965 Axisymmetrical creep buckling of clamped shallow spherical shells. *Journ. Applied Mechanics, Trans. ASME*, **32**, June, 323–330.

HUANG, N. C. 1967 Nonlinear creep buckling of some simple structures. *Journ. Applied Mechanics, Trans. ASME*, **34**, 651.

HUANG, N. C. 1973 Inelastic buckling of eccentrically loaded columns. *AIAA Journal*, **11**, No. 7, 974–979.

HUANG, N. C. 1976 Creep buckling of imperfect columns. *Journ. Applied Mechanics, Trans. ASME*, **43**, March, 131–136.

HUANG, N. C. and FUNK, G. 1974 Inelastic buckling of a deep spherical shell subject to external pressure. *AIAA Journal*, **12**, No. 7, 914–920.

HULT, J. A. H. 1962 Oil Canning Problems in Creep. IUTAM Coll. Creep in Structures, July 11-15, 1960. Academic Press, New York and Springer, Berlin. 161–173.

IJJAS, GY. 1982 Stability of viscoelastic structures. *Acta Techn. Acad. Sci. Hung.*, **95**, 55–61.

IJJAS, GY. 1994*a* Numerical method for the solution of the differential equations describing the stability behaviour of viscoelastic structures of one degree of freedom. *Acta Techn. Acad. Sci. Hung.*, **106**, 43–58.

IJJAS, GY. 1994*b* On the stability of viscoelastic systems with viscosity coefficients varying in time. *Acta Techn. Acad. Sci. Hung.*, **106**, 127–143.

KÁRMÁN T. VON and BIOT, M. A. 1940 Mathematical Methods in Engineering. McGraw–Hill, New York.

KEMPNER, J. 1962 Viscoelastic buckling. In: Handbook of Engineering Mechanics (W. Függe, Ed.). McGraw-Hill, New York, **54**–1 to **54**–16.

KORN, G. A. and KORN, T. M. 1975 Mathematical Handbook for Scientists and Engineers, Definitions, Theorems and Formulas for Reference and Review. Mc Graw-Hill, New York.

OBRECHT, H. 1977 Creep buckling and postbuckling of circular cylindrical shells under axial compression. *Int. Journ. Solids Structures*, **13**, 337–355.

SCHAPERY, R. A. 1964 Application of thermodynamics to thermomechanical, fracture and birefringent phenomena in viscoelastic media. *Journ. Applied Physics*, **35**, No. 5, 1451–1465.

SZALAI, L. 1989 Creep buckling of a three-hinged structure (von Mises Truss) (In Hungarian). *Építési Kutatás, Fejlesztés*, **22**, No. 3. 160–167.

VINOGRADOV, A. M. 1985 Nonlinear effects in creep buckling analysis of columns. *Journ. Eng. Mech.*, **111**, No. 6, 757–767.

ŻYCZKOWSKI, M. 1962 Geometrically Non-Linear Creep Buckling of Bars. IUTAM Colloquium, Creep in Structures. Stanford University, California. July 11-15. 1960, Springer, Berlin. 307–325.

REFERENCES TO CHAPTER 11

DAVIDSON, J. F. 1953 Buckling of struts under dynamic loading. *Journ. Mech. Phys. Solids*, **2**, 54–66.

ELISHAKOFF, H. 1978 Axial impact buckling of a column with random initial imperfections. *Journ. Appl. Mech.*, **45**, 361–365.

HOFF, N. J. 1951 The dynamics of the buckling of elastic columns. *Journ. Appl. Mech.*, **18**, 68–74.

HOFF, N. J. 1965 Dynamic stability of structures (Keynote address). In: Dynamic Stability of Structures (Herrmann, G., Ed.). Pergamon Press, Oxford.

HOLZER, S. M. and EUBANKS, R. A. 1969 Stability of columns subject to impulsive loading. *Proc. ASCE, Journ Eng. Mech. Div.*, **95**, EM 4, 897–920.

KONING, C. and TAUB, J. 1933 Stoßartige Knickbeanspruchung schlanker Stäbe im elastischen Bereich bei beiderseits gelenkiger Lagerung. *Luftfahrtforschung*, **10**, 55–64.

MEIER, J. H. 1945 On the dynamics of elastic buckling. *Journ. Aeronaut. Sci.*, **12**, 433–440.

SEVIN, E. 1960 On the elastic buckling of columns due to dynamic axial forces including effects of axial inertia. *Journ. Appl. Mech.*, **27**, 125–131.

TAUB, J. 1933 Stoßartige Knickbeanspruchung schlanker Stäbe im elastischen Bereich. *Luftfahrtforschung*, **10**, 65–85.

UNGER, B. 1984 Ein Beitrag zur Ermittlung der Tragfähigkeit knickgefährdeter Stäbe unter stoßartiger Belastung. *Der Stahlbau*, **53**, 203–209.

REFERENCES TO CHAPTER 12

BARTA, J. 1970 Über stabilizierende und destabilizierende Wirkungen. *Acta Techn. Acad. Sci. Hung.*, **68**, 311–317.

BIEZENO, C. B. and GRAMMEL, R. 1953 Technische Dynamik Band 1., Springer, Berlin.

CSONKA, P. 1987 Theory and Practice of Membrane Shells. VDI-Verlag, Düsseldorf and Akadémiai Kiadó, Budapest.

CYWIŃSKI, Z. and KOLLBRUNNER, C. F. 1971 Drillknicken dünnwandiger I-Stäbe mit veränderlichen, doppelt-symmetrischen Querschnitten. Institut für bauwissenschaftliche Forschung, Nr. 18, Leemann, Zürich.

CYWIŃSKI, Z. and KOLLBRUNNER, C. F. 1982 Neues zu einem Paradoxon des Drillknickens. Institut für bauwissenschaftliche Forschung, Nr. 50, Schulthess, Zürich.

CYWIŃSKI, Z. 1986 On a certain paradox of torsional buckling: State-of-the-art. *Proc. 2nd Regional Coll. Stability of Steel Structures, Hungary.* (Ed., Halász, O. and Iványi, M.). **1**, 1/51–1/58.

DULÁCSKA, E. 1972 Stability of rubber balloons. *Proc. IASS Symp. Pneumatic Structures*, Delft University of Technology.

HEGEDŰS, I. 1988 The stability of a hinged bar fixed by weightless chords. *News Letter*, Technical University of Budapest, **6**, No. 3, 15–22.

KANOK-NUKULCHAI, W. and SUSUMPOW, T. 1993 False paradox of torsional buckling. *Journ. Struct. Eng.*, **119**, 3670–3679.

KOLLÁR, L. P. 1990 Postbuckling behavior of structures having infinitely great critical loads. *Mech. Struct. Machines*, **18**, 17–31.

LOTTATI, I. and ELISHAKOFF, I. 1987 On a new 'destabilization' phenomenon: effect of rotary damping. *Ingenieur-Archiv*, **57** 413–419.

NEAL, B. G. and MANSELL, D. S. 1963 The effect of restraint upon the collapse loads of mild steel trusses. *Int. Journ. Mech. Sci.*, **5**, 87–97.

PETERSEN, CH. 1980 Statik und Stabilität der Baukonstruktionen. Friedr. Vieweg, Braunschweig.

POPPER, GY. 1976 The Beck stability problem for visco-elastic bars. *Periodica Polytechnica – Civil Engineering*, Budapest, **20** 135–147.

TARNAI, T. and SZALAI, L. 1978 A többletmegtámasztás hatása a tartók kifordulására (The influence of additional restraints on lateral buckling of beams). *Mélyépítéstudományi Szemle*, **28**, 210–215.

TARNAI, T. 1980 Destabilizing effect of additional restraint on elastic bar structures. *Int. Journ. Mech. Sci.*, **22**, 379–390.

TIMOSHENKO S. P. and GERE, J. M. 1961 Theory of Elastic Stability. McGraw-Hill, New York.

TROSTEL, R. 1962 Berechnung der Membranen. In: Zugbeanspruchte Konstruktionen. Band 1 (Ed. F. Otto). Ullstein Fachverlag, Frankfurt.

ZASLAVSKY, A. 1979 Increased length may enhance critical load. *Israel Journ. Technology, Jerusalem*, **17**, No. 3.

ZIEGLER, H. 1952 Die Stabilitätskriterien der Elastomechanik. *Ingenieur-Archiv*, **20**, 49–56.

ZIEGLER, H. 1968 Principles of Structural Stability. Blaisdell, Waltham, Maassachusetts.

Author Index

A

Adams 147
Agent, R. 142, 419
Akesson, B. 259, 429
Akhiezer, N. I. 26, 416
Allen, H. G. 191, 199, 201, 211, 212, 427
Almroth, B. O. 20, 280, 415, 431
Amazigo, J. C. 295, 431
Anderson, D. 158, 423
Appeltauer, J. 129, 131, 137, 138, 140, 142, 145, 153, 155, 160, 419, 420, 423
Armer, G. S. T. 136, 167, 188, 230, 259, 262, 265, 425, 426, 429, 431
Arnol'd, V. U. 91, 418
Asztalos, Z. 230, 234, 427
Austin, W. J. 286, 289, 431

B

Babcock, C. D. 278, 434
Barta, J. 410, 440
Barta, Th. (T. A.) 138, 140, 142, 145, 153, 155, 160, 259, 419, 429
Basler, K. 153, 420
Batoz, J.-L. 277, 431
Bayazid, H. 295, 433
Bažant, Z. 359, 373, 375, 438
Beck, C. F. 144, 145, 235, 260, 353 420, 427, 429, 435
Beedle, L. S. 130, 148, 157, 420, 425

Beer, H. 147, 151, 420, 425
Benson, A. 201, 427
Bernardinis, M. D. 137, 420
Biezeno, C. B. 403, 440
Biot, M. A. 49, 223, 230, 366, 416, 428, 438, 439
Bleich, F. 46, 175, 223, 339, 344, 416, 426, 427, 436
Bodner, S. R. 277, 431
Bolotin, V. V. 127, 201, 418, 427
Bolton, A. 142, 420
Boresi, A. P. 280, 431
Bornscheuer, F. W. 266, 429
Bódi, I. 326, 434, 435
Bowles, R. E. 143, 420
Bölcskei, E. 352, 436
Böröcz, I. 352, 436
Brand, L. 223, 427
Brandt, B. 261, 430
Brazier, L. G. 299, 431
Britvec, S. J. 156, 420
Broughton, P. 295, 432
Bruce, V. G. 291, 432
Brush, D. O. 20, 280, 415, 431
Budiansky, B. 85, 284, 417, 431, 434
Bürgermeister, G. 84, 137, 152, 339, 417, 420, 436

C

Campus, F. 159, 420

Carrier, G. F. 284, 431
Cassens, J. 153, 420
Cedolin, L. 359, 373, 375, 438
Chilver, A. H. 154, 156, 420, 421
Chong, K. P. 201, 427
Chow, H. L. 144, 420
Chu, K. H. 144, 146, 420
Chwalla, E. 130, 132, 133, 154, 155, 280,
    281, 344, 345, 420, 431, 436
Coull, A. 258, 275, 430
Croll, J. G. 1, 2, 8, 79, 415, 417
Crowe, E. 196, 429
Csellár, Ö. 160, 420
Csonka, P. 188, 196, 208, 234, 236, 352,
    352, 398, 427, 436, 440
Cuteanu, E. 131, 142, 153, 420, 423
Cywiński, Z. 412, 440

D

Danay, A. 258, 267, 430
Davidson, J. F. 389, 390, 439
Davies, J. M. 150, 426
Deutsch, E. 289, 432
Dickie, J. F. 295, 432
Dinnik, A. N. 286, 290, 432
Dischinger, F. 285, 286, 287, 313, 358,
    432, 438
Domokos, G. 113, 418
Driscoll, G. C. 130, 159, 421, 425
Dulácska, E. 80, 83, 85, 143, 153, 157, 175,
    177, 180, 185, 244, 302, 313, 316, 334,
    352, 363, 400, 401, 417, 421, 425, 426,
    430, 432, 436, 438, 440
Dunkerley, S. 30, 416
Dutheil, J. 134, 151, 157, 159, 421
Dym, C. L. 281, 282, 285, 295, 432

E

Eibl, J. 318, 320, 321, 432
El Naschie, M. S. 284, 432
Elishakoff, H. 127, 389, 413, 418, 439, 440
Eubanks, R. A. 389, 439

F

Faber, G. 244, 430
Farshad, M. 290, 432
Federhofer, K. 295, 432
Fey, T. 146, 421
Firoozbaksh, K. 290, 432
Fischer, M. 17, 415
Flachsbart, O. 256, 430
Flügge, W. 80, 358, 417, 438
Fok, W. C. 74, 417
Föppl, L. 35, 416
Freihart, G. 144, 421
Freihart, I. 144, 421
Friemann, H. 152, 423
Fung, Y. C. 291, 432
Funk, G. 371, 438

G

Galambos, Th. V. 130, 147, 148, 157, 421,
    425
Gállik, I. 330, 434
Gandhi, S. N. 158, 422
Gárdonyi, Z. 338, 348, 350, 353, 435, 436
Gáspár, Zs. 5, 101, 113, 118, 125, 128, 415,
    418, 419
Gaylord, E. H. 145, 425
Gellért, M. 258, 267, 430
Gent, A. R. 145, 421
Gere, J. M. 7, 18, 19, 20, 23, 24, 28, 39,
    47, 70, 71, 80, 137, 188, 192, 199, 200,
    205, 220, 230, 233, 234, 241, 244, 248,
    249, 250, 252, 257, 265, 270, 278, 280,
    282, 283, 284, 286, 288, 290, 294, 321,
    323, 327, 335, 339, 341, 344, 351, 402,
    403, 412, 415, 416, 418, 425, 429, 431,
    434, 435, 437, 441
Gilmore, R. 89, 418
Gioncu, V. 125, 153, 156, 418, 420, 421
Giri, J. 20, 415
Glazman, I. M. 26, 416
Glück, J. 258, 267, 430
Godden, W. G. 330, 331, 434

Goder, W. 153, 423
Godley, M. H. R. 156, 421
Goldberg, J. E. 145, 421
Goschy, B. 238, 241, 427
Grammel, R. 403, 440
Gyurkó, J. 326, 435

H

Habel, A. 155, 421
Hackl, K. 125, 418
Hall D. B. 352
Halldorsson, O. P. 155, 422
Halász, O. 135, 145, 149, 158, 160, 421, 422
Hangan, M. D. 142, 422
Hartsock, J. A. 201, 427
Haviár, Gy. 330, 434
Hayman, B. 367, 368, 375, 438
Hegedűs, I. 188, 194, 198, 201, 202, 203, 204, 208, 209, 210, 212, 219, 223, 227, 230, 231, 234, 236, 238, 267, 326, 405, 427, 428, 430, 434, 440
Herber, K. H. 153, 422
Heyman 145, 158
Hilman, L. 286, 432
Hoff, N. J. 291, 359, 388, 389, 432, 438, 439
Holmes, M. 145, 158, 422
Holzer, S. M. 389, 439
Horne, M. R. 56, 135, 143, 149, 153, 154, 416, 422
Huang, N. C. 291, 371, 384, 432, 438
Huddleston, J. V. 154, 422
Hult, J. A. H. 364, 438
Hunt, G. W. 1, 20, 68, 106, 117, 122, 125, 156, 352, 415, 417, 418, 419, 425
Hussey H. D. 352
Hutchinson, J. 80, 84, 85, 87, 417

I

Ijjas, Gy. 369, 379, 386, 438
Irwin, A. W. 267, 430

Ivan, M. 125, 156, 418, 421
Iványi, G. 337
Iványi, M. 145, 160, 324, 418, 421, 422, 434

J

Jennings, A. 154, 422
Johnson, D. E. 137, 143, 422
Jokisch, F. R. 132, 420
Jordan, C. (K.) 223, 224, 428
Jung, F. W. 153, 422

K

Kanok-Nukulchai, W. 412, 440
Kaplan, A. 291, 432
Kármán, Th. 49, 223, 230, 299, 366, 428, 432, 439
Kärcher, H. 149, 422
Kazazaev, V. N. 145, 422
Kee, C. F. 332, 434
Kempner, J. 358, 439
Kerek, A. 356, 436
Kerkhofs, W. 143, 155, 422
Kesti, N. E. 280, 283, 434
Ketter, R. L. 153, 142, 353, 353, 422, 423, 436
Kirste, L. 142, 236, 353, 354, 422
Klöppel, K. 147, 148, 152, 153, 423
Knothe, K. 148, 151, 153, 423
Koiter, W. T. 1, 3, 8, 20, 85, 133, 296, 415, 417
Kollbrunner, C. F. 130, 280, 281, 286, 412, 423, 431, 433, 440
Kollár, L. 21, 26, 43, 50, 58, 80, 83, 85, 153, 157, 161, 177 188, 209, 223, 227, 237, 246, 267, 276, 299, 302, 322, 323, 324, 326, 327, 335, 336, 337, 338, 340, 343, 347, 348, 350, 353, 356, 415, 416, 417, 420, 426, 428, 430, 432, 433, 434, 435, 436
Kollár L. P. 194, 198, 201, 202, 203, 204, 208, 212, 214, 215, 216, 220, 222, 230,

231, 234, 236, 238, 404, 427, 428, 430, 440
Komatsu, S. 335, 435
Koning, C. 389, 397, 439
Korda, J. 352, 436
Korn, G. A. 368, 439
Korn, T. M. 368, 439
Kornoukhov, N. V. 143, 423
Korondi, D. 171, 426
Kounadis, A. N. 20, 415
Kovács, I. 313, 314, 315, 316, 331, 432, 433, 435
Kovács, O. 244, 430
Kókai, T. 266, 430
Kretschmar, H. 84, 137, 339, 417, 420, 436

L

Lebelle, P. 353, 436
Lewis, G. M. 79, 417
Ligeti, R. 235, 428
Lightfoot 143
Ligtenberg, I. F. K. 149, 158, 423
Lipták, L. 142, 162, 275, 423, 426, 430
Liska, A. 352, 353, 436
Lo, D. L. C. 295, 296, 433
Lokschin, A. 286, 290, 329, 433
Lottati, I. 413, 440
Lovass-Nagy, V. 340, 352, 437
Low, M. W. 159, 423
Lőrincz, Gy. 349, 437
Lu, L. W. 130, 145, 148, 156, 159, 421, 423, 425
Lusas 258, 430

M

MacLeod, I. A. 257, 430
Magyar, Á. Sz. 330, 434
Majid, K. J. 150, 158, 423
Mansell, D. S. 410, 440
Massonnet, Ch. 146, 150, 159, 420, 423
Masur, E. F. 295, 296, 433, 434
Mateescu, D. 131, 142, 423

Mayer-Mita, R. 313, 433
Mayers, I. 201, 201, 427
Meier, J. H. 389, 439
Meissner, F. 349, 437
Meister, M. 130, 286, 423, 433
Melan, E. 51, 223, 416, 427
Merchant, W. 142, 143, 147, 149, 151, 420, 422, 423
Mikhlin, S. G. 26, 416
Mladenov, K. A. 127, 418
Moses, F. 146, 423
Murasew 144, 423

N

Nagy, J. 297, 298, 433
Neal, B. G. 410, 440
Nemestóthy, É. 165, 426
Neut, A. van der 74, 75, 78, 79, 417

O

Obrecht, H. 371, 439
Ojalvo, M. 330, 331, 332, 435
Opladen, K. 153, 423
Oran, C. 138, 295, 424, 433
Ostapenko, A. 146, 147, 424
Oxfort, J. 146, 151, 424
Östlund, L. 333, 334, 335, 435

P

Pabarcius, A. 146, 420
Pajunen, S. 125, 419
Papkovich, P. F. 35, 416
Parekh, C. J. 258, 431
Pearson, C. E. 266, 280, 430, 433
Penelis, G. 137, 424
Peters, P. 144, 424
Petersen, Chr. 286, 313, 327, 403, 433, 435, 440
Pettersson, O. 353, 437
Pflüger, A. 84, 271, 313, 339, 417, 430, 433, 437

Plantema, F. J. 39, 194, 220, 416, 428
Podolski, D. M. 258, 430
Pomázi, L. 201, 428
Popov, I. G. 143, 424
Popper, Gy. 413, 440
Poston, T. 89, 97, 101, 419
Puwein, M. G. 137, 140, 141, 142, 144, 424

R

Raevski, A. N. 150, 156, 424
Rafla, K. 347, 353, 437
Ramm, E. 17, 415
Rankine, W. J. M. 55, 416
Rapp, I. H. 295, 434
Ratzersdorfer, J. 223, 230, 428
Reay, N. A. 125, 352
Reeh, H. 261, 430
Reis, J. 59, 417
Resinger, F. 153, 424
Rhodes, J. 74, 417
Roorda, J. 59, 130, 417, 424
Rosman, R. 153, 196, 235, 238, 238, 259, 267, 424, 428, 429, 430
Rozenblat, G. I. 142, 424
Rózsa, M. 152, 286, 424, 433

S

Saafan, S. A. 154, 424
Sahmel, P. 137, 138, 424
Sakimoto, T. 335, 435
Salem, A. H. 142, 147, 149, 423
Sanders, J. L. 285, 433
Sandhu, R. S. 327, 435
Sattler, K. 137, 424
Sawko, F. 147, 424
Sättele, J. M. 17, 415
Scarlat, A. 142, 424
Schäfer, H. 260, 261, 430
Schaber, E. 152, 425
Schack, R. 353, 436
Schapery, R. A. 366, 439
Scheer, J. 151, 425

Schibler, W. 313, 433
Schineis, M. 154, 280, 283, 284, 285, 425, 433, 434
Schmidt, R. 280, 283, 284, 285, 433
Schreyer, H. L. 295, 296, 434
Sevin, E. 389, 390, 439
Shukla, S. 330, 331, 332, 435
Sievers, 137
Sills, L. B. 284, 434
Simitses, G. J. 20, 295, 415, 434
Singer, J. 278, 434
Slavin, A. 142, 425
Smith, B. S. 145, 425
Snitko, N. K. 142, 153, 425
Southwell, R. V. 26, 416
Stafford Smith, B. 196, 258, 275, 428, 430
Stein, P. 330, 435
Steiner, H. 153, 424
Steinhardt, O. 151, 425
Steup, H. 84, 137, 152, 339, 417, 420, 436
Stevens, L. K. 138, 152, 155, 158, 425
Stewart, I. N. 89, 97, 101, 419
Strigl, G. 35, 416
Stüssi, F. 266, 286, 287, 327, 339, 430, 434, 435, 437
Susumpow, T. 412, 440
Svensson, S. E. 79, 417
Switzky, H. 157, 425
Szalai, L. 380, 410, 439, 440
Szerémi, L. 196, 235, 236, 238, 429
Szmodits, K. 266, 431

T

Tall, S. 130, 148, 157, 420, 425
Taranath, B. S. 267, 431
Tarnai, T. 26, 30, 35, 352, 410, 411, 416, 417, 436, 437, 440
Taub, J. 389, 397, 439, 440
Teply, B. 258, 431
Thom, R. 89, 419
Thompson J. M. T. 1, 20, 68, 79, 89, 106, 125, 156, 330, 415, 417, 418, 419, 425, 434

Tietze, M. 286, 287, 288, 434
Timoshenko, S. P. 7, 18, 19, 20, 23, 24, 28, 39, 47, 70, 71, 80, 137, 188, 192, 199, 200, 205, 220, 230, 231, 233, 234, 241, 244, 248, 249, 250, 252, 257, 265, 270, 278, 280, 282, 283, 284, 286, 287, 288, 290, 294, 321, 323, 327, 335, 339, 340, 344, 351, 402, 403, 412, 415, 416, 418, 425, 428, 431, 434, 435, 437, 441
Tokarz, F. J. 327, 435
Tomka, P. 355, 437
Trahair, N. S. 339, 340, 352, 437
Trostel, R. 399, 441
Tulk, J. D. 74, 418

U

Uhlmann, W. 147, 148, 423
Unger, B. 391, 440

V

Verőci, B. 418
Vinogradov, A. M. 388, 439
Vlasov, V. Z. 262, 266, 431
Vogel, U. 148, 425
Vries, K. de 352, 437

W

Walker, A. C. 1, 2, 8, 69, 71, 73, 74, 415, 417, 418

Wang, Ch. K. 155, 157, 422, 425
Weinberger, H. 26, 41, 416, 417
Weinel, E. 299, 434
Wempner, G. A. 280, 283, 434
Wilde, A. M. B. 147, 424
Williams, D. G. 71, 418
Wlassow, W. S. (see also Vlasov) 241, 429
Woinowski-Krieger, S. 18, 415
Wood, R. H. 134, 142, 149, 150, 158, 425
Woolcock, S. T. 340, 352, 437
Wright, E. W. 145, 425

Y

Yamao, T. 435
Yarimci, E. 259, 431
Yegian, S. 353, 437
Yen, Y. Ch. 158, 425
Yoshimura, T. 125, 419
Yura 147

Z

Zalka, K. A. 136, 167, 188, 194, 208, 209, 222, 230, 237, 257, 259, 262, 264, 265, 266, 425, 426, 429, 430, 431
Zar, M. 144, 145, 420
Zaslavsky, A. 407, 441
Zeeman, E. C. 89, 127, 419
Ziegler, H. 404, 413, 441
Zienkiewitz, O. C. 258, 431
Życzkowski, M. 388, 439

Subject Index

A

A-orthogonal 36, 37, 40, 41, 57
A-orthogonal subspace 36, 37
accidental imperfections 129
active (unstable) bars 134, 129, 157
active (essential) variables 91, 92, 100, 101
amplification factor 257, 258
anticlinal point of bifurcation 126, 127
arches with hangers or struts 276, 308 to 319, 329 to 333
arches with parabolic axis 285
asymmetric bifurcation 4, 86, 104, 105, 106, 155, 270, 371, 376, 383
asymmetric post-critical behaviour 363, 371
asymptotic critical load 62, 73, 76, 78, 79
axially compressed cylinder 80, 85
Ayrton-Perry approximation 159

B

B-beam 217, 218, 219, 220
beam-column 31, 198
bending deformation 39, 41, 132, 133, 136, 191, 193, 197, 217, 220, 238, 335, 350, 390, 391, 413
bending stiffness 65, 67, 76, 77, 129, 132, 134, 157, 166, 175, 189, 190, 193, 195, 197, 200, 202, 205, 207, 217, 222, 233, 234, 235, 237, 243, 244, 245, 261, 263, 339, 343, 343, 348, 349, 350, 410

bending vibration 390
bifurcation 59, 67, 73, 78, 88, 130, 131, 132, 133, 135, 142, 149, 151, 153, 154, 155, 156, 157, 158, 276, 277, 381
bifurcation model 130, 131, 132, 133, 156
bifurcation set 92, 94, 95, 96, 97, 98, 99, 101, 102, 108, 118, 122
Biot's equation 366, 373, 375, 378, 382
box bar 74, 75, 77, 78, 79
braced column 60 to 68, 74
bracing core 242 to 256, 269, 275
bracing element 242, 259, 260, 261, 266, 268
bracing system 242, 258, 259, 260, 261, 263, 264, 266, 275
buckling factor 155, 156, 157, 159
buckling length 144, 145, 157, 158, 164, 165, 174, 179, 185, 186, 201, 202, 203, 211, 214, 216, 234, 237, 250, 251, 350, 402, 403, 404
butterfly catastrophe 91, 96, 115, 116

C

canonical form 91, 92, 93, 95, 96, 97, 99, 101, 102, 108, 127
catenary arch 289, 290
central load 7, 167
characteristic equation 40, 135, 224, 282, 404, 412
circular axis 277, 281, 284, 321, 328

clamped arch 282, 284, 285, 286, 289, 295,
296, 326, 327, 330
closed hollow section 74
Code of Practice provisions 257
coefficient of resilience 390, 395, 396
column of variable cross section 27
compatibility methods 146
composite stability phenomena 16
conformal mapping 368
conservation of momentum 391
conservative force 57, 359, 376
constant directional load 131, 277, 278, 280
to 286, 290, 327, 327
continuous bar on elastic support 43
core 150, 159, 187 to 235, 242 to 274
coupled shear wall 196, 208, 237, 238, 238,
266
creep buckling 358, 359, 361, 363, 388
critical imperfection territory 101 to 113,
120, 123, 127
critical point 89 to 92, 102, 105, 106, 113
to 118, 122, 126, 127
critical time 358, 359, 362, 363, 386, 388
Csonka-beam 207, 208, 218, 219, 221, 238
cusp 91, 106, 109, 115
cuspoid catastrophe 91, 92, 93, 100, 106,
107, 109, 111, 113, 115, 116, 117
cut-out subsystems 142, 147, 148
cylindrical shell 79 to 87, 210

D

damping 412, 413
deck-type arch bridges 308, 311, 313, 321,
329, 332, 335
deformation modulus 155
deformation of the cross section 299, 306,
337
density of the strain energy 138
destabilizing effect 309, 311, 317, 410 to
413
determinacy 91, 92, 100
diffeomorphism 89, 90

difference equation 223, 224, 225, 227, 228,
231
differential operators 28, 32
direct plastic design 158
Dischinger material 358, 363, 377, 378, 379,
380, 383
discrete spectrum 26, 34
divergence 11, 64, 130, 133, 134, 135, 145
to 151, 153, 157, 158, 159
Donnell's theory 80
double-cusp catastrophe 100, 127
drop test 396
dual cusp catastrophe 93, 95, 106, 107, 109,
110, 113
dual form 91, 101
Dunkerley theorem 30 to 38, 53, 57
dynamic load 389, 390

E

effective buckling length 235, 237
effective moment of inertia 300, 302, 303
eigenelements 34
eigenfunction 82, 83, 84, 130, 135, 149,
257, 258, 274
eigenvalue 26, 27, 28, 30, 31, 34 to 40, 49,
51, 56, 57, 228, 258, 389
elastic foundation 43 to 50, 199, 200, 202,
207, 208, 209, 218 to 221, 319
elastic rigidity matrix 151
elastic support 43 to 50, 161, 177, 272, 273,
280, 281, 326
elastically supported arches 326
elastically supported bars 407, 408
elliptic umbilic catastrophe 98, 91, 97, 100,
118, 126
elongation of the hangers 312, 314, 315,
316, 321
energy method 42, 84, 136, 137, 243, 277,
313, 324, 337, 344, 347, 348, 349, 351
entropy 366
envelope 101, 103, 105, 107, 109, 119, 120,
123, 188

equilibrium path 1 to 14, 68, 88, 100, 101, 104, 110, 111, 112, 116, 117, 127, 135, 146, 151, 153, 362, 366, 367, 371, 373, 375, 385
equilibrium state 104, 105, 399, 402
equilibrium surface 92, 93, 97, 101, 127
equivalent continum 136, 259 to 266
external alien work 30
external co-ordinate 362, 363, 376, 379
external parameter 369, 376

F

falling load 391, 393
families of functions 88, 90
festoon curve 407, 408, 409
flexural-torsional buckling 24, 159, 160, 243, 245, 252, 253, 254, 258, 262, 264, 265,
fluid type material 363, 371, 358, 371,
fold 72, 75, 76, 91, 100, 352
fold catastrophe 92, 93, 100, 102, 103, 105
fork-like support 323, 326, 336, 341, 341, 344, 345, 349, 350, 410
function or map 89
Föppl-Papkovich theorem 35, 37, 39, 41, 68, 149, 194, 219, 220, 254, 255, 267, 315, 316

G

Galerkin's method 84, 85
generalized co-ordinate 88, 100, 102, 105
generalized Southwell plot 20
geometric rigidity matrix 151
global buckling 59, 69, 60, 74, 80, 199, 201, 230, 234, 237, 276, 313, 315, 316, 320, 332, 333
global buckling length 237
global safety factor 257
gradient 86, 92, 100
gradient vector 89

H

Hessian matrix 89, 90, 92, 100, 126, 369
higher order (nonlinear) theory 133
Hilbert space 26, 28, 30, 34, 36, 51
homeoclinal bifurcation 125, 126
honey-comb core 187
hydrostatic pressure 276 to 285
hyperbolic umbilic catastrophe 91, 97 to 100, 118, 120, 125, 126

I

Iasinski formula 157
impact, not perfectly plastic 395
impact, perfectly elastic 391, 394, 395
impact, perfectly plastic 390
imperfection sensitive 7, 9, 10, 12, 15, 16, 59, 61, 63, 74, 86, 99, 110, 111, 115, 155, 156, 157, 284, 297, 405
imperfection-sensitivity surface 101, 102, 108, 118, 119, 122
impulse 392, 395, 396
impulse theorem 391, 392, 393, 394
inert mass 359, 360, 362, 363, 366, 386, 389
inextensional deformation 71, 277, 279, 292, 309, 312, 313, 314, 316, 320, 326, 330, 332, 392
initial crookedness (curvature) 79, 391, 393
initial imperfections 255, 277, 359, 363, 364, 366, 367, 382, 384, 388, 405, 406
initial velocities 392
interaction 59, 60, 69, 74, 79, 109, 156, 160, 199, 201, 245, 258, 259, 264, 265, 275
internal energy 138, 362, 369
internal parameter 366, 375
internal pressure 80 to 86
internal work 29, 243, 244

K

k-determinate 91
Kelvin solid 358
Kollár conjecture 43 to 49, 57
Kollár critical load parameter 45, 47, 49

L

laced (Vierendeel) column 41, 43, 45, 223, 230 to 235
lateral bending rigidity 323
lateral buckling 24, 31, 32, 33, 53, 54, 55, 156, 159, 321, 327 to 338, 339, 340, 343, 344, 350 to 355, 410
level line 103, 105, 109, 120, 123
lift-slab 402
limit curve 134, 135
limit point 11, 22, 100, 104, 105, 130, 276, 371, 376
linear elastic theory 133
linear operator 26, 29, 30
local buckling 59, 60, 65, 69, 74, 80, 199, 228, 230, 274, 276, 315, 320, 332, 333
locus of critical points 367, 374, 375, 382
longitudinal shock wave 390
longitudinal stresses 303, 306, 321, 340

M

mathematical model 88, 130, 133
Maxwell fluid 359, 360, 363, 364, 371, 376, 380, 383
Maxwell-Betti theorem 30
measure of stability 355
mechanical models 130
mechanism 134, 135, 147, 148, 149, 151, 404
Melan theorem 50, 51, 53
membrane action 17, 255
Merchant formula 149, 150, 156
method of split rigidities 39, 41
minimum of the total potential energy 362, 363, 376, 386

mixed mechanism 150
moment increase factor 153, 395
monoclinal bifurcation 125, 126
Morse lemma 89
Morse l-saddle 89
Morse $(n - i)$-saddle 89, 91
movement equations 389
multi-bay frame 235, 238
multi-layer sandwich 209, 216, 222, 235

N

natural circular frequency 51, 52, 53
near-coincidence of critical loads 118, 122
Newtonian fluid 388
non-degenerate critical point 89
nonconservative forces 57, 407, 409, 412
nonequilibrium thermodynamics 362, 366
nonlinear interaction formula 35, 56
nonlinear theory 59
non-sway stability of the braced column 268, 271 to 274
not conservative 277, 402

O

one-hinged arch 282, 284, 288
orthogonal 59, 68, 130, 131, 274
own critical load 179 to 186

P

parabolic umbilic catastrophe 91, 97, 125, 126
paradox of plate buckling 18
parallel connection 217, 218, 219
partial critical load 136, 349, 351
partial rigidizing 36, 37, 40, 42, 57
passive (supporting) bars 129, 134, 157
passive (inessential) variable 101
Plantema paradox 39, 219, 220
plastic critical stress 267
plasticity 13 to 17, 79, 130, 145, 173, 335
positive definite 26, 29, 30, 34, 36, 51

positivity of operator 29, 30
post-buckling load-bearing capacity 59 to 79, 85, 86, 291, 302, 335, 406
post-critical behaviour 1 to 7, 16, 19, 61 to 87, 133, 155, 255, 256, 276, 284, 285, 335, 338, 363, 383, 405, 407
pressure line 148, 285, 309, 311
primary deformation 130, 131, 132
principal axes 243, 252 to 255, 261 to 263
probability density function 127
probability of instability 127
pseudo-equilibrium surface 362, 365 to 369, 373, 374, 378, 381, 382, 384

Q

quadratic Dunkerley formula 35
quadratic eigenvalue problem 30, 35
quadratic functional 31, 34, 36

R

radius of gyration 67, 247, 248, 252, 253, 261, 291, 341
Rankine formula 55, 56, 57, 149
Rankine-Merchant formula 56
Rankine-Merchant load factor 56
Rayleigh quotient 27, 36, 37
reduced critical load 72, 158
regular structures 188, 223, 259
reinforced concrete bracing element 268
replacement continuum 188, 223, 226, 229, 231 to 238
ribbed plate 69, 72
rigid-body displacement 278, 279
rubber balloons 398 to 402

S

safety 1, 244, 257, 267, 268
safety factor 20, 154, 155, 156, 185
Saint-Venant torsional stiffness 23, 246 to 251, 260, 264, 267
sandwich bar with thick faces 232, 235, 238

sandwich beam with thick faces 190, 194, 195 to 204, 205 to 217, 218, 219, 220, 222, 232, 233
sandwich beam with thin faces 190, 192, 206, 207, 219, 229, 230, 233, 237
sandwiches 187, 188, 195, 202, 203, 209, 235, 238, 241
secant geometric matrix 151
second-order elastic computation 148, 151, 152, 160
second-order elasto-plastic method 146, 147
second-order rigid-plastic computation 147, 148
second-order theory 135
secondary deformation 130, 132
secondary path 117, 155
serial connection 217, 218, 219
shallow sandwich shell 215
shallow shell 80, 81, 215
shear centre 31, 33, 34, 239, 246 to 249, 252 to 266, 321, 323, 339, to 345, 349 to 351
shear deformation 39 to 43, 136, 188 to 193, 197, 201, 205 to 209, 213, 214, 217, 221, 230, 237, 238, 241, 266, 276, 294 to 298, 333
shear stiffness 250, 251, 233
shear wall 235, 237, 238, 242, 255, 258, 259, 266, 267
shell arch 298, 299, 302, 303, 308, 324, 335, 336, 337, 357
shell-beam row 356, 357
shifting operator 224
smooth function 89, 90, 91
snapping through 11, 12, 22, 130, 153, 154, 276, 277, 290, 291, 293, 295, 297, 339, 355, 356, 357
soft rubber material 401
softening 129, 400, 407
solid-type material 363, 371
Southwell theorem 26, 27, 28, 29, 30, 56, 57, 217, 219, 221, 250, 251, 264
Southwell's plot 20
spatial behaviour 251, 253, 255

spectrum of a linear operator 26
spherical shell 87, 125
splitting lemma 90, 101
splitting parameter 117
spring rigidity 43, 44, 45, 47
stability domain 24, 53, 54, 55
standard cusp 93, 94, 106, 109, 111, 112, 113
standard forms 91
statically indeterminate structure 363, 148
storey mechanism 150
strain energy 84, 281, 344, 394
strength of the bracing core 242, 256, 257
stress-softening material 400
Strigl formula 34, 56
structurally stable 90, 91, 100
summation theorem 23, 25, 26, 57, 349, 351
suspended beam 344, 351, 352
swallowtail catastrophe 91, 95, 96
sway critical load of columns 269, 270, 272, 273
symbolic umbilic catastrophe 91, 125, 126
symmetric linear operator 26, 34, 36, 51, 57
symmetric stable bifurcation 2, 110, 111, 113, 114, 363, 380, 383
symmetric unstable bifurcation 3, 109, 112, 115, 363, 366, 376
symmetry of operator 29, 30

T

tangent geometric matrix 151
tensile stiffness 65, 210, 215, 216
theorem of conservation of energy 394
third-order theory 135
Thom's theorem 89, 90, 91, 100, 117, 127
three-hinged arch 283, 285, 287, 289
three-parameter solid 363, 371, 376, 377, 379, 380, 384
through-type arch bridge 308, 314, 321, 329, 333

Timoshenko-beam 190, 191, 193, 207, 218, 219, 220, 237, 238
torsional buckling 23, 24, 32, 38, 50, 74, 241, 243 to 267, 276, 321, 332, 335, 340, 350, 351, 411, 412
torsional stiffness 243, 245, 247, 249, 260, 261, 264, 268, 323
total potential energy 42, 57, 88, 88, 100, 105, 106, 113, 114, 115, 116, 126, 368, 369, 375, 379, 382, 383, 386
transverse bending moment 303, 308
transverse vibration of the bar 51, 53
triple root 125, 126
truss 174, 188, 223, 226 to 230, 309, 339, 352 to 355, 379, 380
twin arches 321, 333, 334, 335
two-hinged arch 279, 280, 281, 282, 283, 285, 287, 290, 294, 295, 296, 318
two-legged plane frame 48
two-parameter eigenvalue problem 51

U

umbilic bracelet 97
umbilic catastrophe 91, 97, 100, 117, 119, 121, 123, 125, 126
unfolding 90, 91

V

Vierendeel columns see: laced columns
von Mises truss 364, 371, 380

W

warping stiffness 23, 246, 248, 250, 251, 261, 266

Milton Keynes UK
Ingram Content Group UK Ltd.
UKHW021900071024
449327UK00021B/1593